Perspectives on
Food-Safety Issues
of Animal-Derived Foods

Perspectives on Food-Safety Issues of Animal-Derived Foods

■ ■ ■

Edited by Steven C. Ricke and Frank T. Jones

The University of Arkansas Press
Fayetteville
2010

ISBN-10: 1-55728-919-0
ISBN-13: 978-1-55728-919-3

14   13   12   11   10   5   4   3   2   1

Text design by Ellen Beeler

⊗ The paper used in this publication meets the minimum requirements of the American
National Standard for Permanence of Paper for Printed Library Materials Z39.48-1984.

Library of Congress Cataloging-in-Publication Data

Perspectives on food-safety issues of animal-derived foods / edited by Steven C. Ricke and
Frank T. Jones.
      p. cm.
   Includes bibliographical references and index.
   ISBN 978-1-55728-919-3 (hardback : alk. paper)
   1. Food—Microbiology. 2. Food of animal origin—United States—Safety measures.
   I. Ricke, Steven C., 1957– II. Jones, Frank T.
   QR115.P47 2010
   664.001'579--dc22

                                                                          2009041590

This publication was made possible by the support of the Food Safety Consortium
through the U.S. Department of Agriculture Cooperative State Research, Education,
and Extension Service.

Contents

Emerging Issues in Food Safety

∎ 1 ∎

A Brief History of the
Food Safety Consortium

James H. Denton

Origins and Development

The Food Safety Consortium (FSC) was established by Congress through a Cooperative Research Service grant in 1988 to conduct scientific investigation in four areas of immediate concern to consumers, growers, and processors:

- To determine the most effective intervention points to control microbiological or chemical hazards.
- To develop technology for rapid identification of infectious agents and toxins.
- To develop risk-monitoring techniques to detect hazards in the distribution chain.
- To develop a statistical framework to evaluate the health risks posed by contamination of the food chain by infectious agents and toxins.

After the FSC was established, it expanded its mission to include three additional tasks:

- To develop risk assessment and interdiction actions in hazard reduction and control.
- To develop technology to reduce the hazards and improve the quality of animal food products, which will complement the development of

Hazard Analysis Critical Control Point (HACCP) programs by the USDA.

• To develop, complement, and maintain an aggressive technology transfer system that effectively communicates the work of the Consortium to consumers, industry, government, and other scientific investigators.

The research work was divided to correspond to each university's area of strength: poultry at Arkansas, pork at Iowa State, and beef at Kansas State.

How the FSC Matured

In 1986, after the CBS newsmagazine "60 Minutes" broadcast a program about food safety, Arkansas poultry industry leaders were concerned that more work needed to be done on at least two fronts: to inform the public of industry's efforts toward safer food and to support a cohesive national food-safety research effort.

The poultry leaders contacted Senator Dale Bumpers of Arkansas and soon efforts were under way to establish a food-safety research organization that would explore scientific solutions to problems in animal meats. In 1988, Congress approved the establishment of the Food Safety Consortium with the University of Arkansas, Iowa State University, and Kansas State University as members. The first research coordinator, serving from 1988 to 1989, was Bernie Daniels, with the first funding received in mid- to late 1989.

One of the poultry industry leaders who met with Bumpers was Richard Forsythe, then an executive with Campbell Soup in Fayetteville, Arkansas, and chair of the Arkansas Poultry Federation's technical advisory committee. By 1989, Forsythe had joined the faculty of the University of Arkansas poultry science department and assumed duties as coordinator of the Food Safety Consortium. He began working with colleagues Curtis Kastner of Kansas State and George Beran of Iowa State to advance the FSC from an entity that existed only on paper to a working research organization.

Forsythe, Kastner, and Beran met with Clark Burbee of the U.S. Department of Agriculture Cooperative State Research, Education and Extension Service to prepare a budget request for allocation of the congressional appropriation for fiscal 1990–1991.

As this process was under way in 1990, the FSC selected a steering committee and a technical advisory committee to oversee operations. The three universities' program leaders asked faculty at their respective campuses to submit proposals for FSC research projects. A formal review process was established by each campus for their research proposals and in 1997, under the leadership of Charles Scifres, the

University of Arkansas initiated an off-campus peer review process based on the USDA NRI format with experts in microbiology and engineering food safety.

During the early years, the program leaders at the campuses knew which faculty members in various departments concentrated on particular areas of research that would be relevant to food safety. Those professors were asked to develop proposals for FSC funding. The FSC grants also provided many faculty members their first opportunity to hire postdoctoral students to assist in research projects.

During the second year of the FSC grant, Forsythe contacted the University of Arkansas for Medical Sciences in Little Rock (UAMS), a separate institution from the Fayetteville campus, and Arkansas Children's Hospital (ACH) in Little Rock. UAMS and ACH researchers were examining poultry-related projects such as studies of the prevalence and control of *Salmonella*. UAMS and ACH then began submitting proposals as partners under the University of Arkansas segment of the annual grant.

Forsythe continued as coordinator of the FSC until his retirement in 1995. Gerald Musick, the University of Arkansas associate vice president for agriculture research, was the first chair of the FSC Steering Committee until 1993. Musick was succeeded as steering committee chair by L. B. (Bernie) Daniels, the associate director of the U of A Agricultural Experiment Station. In 1995, Daniels also assumed the coordinator's duties upon Forsythe's retirement. Daniels returned to full-time faculty research the next year and was succeeded as coordinator and steering committee chair by Charles J. Scifres, the dean of the Dale Bumpers College of Agricultural, Food and Life Sciences at the University of Arkansas. Michael Johnson became technical coordinator in 1996 until 2006, when he was succeeded by Steven Ricke. Today the leadership team consists of campus program directors Curtis Kastner at Kansas State, James Dickson at Iowa State, and Steven Ricke at Arkansas.

Over the years the annual fall research meeting was conducted in Kansas City, Missouri, and then shifted to the current format of a three-year rotation among the three contributing university campuses. Each university's program leader has delivered presentations at the meeting reporting the year's research accomplishments. A written summary of the research highlights and a complete listing of each university's published articles and scholarly presentations have been published annually by the FSC (Food Safety Consortium 1998).

Establishing Focus

The focus of the FSC in the initial stages was to develop capacity in the scientific cadre at the University of Arkansas, Iowa State University, and Kansas State University required to form the basis for future work in addressing the stated objectives.

This included the establishment of laboratories with state-of-the-art equipment as well as identification of scientists possessing the required skills and experience to accept the challenge. The key to optimizing the effectiveness of the FSC was forging trusting relationships between and among the scientists of the FSC and beyond.

The early leadership in the FSC provided by Dick Forsythe (AR), George Beran (IA), and Curtis Kastner (KS) led to the initial focus of the FSC in 1991–1992 to define the prevalence of foodborne pathogenic bacteria associated with pork (*Salmonella, Staphylococcus aureus, Listeria monocytogenes*) and beef products (*L. monocytogenes*).

It is important to note that some of the early investigations at Iowa State also included work on the characterization of viral organisms. Limited work on the development of rapid identification techniques was initiated at the University of Arkansas (*L. monocytogenes* and *Salmonella*). The use of electronic cattle identification systems for tracking cattle through all stages of production, processing, and marketing was evaluated at Kansas State University. The work on intervention strategies for poultry processing was initiated at the University of Arkansas in cooperation with the Southeastern Poultry and Egg Association and the National Broiler Council.

Following the initial organizational efforts the FSC began to focus on meeting the targeted objectives. Considerable effort and resources were expended on the development of rapid detection and identification methodologies. Leadership in these efforts was provided by Dan Fung (KS), Mike Johnson and Mike Slavik (AR), and Irene Wesley (IA-NADC). Later efforts in the development of biosensors was largely led by Yanbin Li (AR). Leadership in applying these detection technologies to the assessment of intervention strategies was provided by Curtis Kastner and Randy Phebus (KS), Amy Waldroup (AR), and Jim Dickson (IA). Application of these methodologies in conducting early epidemiological research regarding the prevalence of *Salmonella* in children in Arkansas, which was higher than the national average, was provided by Gordon Schutze with the University of Arkansas Medical Sciences Pediatric Unit. Early work in risk assessment and analysis was conducted by Yanbin Li (AR). These efforts led to work in preharvest intervention strategies through the efforts of Dayton Steelman on insect transmittal of pathogens and Billy Hargis on probiotics to control *Salmonella* in broilers and turkeys.

Building Our Reputation

The worth of any organization is measured by its effectiveness in meeting the needs of their clientele or constituencies. The FSC was challenged in meeting the needs of four primary groups: meat and poultry processors, regulatory personnel, meat and poultry producers, and most important, consumers of meat and poultry products. In meeting the needs of processors it was imperative to develop and evaluate new technologies designed to reduce or eliminate pathogenic organisms. These technologies included chlorine, ozone, CPC (cetyl pyridinium chloride), steam pasteurization, carcass rinsing, and irradiation. These not only form the basis for effective interventions in processing, they also provide the basis for effective interventions as defined in HACCP systems for meeting regulatory requirements. In addition to the needs of processors and regulators it is imperative that the needs of producers are also addressed. Identification of vectors by which pathogens are spread, and other intervention strategies such as competitive exclusion technologies, provide real-world solutions to these producers in meeting the goal of a safe meat and poultry supply. Finally, in meeting the needs of consumers of meat and poultry products, not only is the FSC challenged with developing detection methods, intervention strategies, assisting in epidemiological investigations, risk assessment and analysis, and preharvest interventions, it also has the obligation for contributing to the effective education of consumers regarding the safe handling of poultry and meat products.

Expanding Our Influence

The primary objectives of the FSC are research driven for the purpose of maintaining relevance in providing solutions to existing as well as new and emerging challenges to the safety of the meat and poultry supply. The most important added responsibility for the FSC was the incorporation of education and outreach as part of its mission. The importance of consumer education can never be underestimated, and it is imperative that the information be science based and not driven by emotion or sensationalism.

The FSC recognized the need for a multifaceted approach to providing education and outreach and had the capacity to meet this challenge. The greatest challenge was the development of HACCP workshops for the meat- and poultry-processing industries as a means to assist in implementing the 1996 Pathogen Reduction Rule and HACCP implementation legislation as required by USDA-FSIS (Food Safety Inspection Service). Arkansas scientists became involved in HACCP workshops in 1993 and later became founding members of the International HACCP Alliance for addressing poultry industry training and education needs. A team of investigators from Arkansas in the FSC in collaboration with a multiagency task force developed the Operation Food Safety public school curriculum for grades pre-K through 12 as part of the Public Health Framework. In addition collaborations were developed with the National Restaurant Association for the delivery of the ServSafe food-safety management program to food service establishments. A similar collaboration was developed with the retail supermarket industry in providing the Super Safe Mark food-safety management program to retail establishments. The National Alliance for Food Safety and Security, which formed in 1998 with funding provided by the USDA-ARS under the leadership of Floyd Horn, originated from the FSC and includes 20 land-grant institutions in partnership addressing needs in food safety and security.

Bridging the Gap

The approach in addressing the issue of food safety is best described as the Food Safety Consortium. It is imperative that each segment of the food marketing chain understands and addresses the responsibilities and needs inherent in its part of the system. Beginning with the food producer, and including the food processor, food service, food retail, food regulatory, and food consumer, each segment is unique with regard to its responsibilities and needs. Only when each segment is viewed as part of a larger system can the needs of the entire system be met.

Balancing Basic and Applied Research Needs

The primary focus of basic research in food safety has been traditionally on detection methodologies that are more rapid and more accurate and more sensitive. An example is the work on biosensors for real time detection of human pathogens. The application of these technologies has been primarily for monitoring of food production, processing and marketing systems. The age-old question, however, is "now that we have detected the pathogen, what are we going to do about it?" This leads us to the inescapable conclusion that we must apply the basic research approach to developing effective intervention strategies for the reduction or elimination of the subject pathogen. This requires that greater research commitment be made to gaining a better understanding of microbial physiology and ecology of human pathogens. We must make use of the knowledge we have concerning the niches and susceptibility of the organisms to novel interventions.

Recent Successes

The FSC scientists are rising to the challenge in many ways that offer encouraging results. The advancements are novel applications of basic tools that we have gleaned from earlier research. Competitive exclusion technologies have been attempted in several venues and using a novel screening method is demonstrating effectiveness in pathogen control that is economically feasible. The utilization of bacteriophages (viruses specific to a known pathogen) is being demonstrated as an effective pathogen control method. The use of various bacteriocins from sources like *Bacillus* and food fermentation bacteria, including nisin and pediocin, have shown excellent promise in controlling specific pathogens under test conditions.

FSC Future Directions

As the FSC continues to build on its proud tradition there are many new and exciting areas that show great promise in advancing improvements in food safety. The advancements in cell and molecular knowledge and innovative applications of these technologies are extremely encouraging and show considerable promise in resolving issues of long standing. The potential for novel intervention strategies and novel vaccine development are two of the primary areas in which advancement can be very rapid. The efforts of the USDA-FSIS to move to the next generation of improved meat and poultry inspection—risk-based inspection—will rely heavily on improved detection and monitoring systems, and food-processing establishments will rely on more sophisticated intervention techniques. Heightened awareness of food security issues will also require increased attention to strategies for assuring these concerns are addressed. The education and outreach responsibilities of the FSC mission will require consideration and adoption of distance education approaches.

Consortium Leadership

In addition to the coordinators and chairs listed previously, the Consortium has been the beneficiary of guidance provided by the other members of the steering committee. The committee consists of representatives of each member university, the stakeholder industries, and the USDA. Listed below are the names of the steering committee members, their affiliation, and their years of service:

- Gerald Musick, University of Arkansas, 1991–1993
- Dell Allen, Excel Corporation, 1991–1994
- Clark Burbee, USDA Cooperative State Research, Education and Extension Service, 1991–1995
- L. B. Daniels, University of Arkansas, 1991–1995
- Donald Derr, USDA Food Safety and Inspection Service, 1991–1994
- Thomas A. Fretz, Iowa State University, 1991–1994
- George Ham, Kansas State University, 1991–2000
- Stan Harris, National Veterinary Services Laboratory, 1991–1994
- Mike Telford, Iowa Pork Producers Association, 1991–1997
- James Whitmore, Tyson Foods, 1991–1992
- Ellis Brunton, Tyson Foods, 1992–2003
- Jack Riley, Kansas State University, 1993–2006
- James Denton, University of Arkansas, 1995–present

- Frank Flora, USDA Cooperative State Research, Education and Extension Service, 1995–1998
- Richard Forsythe, University of Arkansas, 1994–1996
- Brenda Halbrook, USDA Food Safety and Inspection Service, 1994–1997
- Jim Riemann, Excel Country Fresh Meats/Certified Angus Beef Program, 1994–present
- Colin Scanes, Iowa State University, 1994–2000
- Lee Ann Thomas, National Veterinary Services Lab, 1994–1995
- Charles Scifres, University of Arkansas, 1996–1999
- Irene Wesley, National Animal Disease Center, 1996–present
- Richard Ellis, USDA Food Safety and Inspection Service, 1997–2002
- Rich Degner, Iowa Pork Producers Association, 1997–present
- Gregory J. Weidemann, University of Arkansas, 1997–2006
- Craig Wilson, Costco Wholesale, 1998–present
- Marc Johnson, Kansas State University, 2000–2002
- Donald L. Reynolds, Iowa State University, 2000–present
- Robert O. (Neal) Apple, Tyson Foods, 2003–2004
- Fred Cholick, Kansas State University, 2003–present
- Lynda L. Kelley, USDA Food Safety and Inspection Service, 2003–present
- Sharon Beals, Tyson Foods, 2004–2007
- Mark J. Cochran, University of Arkansas, 2006–present
- Ed Moix, Tyson Foods, 2008–present

Summary

In summary, the FSC has a proud history of accomplishment and success stories relating to their contributions to improved food safety in the United States. For future success, the FSC must build on its strengths. In addition it must continue to provide leadership in accepting the challenges for improving food-safety systems. The scientists must continue to anticipate the needs for the food-safety system in the United States. Innovation in the application of new knowledge and new scientific information will be key to providing the necessary leadership. FSC scientists will be involved in cutting-edge research, which is, by its nature, high-risk research and not always successful in the beginning. However, the courage to accept this leadership role is the expectation that has been placed on the FSC and one that the FSC has accepted before. The future for the FSC is very bright indeed.

References

Food Safety Consortium: origins and development. Food Safety Consortium–10 years of service, 1988–1998. Food Safety Consortium, Fayetteville, AR.

■ ■ ■

Preharvest Foodborne Pathogen Ecology and Intervention Strategies

■ ■ ■

▌2▐

Novel Strategies for the Preharvest Control of *Campylobacter* in Poultry

Ixchel Reyes-Herrera, Ann M. Donoghue,
and Dan J. Donoghue

Introduction

Ingestion of food products contaminated with *Campylobacter jejuni* or *coli* is one of the main causes of foodborne infections. These pathogens are usually reported among the leading causes of laboratory-confirmed foodborne illnesses in the United States and in many other countries around the world (Coker et al. 2002; Moore et al. 2005; CDC 2006). The latest report from the CDC indicated that these organisms accounted for approximately 28% of the laboratory-confirmed food-borne-related illnesses in 2005 (CDC 2006). Epidemiological evidence indicates that a large number of human *Campylobacter* infections result from the improper preparation and consumption of contaminated poultry and poultry products (Jacobs-Reitsma 2000; Corry and Atabay 2001; Lee and Newell 2006). Despite extensive research in this area, reducing or preventing *Campylobacter* colonization in poultry and subsequent human exposure to contaminated poultry products remains a challenge. This chapter briefly reviews the importance of this pathogen for food safety and presents some of the work conducted in our laboratory to understand and reduce the incidence of these organisms in poultry.

Campylobacter

Organisms of the genus *Campylobacter* are small, gram-negative, spiral curved rods. *Campylobacter* spp. are microaerophilic and nutritionally fastidious organisms that grow at temperatures between 35–44°C, with an optimal growth temperature of

42°C (van Vliet and Ketley 2001). They are motile due to the presence of polar flagella at one or both ends of the cell. They can undergo a morphological change from a spiral to a coccoid form, as a defense mechanism during stressful environmental conditions or during prolonged culture. However, the true significance of this coccoid form is controversial (Ziprin and Harvey 2004). The mechanisms of *Campylobacter* pathogenesis are still poorly understood but they are thought to be mediated by a number of virulence factors including chemotaxis, motility, adhesion, invasion, and the production of toxins (van Vliet and Ketley 2001).

Campylobacteriosis, or human enteritis due to *Campylobacter* spp., is the most common illness associated with human *Campylobacter* infection (Coker et al. 2002; Moore et al. 2005). Ingestion of improperly prepared or cross-contaminated foodstuffs is thought to be one of the main routes of infection for humans. Infection can occur after ingestion of as little as 500 organisms (Black et al. 1988) and clinical symptoms appearing one to seven days later. The disease is characterized by acute abdominal pain, fever, malaise, vomiting (in roughly 15% of the afflicted patients), and diarrhea. The diarrhea is usually self limiting, lasting about five to eight days, and may occur as a mild watery form, most common in developing countries, to a more severe and bloody diarrhea, common in industrialized nations (Coker et al. 2002; Moore et al. 2005). Most victims of campylobacteriosis do not seek, and seldom require, professional medical care or hospitalization; therefore, it is probable that official reports underestimate the incidence of the disease. The infection can predispose to complications such as appendicitis, colitis, toxic megacolon, intestinal hemorrhage, and peri-rectal abscess (Skirrow and Blaser 2000). Aside from enteric symptoms caused by *Campylobacter,* exposure to the organism has been highly correlated to systemic sequelae such as the Guillain-Barré syndrome (Komagamine and Yuki 2006), Reiter's syndrome (Nachamkin 2002), myocarditis (Uzoigwe 2005), and even dental and skin lesions (Ihara et al. 2003).

Sources of Exposure to *Campylobacter*

The main sources of human infection are direct exposure to infected animals and ingestion of improperly prepared contaminated foods, raw milk, and water. Shedding of the bacteria through feces is a common pathway of the disease, therefore direct exposure to infected pets (Lee et al. 2004) or infected farm and wild animals (French et al. 2005) is a potential source of exposure to the pathogen. However, the ingestion of *Campylobacter*-contaminated foods has been associated with a majority of the reported cases of the disease (Jacobs-Reitsma 2000). Some food products that are commonly mentioned are raw vegetables, such as spinach, mushrooms, fenugreek (see, e.g., Kumar et al. 2001), unpasteurized raw milk (Moore et al. 2005), and improperly prepared meat products from cattle, shellfish, and poultry (Jacobs-Reitsma 1997; Moore et al. 2005). Waterborne outbreaks have been reported in Europe and in developing countries (Oberhelman and Taylor 2000; Clark 2003).

Campylobacter and Poultry

Prevalence of *Campylobacter* species in retail chicken meat as high as 75–100% have been documented in the United States (Stern et al. 1995, 2001; Son et al. 2007). Similar isolation rates from retail poultry products have been reported for Hungary, Wales, and the U.K. (Meldrum et al. 2005; Jozwiak et al. 2006). The high prevalence of the organism in commercial poultry may be explained by the fact that *Campylobacter* grows best at temperatures of 42°C, which is similar to the normal body temperature of chickens (Jacobs-Reitsma 1997; van Vliet and Ketley 2001). However, the actual prevalence in a given flock can be affected by different factors, such as the age of the birds, season of the year (Newell and Wagenaar 2000), type of production system (El-Shibiny et al. 2005), and even the combination of multiple strains of *Campylobacter* that can be isolated in the same flock (Newell and Wagenaar 2000).

Campylobacter species tend to be commensals in the avian host (Stern et al. 1995; Hendrixson and DiRita 2004; Lee and Newell 2006) and are considered a common resident in the gastrointestinal tract of several avian species, including intensively raised birds such as chickens, turkeys, ducks, quails, and ostriches (Risdale et al. 1998; van Vliet and Ketley 2001; Ley et al. 2001). In the case of chickens, *Campylobacter* spp. colonization occurs primarily in the lower intestines, particularly in the cecal and cloacal crypts (Meinersmann et al. 1991). It can also be recovered from the crop (Berrang et al. 2000), the gallbladder, and the liver (Fernandez and Pison 1996; Cox et al. 2005). The isolation rate in the gastrointestinal tract can be as high as 10^5–10^9 cfu (Corry and Atabay 2001). This organism can also be isolated from the reproductive tract of both males and females (Cole et al. 2004; Donoghue, Blore et al. 2004), the respiratory tract (Berrang et al. 2003), and several lymphoid organs (Cox et al. 2005). Thus, *Campylobacter* is found in high numbers in avian species and does not cause apparent adverse affects on their health, unlike the situation in humans (see previous section).

Sources of Contamination

Newell and Fearnley (2003), Jacobs-Reitsma (1997), and Jacobs-Reitsma and coworkers (2001) have comprehensively reviewed the potential routes of transmission and sources of infection. Multiple sources of possible horizontal contamination have been documented, including exposure to used litter (Montrose et al. 1985), contaminated water sources (Zimmer et al. 2003), insects (e.g., Skov et al.

2004; Nichols 2005), rodents (Annan-Prah and Janc 1988), wild birds (Newell and Fearnley 2003), other animals (Lee et al. 2004), and human activity (Stern et al. 2001). Yet, the importance of these sources is not completely understood. The presence of *Campylobacter* spp. in the reproductive tracts of both male and female poultry (Newell and Fearnley 2003; Cole et al. 2004; Donoghue, Blore et al. 2004) indicates the possibility of vertical transmission of the pathogen from breeder flocks to the commercial broiler flocks (Cole et al. 2004; Vizzier-Thaxton et al. 2006). However, Callicot and coworkers (2006) examined the possibility of vertical transmission in over 60,000 commercial breeder birds kept in quarantine conditions and found no evidence to support this mechanism of transmission. But, regardless of the actual source of infection, it has been reported that once some birds in the flock become positive, the whole flock will rapidly become colonized and the organism will persist in the flock until slaughter, leading to potential carcass contamination at the processing plant (Corry and Atabay 2001; Hargis et al. 2001).

Strategies to Reduce and/or Eliminate *Campylobacter* Colonization in Poultry

Numerous intervention strategies have focused on reducing environmental sources of the organism in the environment and modifying host-pathogen interactions. However, consistent intervention strategies remain a challenge (Newell and Wagenaar 2000; Mead 2000, 2002; Newell and Fearnley 2003; Donoghue, Hargis et al. 2004; Anderson et al. 2005; Stern et al. 2005; Lee and Newell 2006; Wagenaar et al. 2006). Many laboratories have focused on understanding how *Campylobacter* interacts with the intestine and its microflora and have worked toward developing microbial cultures antagonistic to *Campylobacter* (probiotics).

Use and Development of Probiotic Cultures

The indigenous microflora play an important role in the mechanisms of immune modulation in the gastrointestinal tract (Kelly et al. 2005; Ismail et al. 2005) The gastrointestinal system of the newly hatched neonate is sterile and highly susceptible to pathogen colonization whereas the mature bird can be resilient to colonization (Nurmi and Rantala 1973). Several studies have demonstrated that very few *Salmonella* cells can infect a newly hatched chick whereas the older bird has a mature microflora capable of resisting pathogens (see review, Barrow and Page 2000). The protective influence of maternal transfer of enteric microflora has a demonstrated benefit in many species, including humans. Unfortunately in many poultry operations, transfer of microflora from the hen to offspring no longer occurs because chicks are raised separately from parent flocks.

The concept of accelerating the development of normal enteric microflora, thereby increasing the resistance of young poultry to infection, was first described by Nurmi and Rantala (1973). These researchers collected microflora from mature chickens and inoculated newly hatched chicks, thereby significantly reducing *Salmonella* colonization. This strategy has been called "competitive exclusion," the "Nurmi effect," or "probiotic supplementation" and numerous studies have demonstrated reduction in *Salmonella* colonization in poultry using mixed undefined enteric cultures (Mead and Impey 1986; Stavric and D'Aoust 1993; Mead 2000, 2002; Nisbet 2002).

There are several proposed mechanisms of protection provided by the enteric microflora and by effective competitive exclusion cultures (Nurmi et al. 1992; Hollister et al. 1999), including competition for binding sites, competition for nutrients, production of antibacterial substances and immunostimulation. Currently, several probiotic preparations are successfully being used to reduce *Salmonella* infections in poultry production (Mead 2002; Andreatti Filho et al. 2003; Hargis et al. 2003). However, this approach is not as efficacious in controlling *Campylobacter* colonization in poultry (Mead 2000; Stern et al. 2001). To improve the reliability of these cultures against *Campylobacter*, researchers focused on the observation that the organism preferentially colonizes the intestinal crypts (Beery et al. 1988; Meinersmann et al. 1991). Schoeni and Doyle (1992) developed a defined probiotic culture isolated from the intestinal microflora of *Campylobacter*-free hens that reportedly produced anti-*Campylobacter* metabolites *in vitro*. Stern and coworkers (1994) reported the development of a probiotic culture obtained from deep tissue scrapings of intestinal mucosa, the same niche that *Campylobacter* occupies. Unfortunately, these cultures have shown inconsistent results against *Campylobacter*.

Our laboratory attempted to develop improved probiotic cultures using methods to prescreen individual bacterial isolates for *in vitro* efficacy against *Campylobacter*. Initially, hundreds of individual enteric isolates collected from healthy poultry were co-incubated with various concentrations of *Campylobacter* in 96 well plates (Donoghue, Hargis et al. 2004). With this method we were successfully able to detect a number of isolates with the ability to inhibit *Campylobacter, in vitro*. Although effective *in vitro*, most combinations when tested against young poultry were not consistently effective against *in vivo Campylobacter* colonization.

In follow-up experiments, isolates demonstrating *in vitro* efficacy were tested against *Campylobacter* using a soft-agar overlay technique, a modified procedure from Miyamoto and coworkers (2000). The isolates that successfully restricted the growth of *Campylobacter* in the area surrounding the isolate (zone of inhibition) were selected for an *in vivo* challenge (Figure 2.1).

Figure 2.1. Example of zone of inhibition produced by a cecal isolate able to inhibit the *in vitro* growth of *Campylobacter.* Cecal isolates were collected from young poults by diluting cecal contents, growing on standard MRS, and selecting individual colonies for testing. Ability to inhibit *Campylo - bacter* growth is evaluated using an agar overlay technique modified for *Campylo - bacter* from Miyamoto et al. 2000. Used with permission of *Poultry Science.*

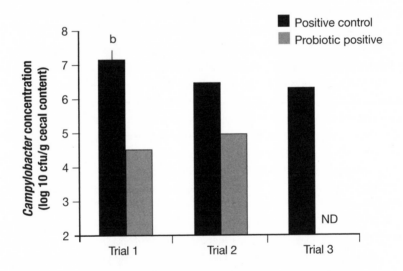

Figure 2.2. Mean effects of *Campylobacter* colonization in young poults after dosing with a probiotic isolate with previously detected anti-*Campylobacter* properties (Holliman et al. 2003). Different probiotic cultures were previously tested and selected based on their efficacy against *Campylobacter.* One of those cultures produced these results in an *in vivo* challenge. Young poults were orally challenged 3 days posthatch, with approximately 10^4 cfu of a mixture of wild-type *Campylobacter coli.* On day 10, poults were euthanized and cecal contents were aseptically collected and enumerated for *Campylobacter.* Each value represents at least 10 poults/treatment group.

These isolates were dosed separately or in combination, via oral gavage, to commercially obtained poults at day of hatch. At day 3 of age poults were challenged with a combination of different *Campylobacter jejuni* isolates. Cecal samples were collected at day 10 of age and results were compared with our positive controls. Of all the applied treatments, one particular isolate combination consistently produced significant reductions of *Campylobacter* colonization sites in three separate trials (Figure 2.2; Holliman et al. 2003).

Further work is needed to identify the specific concentrations of the probiotic that produces the highest reduction of the pathogen, as well as understanding the interaction between the different isolates and the native gastrointestinal microbiota. In addition, it is probable that a number of different factors in the gastrointestinal environment can interfere with the efficacy of probiotic cultures.

Bacteriocins

Through efforts to improve the effectiveness of probiotic cultures against *Campylobacter,* researchers have observed that certain bacteria produce metabolites that are inhibitory to *Campylobacter* growth *in vitro* (Svetoch et al. 2005). These metabolites, identified as bacteriocins, are proteins naturally produced by bacteria that kill or inhibit the growth of other bacteria, particularly gram-negatives (Cleveland et al. 2001). Unlike antibiotics, bacteriocins have no known toxic effects and have a narrow killing spectrum (Riley and Wertz 2002). Bacteriocins have been demonstrated to inhibit or kill other foodborne pathogens, such as *Listeria, Clostridium,* and *Salmonella,* and are used in food processing and preservation. For example, the bacteriocin nisin is considered a generally recognized safe compound and is approved for use in foods (Natrajan and Sheldon 2000). Bacteriocin-like compounds have also been shown to have direct antimicrobial activity, *in vitro,* against *Campylobacter* (Chaveerach et al. 2004). It was recently reported that purified bacteriocins produced by certain strains of *Lactobacillus salivarius (B-OR7)* and *Paenibacillus polymyxa (B-602)* were inhibitory to *Campylobacter jejuni* growth *in vitro* and had efficacy in chickens after challenge (Svetoch et al. 2005; Stern et al. 2005). As a follow-up study we demonstrated that these bacteriocins reduced *Campylobacter coli* cecal colonization to undetectable levels in turkeys in three separate trials (Cole et al. 2006) whereas approximately 10^6 cfu/gram cecal contents of *Campylobacter* was detected in positive control birds (Table 2.1). We also observed that administration of the bacteriocins significantly reduced crypt depth (Figure 2.3) and goblet cell density (Figure 2.4) in the duodenum of turkey poults (Cole et al. 2006). As *Campylobacter* preferentially colonizes mucin in the cecal crypts of poultry (Beery et al. 1988; Meinersmann et al. 1991), this may provide clues as to how these peptides alter the preferential colonization sites for *Campylobacter.* This result suggests that in addition to the direct bactericidal or bacterio-

Table 2.1. Reduction of Cecal Campylobacter Concentrations and Incidence in Commercial Turkey Poults Treated with Bacteriocins[1]

Trial	Positive control B602	Bacteriocin OR7	Bacteriocin
1	$2.6 \times 10^{6,a}$ (10/10)	ND[b] (0/10)	ND[b] (0/10)
2	$3.6 \times 10^{5,a}$ (10/10)	ND[b] (0/10)	ND[b] (0/10)
3	$3.4 \times 10^{5,a}$ (10/10)	ND[b] (0/10)	ND[b] (0/10)

[a,b] Means within rows with no common superscript differ significantly ($P < 0.01$)

[1] Data represent log-10 colony-forming units of *Campylobacter* per gram of cecal contents collected from 3 separate trials (n = 10 poults/treatment per trial; total 30 poults/trial; Cole et al. 2006). Incidence of *Campylobacter* is represented as the number of positive ceca out of 10 birds. In each trial, poults were orally challenged 3 days posthatch with approximately 10^6 cfu of a mixture of 3 *Campylobacter coli* isolates. On day 10 to 12 posthatch, the 2 treatment groups were fed a diet containing purified bacteriocins at a dose of 250 mg/kg of feed, and the positive control group was fed the same commercial diet without bacteriocins. After 72 hours of treatment with bacteriocins, turkeys were euthanized and ceca were collected for enumeration of *Campylobacter*. ND = the concentration of bacteria was below detectable levels ($<10^2$ cfu/g of cecal contents).

static activity, the oral treatment of bacteriocins in poultry altered the physical or functional characteristics of the preferential sites of colonization of *Campylobacter*. The reduction in crypt size could expose *Campylobacter* to different nutrient or chemical environments (e.g., increased oxygen tension) and consequently limit its growth and colonization. It is also possible that different microbiota will colonize these smaller crypts, with the ability to outcompete *Campylobacter*. In addition, the reduction in goblet cell numbers and subsequent mucin production (Geyra et al. 2001) may further limit *Campylobacter* colonization, as it has been demonstrated that *Campylobacter* can use mucin as a nutrient source for growth (Fernandez et al. 2000). To our knowledge, this is the first study showing that an alteration in the gastrointestinal (GI) tract may be associated with the elimination of *Campylobacter* colonization.

Although the administration of bacteriocins eliminated detectable enteric *Campylobacter* in these studies, it is still possible *Campylobacter* may reside in the GI tract at very low levels. Previous research from our laboratory has demonstrated that even if *Campylobacter* is eliminated from most, but not all enteric locations, the remaining enteric *Campylobacter* can recolonize the gut within a few

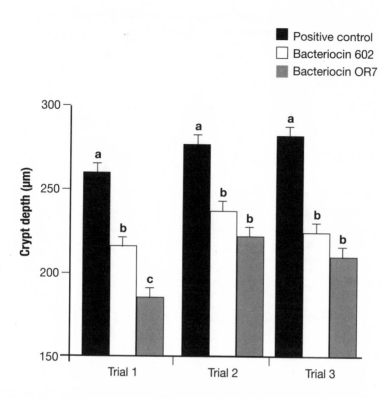

Figure 2.3. Effect of bacteriocins on duodenal crypt depth in turkey poults after oral challenge with *Campylobacter*. Values are means ± SEM, representing 10 birds/group and 10 measurements/parameter per bird from 3 separate trials (Cole et al. 2006). In each trial, poults were orally challenged 3 days posthatch with approximately 10^6 cfu of a mixture of 3 *Campylobacter coli* isolates. On day 10 to 12 posthatch, the 2 treatment groups were fed a commercial diet containing purified bacteriocins at a dose of 250 mg/kg of feed, and the positive control group was fed the same commercial diet without bacteriocins. On day 13 posthatch (10 day postchallenge), turkeys were euthanized and duodenal loops were collected for morphometric analysis. Means with no common superscripts differ significantly ($P < 0.05$) between treatments within trials. Used with permission of *Poultry Science*.

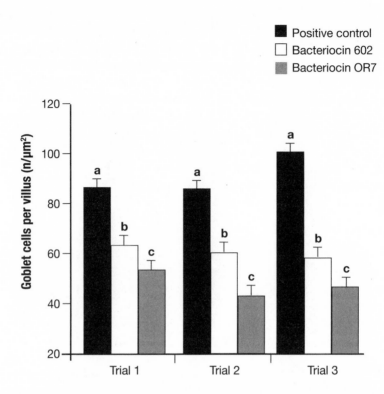

Figure 2.4. Effect of bacteriocins on duodenal goblet cell density in turkey poults after oral challenge with *Campylobacter.* Values are mean ± SEM, representing 10 birds/treatment group and 10 measurements/ parameter per bird from 3 separate trials (Cole et al. 2006). In each trial, poults were orally challenged 3 d posthatch with approximately 106 cfu of a mixture of 3 *Campylobacter coli* isolates. On d 10 to 12 posthatch, the 2 treatment groups were fed a commercial diet containing purified bacteriocins at a dose of 250 mg/kg of feed, and the positive control group was fed the same commercial diet without bacteriocins. On d 13 posthatch (10 d postchallenge), turkeys were euthanized and duodenal loops were collected for morphometric analysis. Means with no common superscript differ significantly ($P < 0.05$) between treatments within trials.

days following antibiotic treatment (Farnell et al. 2005). Therefore, bacteriocins should be dosed just before marketing to reduce any potential recolonization of the gastrointestinal tract by this organism. Furthermore, even if bacteriocin treatment did not totally eliminate *Campylobacter*, the approximately 4-log reduction in *Campylobacter* concentrations obtained in these studies would provide a significant benefit to human food safety. Research by Rosenquist and coworkers (2003) reported that even a 2-log reduction in carcass contamination would reduce the human incidence of campylobacteriosis in humans by 30-fold. Although the use of bacteriocins to reduce the incidence of *Campylobacter* appears promising, regulatory approval will be required before these compounds can be used by the poultry industry. Approval for these compounds is currently being sought from the Food and Drug Administration (FDA).

Summary

Campylobacter remains an important challenge for controlling foodborne illness associated with poultry products. There is extensive literature on the prevalence, source of contamination, and colonization of *Campylobacter* in poultry. Determining how *Campylobacter* coexists in the poultry host is providing keys toward developing methods to reduce or eliminate this elusive organism prior to slaughter.

References

Anderson, R. C., R. B. Harvey, J. A. Byrd, T. R. Callaway, K. J. Genovese, T. S. Edrington, Y. S. Jung, J. L. McReynolds, and D. J. Nisbet. 2005. Novel preharvest strategies involving the use of experimental chlorate preparations and nitro-based compounds to prevent colonization of food-producing animals by foodborne pathogens. Poult. Sci. 84:649–654.

Andreatti Filho, R. L., H. Marcos Sampaio, M. Rodriques Barros, and P. Roberto Gratao. 2003. Use of cecal microbiota cultured under aerobic or anaerobic conditions in the control of experimental infection of chicks with *Salmonella enteritidis*. Vet. Microbiol. 92(3):237–244.

Annan-Prah, A., and M. Janc. 1988. The mode of spread of *Campylobacter jejuni/coli* to broiler flocks. J. Vet. Med. B. 35:11–18.

Barrow, P. A., and K. Page. 2000. Inhibition of colonisation of the alimentary tract in young chickens with *Campylobacter jejuni* by pre-colonisation with strains of *C. jejuni*. FEMS Microbiol. Lett. 182:87–91.

Beery, J. T., M. B. Hugdahl, and M. P. Doyle. 1988. Colonization of gastrointestinal tracts of chicks by *Campylobacter jejuni*. Appl. Environ. Microbiol. 54:2365–2370.

Berrang, M. E., R. J. Buhr, and J. A. Cason. 2000. *Campylobacter* recovery from external and internal organs of commercial broiler carcass prior to scalding. Poult. Sci. 79:286–290.

Berrang, M. E., R. J. Meinersmann, R. J. Buhr, N. A. Reimer, R. W. Phillips, and M. A. Harrison. 2003. Presence of *Campylobacter* in the respiratory tract of broiler carcasses before and after commercial scalding. Poult. Sci. 82:1995–1999.

Black, R. E., M. M. Levine, M. L. Clements, T. P. Hughes, and M. J. Blaser. 1988. Experimental *Campylobacter jejuni* infection in humans. J. Infect. Dis. 157:472–479.

Callicot, K. A., V. Friethriksdottir, J. Reiersen, R. Lowman, J. R. Bisaillon, E. Gunnarsson, E. Berndtson, K. L. Hiett, D. S. Needleman, and N. J. Stern. 2006. Lack of evidence for vertical transmission of *Campylobacter* spp. in chickens. Appl. Environ. Microbiol. 72(9):5794–5798.

Centers for Disease Control and Prevention (CDC). 2006. Preliminary FoodNet data on the incidence of infection with pathogens transmitted through food—10 states, United States, 2005. Morb. Mortal. Wkly. Rep. 14(55):392–395.

Chaveerach, P., L. J. Lipman, and F. van Knapen. 2004. Antagonistic activities of several bacteria on *in vitro* growth of 10 strains of *Campylobacter jejuni/coli*. Int. J. Food Microbiol. 90:43–50.

Clark, C. G. 2003. Characterization of waterborne outbreak-associated *Campylobacter jejuni*, Walkerton, Ontario. Emerg. Infect. Dis. 9(10):1232–1241.

Cleveland, J., T. J. Montville, I. F. Nes, and M. L. Chikindas. 2001. Bacteriocins: safe, natural antimicrobials for food preservation. Int. J. Food Microbiol. 71:1–20.

Coker, A. O., R. D. Isokpehi, B. N. Thomas, K. O. Amisu, and C. L. Obi. 2002. Human campylobacteriosis in developing countries. Emerg. Infect. Dis. 8:237–243.

Cole, K., A. M. Donoghue, P. J. Blore, and D. J. Donoghue. 2004. Isolation and prevalence of *Campylobacter* in the reproductive tracts and semen of commercial turkeys. Avian Dis. 48:625–630.

Cole, K., M. B., Farnell, A. M., Donoghue, N. J. Stern, E. A. Svetoch, B. N., Eruslanov, L. I. Volodina, Y. N. Kovalev, V. V. Perelygin, E. V. Mitsevich, I. P. Mitsevich, V. P. Levchuk, V. D. Pokhilenko, V. N. Borzenkov, O. E. Svetoch, T. Y. Kudryavtseva, I. Reyes-Herrera, P. J. Blore, F. Solis de los Santos, and D. J. Donoghue. 2006. Bacteriocins reduce *Campylobacter* colonization and alter gut morphology in turkey poults. Poult. Sci. 85:1570–1575.

Corry, J. E. L., and H. I. Atabay. 2001. Poultry as a source of *Campylobacter* and related organisms. J. Appl. Microbiol. 90:96S–114S.

Cox, N. A., C. L. Hofacre, J. S. Bailey, R. J. Buhr, J. L. Wilson, K. L. Hiett, L. J. Richardson, M. T. Musgrove, D. E. Cosby, J. D. Tankson, Y. L. Vizzier, P. F. Cray, L. E. Vaughn, P. S. Holt, and D. V. Bourassaa. 2005. Presence of *Campylobacter jejuni* in various organs one hour, one day, and one week following oral or intracloacal inoculation of broiler chicks. Avian Dis. 49:155–158.

Donoghue, A. M., P. J. Blore, K. Cole, N. M. Loskutoff, and D. J. Donoghue. 2004. Detection of *Campylobacter* or *Salmonella* in turkey semen and the ability of poultry semen extenders to reduce their concentrations. Poult. Sci. 83:1728–1733.

Donoghue, D. J., B. M. Hargis, G. Tellez, and A. M. Donoghue. 2004. Competitive exclusion as a means of controlling *Campylobacter* in poultry. 5th Asia-Pacific Poultry Health Conference. April. Australia.

El-Shibiny, A., P. L. Connerton, and I. F. Connerton. 2005. Enumeration and diversity of *Campylobacters* and bacteriophages isolated during the rearing cycles of free-range and organic chickens. Appl. Environ. Microbial. 71:1259–1266.

Farnell, M. B., A. M. Donoghue, K. Cole, I. Reyes-Herrera, P. J. Blore, and D. J. Donoghue. 2005. *Campylobacter* susceptibility to ciprofloxacin and corresponding fluoroquinolone concentrations within the gastrointestinal tracts of chickens. J. Appl. Microbiol. 99:1043–1050.

Fernandez, F., R. Sharma, M. Hinton, and M. R. Bedford. 2000. Diet influences the colonization of *Campylobacter jejuni* and distribution of mucin carbohydrates in the chick intestinal tract. Cell. Mol. Life Sci. 57:1793–1801.

Fernandez, H., and V. Pison. 1996. Isolation of thermotolerant species of *Campylobacter* from commercial chicken livers. Int. J. Food Microbiol. 29:75–80.

French, N., M. Barrigas, P. Brown, P. Riviero, N. M. Williams, H. Leatherbarrow, R. Birtles, P. Fearnhead, and A. Fox. 2005. Spatial epidemiology and natural population structure of *Campylobacter jejuni* colonizing a farmland ecosystem. Environ. Microbiol. 7(8):1116–1126.

Geyra, A., Z. Uni, and D. Sklan. 2001. Enterocyte dynamics and mucosal development in the posthatch chick. Poult. Sci. 80:776–782.

Hargis, B. M., D. J. Caldwell, and J. A. Byrd. 2001. Microbiological pathogens: live poultry considerations. Pages 121–153 in Poultry meat processing. Casey M. Owens, Christine Alvarado, and Alan R. Sams (eds.) CRC Press.

Hargis, B. M., G. I. Tellez, G. Nava, A. M. Donoghue, J. L. Vicente, S. E. Higgins, D. J. Donoghue, and A. D. Wolfenden. 2003. Pages 109–118 in The role of beneficial micróflora in controlling enteric bacterial diseases: probiotics, prebiotics, and competitive exclusion. Cornell Nutrition Conference for Feed Manufactures.

Hendrixson, D. R., and V. J. DiRita. 2004. Identification of *Campylobacter jejuni* genes involved in commensal colonization in the chick gastrointestinal tract. Mol. Microbiol. 52(2):471–484.

Holliman, J., G. Nava, J. Vicente, L. Bielke, K. Cole, P. Blore, A. Donoghue, J. A. Byrd, B. Hargis, and G. Tellez. 2003. Competitive exclusion cultures consisting of lactic acid bacteria plus organic acid treatment may reduce *Campylobacter* colonization in turkeys. Proceedings of the 12th International Workshop on *Campylobacter, Helicobacter* and Related Organisms. Denmark, September 6–10, 2003.

Hollister, A. G., D. E. Corrier, D. J. Nisbet, and J. R. DeLoach. 1999. Effects of chicken-derived cecal microorganisms maintained in continuous culture on cecal colonization by *Salmonella typhimurium* in turkey poults. Poult. Sci. 78(4):546–549.

Ihara, H., T. Miura, T. Kato, K. Ishihara, T. Nakagawa, S. Yamada, and K. Okuda. 2003. Detection of *Campylobacter rectus* in periodontitis sites by monoclonal antibodies. J. Periodontal Res. 38:64–72.

Ismail, A. S., and L. V. Hooper. 2005. Epithelial cells and their neighbors. IV. Bacterial contributions to intestinal epithelial barrier integrity. Am. J. Physiol. Gastrointest. Liver Physiol. 289:G779–G784.

Jacobs-Reitsma, W. F. 1997. Aspects of epidemiology of *Campylobacter* in poultry. Vet. Q. 19(3):113–117.

Jacobs-Reitsma, W. 2000. *Campylobacter* in the food supply. Pages 467–481 in *Campylobacter.* I. Nachamkin and M. J. Blaser, ed. ASM Press. Washington, DC.

Jacobs-Reitsma, W., C. Becht, T. De Vries, J. Van der Plas, B. Duim, and J. Wagenaar. 2001. No evidence for vertical transmission of *Campylobacter* in a study on Dutch breeder and broiler farms. Int. J. Med. Microbiol. 291:39.

Jozwiak, A., O. Reichart, and P. Laczay 2006. The occurrence of *Campylobacter* species in Hungarian broiler chickens from farm to slaughter. J. Vet. Med. B Infec. Dis. Vet. Public Health 53(6):291–294.

Kelly, D., S. Conway, and R. Aminov. 2005. Commensal gut bacteria: mechanisms of immune modulation. Trends in Immunol. 26(6):326–333.

Komagamine, T., and N. Yuki. 2006. Ganglioside mimicry as a cause of Guillain-Barré syndrome. CNS. Neurol. Disord. Drug Targets 5(4):391–400.

Kumar, A., R. K. Agarwal, K. N. Bhilegaonkar, B. R. Shome, and V. N. Bachhil. 2001. Occurrence of *Campylobacter jejuni* in vegetables. Int. J. Food Microbiol. 67:153–155.

Lee, M. D., and D. G. Newell. 2006. *Campylobacter* in poultry: Filling an ecological niche. Avian Dis. 50:1–9.

Lee, M. K., S. J. Billington, and L. A. Joens. 2004. Potential virulence and antimicrobial susceptibility of *Campylobacter jejuni* isolates from food and companion animals. Foodborne Pathog. Dis. 1:223–230.

Ley, E. C., T. Y. Morishita, T. Brisker, and B. S. Harr. 2001. Prevalence of *Salmonella, Campylobacter* and *Escherichia coli* on ostrich carcasses and the susceptibility of ostrich-origin *E. coli* isolates to various antibiotics. Avian Dis. 45(3):696–700.

Mead, G. C. 2000. Prospects for "competitive exclusion" treatment to control *salmonellas* and other foodborne pathogens in poultry. Vet. J. 159:111–123.

Mead, G. C. 2002. Factors affecting intestinal colonisation of poultry by *Campylobacter* and role of microflora in control. World's Poult. Sci. J. 58:169–178.

Mead, G. C., and C. S. Impey. 1986. Current progress in reducing *Salmonella* colonization of poultry by "competitive exclusion." Soc. Appl. Bacteriol. Symp. Ser. 15:67S–75S.

Meinersmann, R. J., W. E. Rigsby, N. J. Stern, L. C. Kelley, J. E. Hill, and M. P. Doyle, 1991. Comparative study of colonizing and noncolonizing *Campylobacter jejuni*. Am. J. Vet. Res. 52:1518–1522.

Meldrum, R. J., I. D. Tucker, R. M. Smith, and C. Edwards. 2005. Survey of *Salmonella* and *Campylobacter* contamination of whole, raw poultry on retail sale in Wales in 2003. J. Food Prot. 68:1447–1449.

Miyamoto, T., T. Horie, T. Fujiwara, T. Fukata, K. Sasai, and E. Baba. 2000. *Lactobacillus* flora in the cloaca and vagina of hens and its inhibitory activity against *Salmonella enteritidis* in vitro. Poult. Sci. 79:7–11.

Montrose, M. S., S. M. Shane, and K. S. Harrington. 1985. Role of litter in the transmission of *Campylobacter jejuni*. Avian Dis. 29:392–399.

Moore, J. E., D. Corcoran, J. S. G. Dooley, S. Fanning, B. Lucey, M. Matsuda, D. A. McDowell, F. Megraud, B. C. Millar, R. O'Mahony, L. O'Riordan, M. O'Rourke, J. R. Rao, P. J. Rooney, A. Sails, and P. Whyte. 2005. *Campylobacter*. Vet. Res. 36:351–382.

Nachamkin, I. 2002. Chronic effects of *Campylobacter* infection. Microbes Infect. 4:399–403.

Natrajan, N., and B. W. Sheldon. 2000. Inhibition of *Salmonella* on poultry skin using protein- and polysaccharidae-based films containing a nisin formulation. J. Food Prot. 63:1268–1272.

Newell, D. G., and C. Fearnley. 2003. Sources of *Campylobacter* colonization in broiler chickens. Appl. Environ. Microbiol. 69:4343–4351.

Newell, D. G., and J. A. Wagenaar. 2000. Poultry infections and their control at the farm level. Pages 497–509 in *Campylobacter*. I. Nachamkin and M. J. Blaser, eds. 2nd ed. ASM Press, Washington, DC.

Nichols, G. L. 2005. Fly transmission of *Campylobacter*. Emerg. Infect. Dis. 11:361–364.

Nisbet, D. J. 2002. Defined competitive exclusion cultures in the prevention of entero-pathogen colonisation in poultry and swine. Antonie van Leeuwenhoek 81:481–486.

Nurmi, E., and M. Rantala. 1973. New aspects of *Salmonella* infection in broiler production. Nature 241:210–211.

Nurmi, E., L. Nuotio, and C. Schneitz. 1992. The competitive exclusion concept: development and future. Int. J. Food Microbiol. 15(3–4):237–240.

Oberhelman, R. A., and D. N. Taylor. *Campylobacter* infections in developing countries. Pages 139–154 in *Campylobacter*. I. Nachamkin and M. J. Blaser, eds. 2nd ed.

Riley, M., and J. E. Wertz. 2002. Bacteriocins: evolution, ecology, and application. Annu. Rev. Microbiol. 56:117–137.

Risdale, J. A., H. I. Atabay, and J. E. L. Corry. 1998. Prevalence of *Campylobacters* and *Arcobacters* in ducks at the abbatoir. J. Appl. Microbiol. 85:567–573.

Rosenquist, H., N. L. Nielsen, H. M. Sommer, B. Norrung, and B. B. Christensen. 2003. Quantitative assessment of human campylobacteriosis associated with thermophilic *Campylobacter* species in chickens. Int. J. Food Microbiol. 83:87–103.

Schoeni, J. L., and M. P. Doyle. 1992. Reduction of *Campylobacter jejuni* colonization of chicks by cecum-colonizing bacteria producing anti-*C. jejuni* metabolites. Appl. Environ. Micro. 58:667–670.

Skirrow, M. B., and M. J. Blaser. 2000. Clinical aspects of *Campylobacter* infection. Pages 69–88 in *Campylobacter*. 2nd ed. I. Nachamkin and M. J. Blaser, eds. ASM Press, Washington, DC.

Skov, M. N., A. G. Spencer, B. Hald, L. Petersen, B. Nauerby, B. Carstensen, and M. Madsen. 2004. The role of litter beetles as potential reservoir for *Salmonella enterica* and thermophilic *Campylobacter* spp. between broiler flocks. Avian Dis. 48:9–18.

Son, I., M. D. Englen, M. E. Berrang, P. J. Fedorka-Cray, and M. A. Harrison. 2007. Prevalence of *Arcobacter* and *Campylobacter* on broiler carcasses during processing. Int. J. Food Microbiol. 113(1):16–22.

Stavric, S., and J. Y. D'Aoust. 1993. Undefined and defined bacterial preparations for the competitive exclusion of *Salmonella* in poultry—a review. J. Food Prot. 56:173–180.

Stern, N. J., D. M. Jones, I. V. Wesley, and D. M. Rollins. 1994. Colonization of chicks by non-culturable *Campylobacter* spp. Lett. Appl. Microbiol. 18:333–336.

Stern, N. J., M. R. Clavero, J. S. Bailey, N. A. Cox, and M. C. Robach. 1995. *Campylobacter* spp. in broilers on the farm and after transport. Poult. Sci. 74:937–941.

Stern, N. J., P. Fedorka-Cray, J. S. Bailey, N. A. Cox, S. E. Craven, K. L. Hiett, M. T. Musgrove, S. Ladely, D. Cosby, and G. C. Mead. 2001. Distribution of *Campylobacter* spp. in selected U.S. poultry production and processing operations. J. Food Prot. 64:1705–1710.

Stern, N. J., E. A. Svetoch, B. V. Eruslanov, Y. N. Kovalev, L. I. Volodina, V. V. Perelygin, E. V. Mitsevich, I. P. Mitsevich, and V. P. Levchuk. 2005. *Paenibacillus polyxma* purified bacteriocin to control *Campylobacter jejuni* in chickens. J. Food Prot. 68:1450–1453.

Svetoch, E. A., N. J. Stern, B. V. Eruslanov, Y. N. Kovalev, L. I. Volodina, V. V. Perelygin, E. V. Mitsevich, I. P. Mitsevich, V. D. Pokhilenko, V. N. Borzenkov, V. P. Levchuk,

O. E. Svetoch, and T. Y. Kudriavtseva. 2005. Isolation of *Bacillus circulans* and *Paenibacillus polymyxa* strains inhibitory to *Campylobacter jejuni* and characterization of associated bacteriocins. J. Food Prot. 68:11–17.

Uzoigwe, C. 2005. *Campylobacter* infections of the pericardium and myocardium. Clin. Microbiol. Infect. 11(4):253–255.

van Vliet, A. H. M., and J. M. Ketley. 2001. Pathogenesis of enteric *Campylobacter* infection. J. Appl. Microbiol. 90:45S–56S.

Vizzier-Thaxton, Y., N. A. Cox, L. J. Richardson, R. J. Buhr, C. D. McDaniel, D. E. Cosby, J. L. Wilson, D. V. Bourassa, and M. B. Ard. 2006. Apparent attachment of *Campylobacter* and *Salmonella* to broiler breeder rooster spermatozoa. Poult. Sci. 85(4):619–624.

Wagenaar, J. A., D. J. Mevius, and A. H. Havelaar. 2006. *Campylobacter* in primary animal production and control strategies to reduce the burden of human campylobacteriosis. Rev. Sci. Tech. 25(2):581–594.

Zimmer, M., H. Barnhart, U. Idris, and M. D. Lee. 2003. Detection of *Campylobacter jejuni* strains in the water lines of a commercial broiler house and their relationship to the strains that colonized the chickens. Avian Dis. 47:101–107.

Ziprin, R. L., and R. B. Harvey. 2004. Inability of cecal microflora to promote reversion of viable nonculturable *Campylobacter jejuni*. Avian Dis. 48:647–650.

3

Colonization and Transmission of *Escherichia coli* O157:H7 in Swine

Nancy A. Cornick, Dianna M. Jordan,
Sheridan L. Booher, and Harley W. Moon

Introduction

Escherichia coli O157:H7 and other serogroups of Shiga toxin-producing *E. coli* (STEC) have emerged over the last several decades as a significant cause of food-borne illness in the United States. Approximately 5–10% of people clinically infected by these bacteria develop a systemic disease, hemolytic uremic syndrome, which has a fatality rate of approximately 5%. The Centers for Disease Control estimates that STEC cause some 110,000 illnesses and 90 deaths annually in the United States (Mead et al. 1999). In addition, the economic consequences of recalling large lots of food for public health reasons are significant. Cattle are considered to be the primary reservoir for STEC. Depending on the season, the methods used for bacterial culture and the age of the animals, the prevalence of *E. coli* O157:H7 in U.S. cattle ranges from 2–28% (Hancock et al. 1994; Elder et al. 2000). *E. coli* O157:H7 has also been recovered from other ruminants such as sheep (Kudva et al. 1996) and deer (Keene et al. 1997; Sargeant et al. 1999).

Epidemiology of *E. coli* O157:H7 in Swine

In contrast to ruminants, STEC are only occasionally recovered from nonruminant animals such as dogs, birds, and raccoons (Beutin et al. 1993; Wallace et al. 1997; Hancock et al. 1998). A 1995 USDA swine survey involving 4,200 head reported the prevalence of *E. coli* O157:H7 was <0.07% (Bush 1997) and the 2000 USDA survey did not recover any *E. coli* O157:H7 from 2,526 animals (APHIS 2001). However, *E. coli* O157:H7 has been recovered from the colon contents of

6/305 (2%) pigs at a U.S. slaughter facility (Feder et al. 2003) and from 13/1,102 swine fecal samples collected at agricultural fairs (Keen et al. 2006). The organism has also been recovered from healthy swine in Japan (3/221), The Netherlands (1/145), Norway (2/1,976), Chile (13/120), and Canada (40/660) (Heuvelink et al. 1999; Nakazawa et al. 1999; Johnsen et al. 2001; Gyles et al. 2002; Borie et al. 1997). During the summer of 2006, spinach potentially contaminated by both feral swine and cattle manure caused a large outbreak of human illness in the United States (Jay et al. 2007). In addition, a small family cluster of *E. coli* O157:H7 infections was traced back to dry pork salami (Conedera et al. 2007).

Experimental Infections

Experimentally, we have shown that *E. coli* O157:H7 can establish and maintain a population in the intestinal tract of some market-weight pigs for at least two months (Booher et al. 2002). In that experiment two different strains of *E. coli* O157:H7 were included in a cocktail inoculum along with three other pathogenic

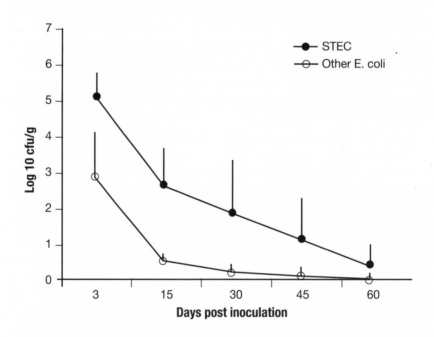

Figure 3.1. Mean fecal shedding of *E. coli* by pigs inoculated with 5 strain cocktail. The inoculum consisted of 2 *E. coli* O157:H7 strains (STEC), 2 enterotoxigenic *E. coli* strains and an enteropathogenic *E. coli* strain.

E. coli strains (two enterotoxigenic porcine pathogens and one enteropathogenic human pathogen). All pigs were fed a commercially available antibiotic-free diet for two weeks prior to inoculation. The geometric mean fecal shedding of the *E. coli* O157:H7 strains was of a higher magnitude and occurred over a longer time period than shedding of the other *E. coli* strains (Figure 3.1). At necropsy, *E. coli* O157:H7 was recovered primarily from tissues of the lower gastrointestinal tract, as it is from experimentally inoculated ruminants (Cornick et al. 2000; Booher et al. 2002; Cornick and Helgerson 2004). In addition, significant numbers of *E. coli* O157:H7 may be recovered from the tonsils of some pigs (Booher et al. 2002; Cornick and Helgerson 2004). This is in contrast to sheep in which the organism is rarely, if ever recovered from the tonsils. Taken together these studies demonstrate that the magnitude and persistence of fecal shedding of *E. coli* O157:H7 by experimentally inoculated swine is similar to that which occurs in ruminants (Figure 3.2) and suggests that there is not an absolute biological barrier to the colonization of swine by *E. coli* O157:H7.

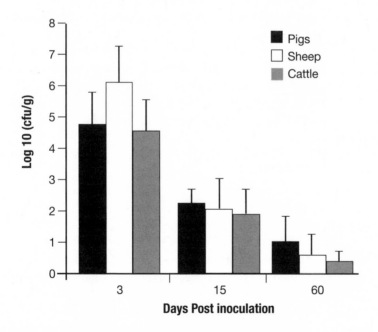

Figure 3.2. Mean fecal shedding of *E. coli* O157:H7 by pigs (n = 15), sheep (n = 39), and cattle (n = 8) inoculated with 10^{10} cfu.

Colonization Mechanisms

The mechanisms utilized by *E. coli* O157:H7 to colonize swine and ruminants are not completely understood. However, some of our work with isogenic mutants suggests that there are differences in how the organism interacts with the gastrointestinal epithelium in these two animal species. Intimin, an outer membrane protein that is required for colonization in some animal models of pathogenesis, and the intimin receptor, Tir, have been shown to be important in the colonization of both sheep and cattle by *E. coli* O157:H7 (Dean-Nystrom et al. 1998; Cornick et al. 2002; Woodward et al. 2003; Sheng et al. 2006; Vlisidou et al. 2006). However, when 12-week old pigs were dually inoculated with the wild-type strain and an isogenic Δ*eae* (intimin) mutant, similar numbers of both strains were recovered from fecal samples and from tissues throughout the alimentary tract for up to five weeks post inoculation (Jordan et al. 2005). This was confirmed by the work of Best et al. (2006) using a Shiga toxin negative parent strain and an isogenic intimin mutant. In addition, their work demonstrated that an aflagellar mutant also colonized pigs to a similar extent as the wild-type parent. The *E. coli* O157:H7 genome contains two homologous operons of long polar fimbriae (*lpf*), an important colonization factor of *Salmonella enterica* serovar Typhimurium. When a *lpf* double mutant was inoculated into pigs it was recovered in lower numbers than the isogenic parent, but was recovered intermittently for two months from some animals (Jordan et al. 2004). In contrast, mutations in the *lpf* operons did not appear to have a significant effect on the magnitude and persistence of *E. coli* O157:H7 in sheep. It is likely that *E. coli* O157:H7 contains redundant adherence mechanisms and the influence of any one factor may depend on the host species.

Transmission among Swine

Experimentally infected swine can transmit *E. coli* O157:H7 to naïve pigs when the animals are housed together in close contact and share water and food sources (Cornick and Helgerson 2004). Donor animals shedding $<10^4$ cfu/g of *E. coli* O157:H7 transmitted the organism to >50% of the exposed pigs housed in close contact (Table 3.1). In some cases the donor was shedding $<10^2$ cfu/g at the time the naïve pigs were exposed. On farms where both ruminants and swine are raised in close contact, transmission of *E. coli* O157:H7 between these animal species has been documented (Erikkson et al. 2003). Potential routes of transmission in this study were traced back to animal management practices on these farms. Transmission of STEC experimentally has been reported in both calves (Besser et al. 2001; Cobbold and Desmarchelier 2002) and sheep (Kudva et al. 1997; Cornick et al. 2000). When the naïve animals shared a pen with the inoculated donor, the

Table 3.1. Transmission of E. coli O157:H7 from Inoculated Donor Pigs to Naïve Pen Mates

Time post exposure	Number of pigs shedding/number exposed	
	≤10[4a]	≥10[5]
3 days	10/17	5/5
2 weeks	12/17	5/5

[a]cfu/g of *E. coli* O157:H7 shed by the donor at the time it was moved in with the naïve pigs

transmission of STEC between calves was more efficient than it was when calves were confined in individual pens (Cobbold and Desmarchelier 2002). However, transmission has been documented between calves that did not have nose-to-nose contact with the inoculated donor (Besser et al. 2001). Horizontal transmission between individuals within a herd is likely to be an important component in the establishment and maintenance of an animal reservoir.

While much is known regarding the pathogenesis and virulence of STEC in human disease, much less is known about when and how animals become colonized by STEC. Understanding the factors that contribute to STEC colonization of animals and resolving the discrepancy between experimental *E. coli* O157:H7 infections in swine and the low prevalence of the organism in U.S. herds may suggest management strategies that would potentially decrease or eliminate the colonization and/or transmission of *E. coli* O157:H7. Such information may also be useful to prevent the emergence of swine as a reservoir of *E. coli* O157:H7 in the United States.

References

APHIS. 2001. Swine 2000. 2001.

Besser, T. E., B. L. Richards, D. H. Rice, and D. D. Hancock. 2001. *Escherichia coli* O157:H7 infection of calves: infectious dose and direct contact transmission. Epidemiol. Infect. 127:555–560.

Best, A., R. M. La Ragione, D. Clifford, W. A. Cooley, A. R. Sayers, and M. J. Woodward. 2006. A comparison of Shiga-toxin negative *Escherichia coli* O157 aflagellate and intimin deficient mutants in porcine in vitro and in vivo models of infection. Vet. Microbiol. 113:63–72.

Beutin, L., D. Geier, H. Steinruck, S. Zimmermann, and F. Scheutz. 1993. Prevalence and some properties of verotoxin (shiga-like toxin) producing *Escherichia coli* in seven different species of healthy domestic animals. J. Clin. Microbiol. 31:2483–2488.

Booher, S., N. A. Cornick, and H. W. Moon. 2002. Persistence of *Escherichia coli* O157:H7 in experimentally infected swine. Vet. Microbiol. 89:69–81.

Borie, C., Z. Monreal, P. Guerrero, M. L. Sanchez, J. Martinez, C. Arellano, and V. Prado. 1997. Prevalencia y caracterizacion de *Escherichia coli* enterohemorragica aisladas de bovinos y cerdos sanos faenados en Santiago, Chile. Archives Med. Vet. 29:205–212.

Bush, E. 1997. US swine herd appears free of *Escherichia coli* O157:H7. Food Safety Digest: 4.

Cobbold, R., and P. Desmarchelier. 2002. Horizontal transmission of shiga toxin-producing *Escherichia coli* within groups of dairy calves. Appl. Environ. Microbiol. 68:4148–4152.

Conedera, G., E. Mattiazzi, F. Russo, E. Chiesa, I. Scorzato, S. Grandesso, A. Bessegato, A. Fioravanti, and A. Caprioli. 2007. A family outbreak of *Escherichia coli* O157 haemorrhagic colitis caused by pork meat salami. Epidemiol. Infect. 135:311–314.

Cornick, N. A., S. Booher, and H. W. Moon. 2002. Intimin facilitates colonization by *Escherichia coli* O157:H7 in ruminants. Infect. Immun. 70:2704–2707.

Cornick, N. A., S. L. Booher, T. A. Casey, and H. W. Moon. 2000. Persistent colonization of sheep by *E. coli* O157:H7 and other pathotypes of *E. coli.* Appl. Environ. Microbiol. 66:4926–4934.

Cornick, N. A., and A. F. Helgerson. 2004. Transmission and infectious dose of *Escherichia coli* O157:H7 in swine. Appl. Environ. Microbiol. 70:5331–5335.

Dean-Nystrom, E. A., B. T. Bosworth, H. W. Moon, and A. D. O'Brien. 1998. *Escherichia coli* O157:H7 requires intimin for enteropathogenicity in calves. Infect. Immun. 66:4560–4563.

Elder, R. O., J. E. Keen, G. R. Siragusa, G. A. Barkocy-Gallagher, M. Koohmaraie, and W. W. Laegreid. 2000. Correlation of enterohemorrhagic *Escherichia coli* O157 prevalence in feces, hides, and carcasses of beef cattle during processing. Proc. Natl. Acad. Sci. USA 97:2999–3003.

Erikkson, E., E. Nerbrink, E. Borch, A. Aspan, and A. Gunnarsson. 2003. Verocytotoxin-producing *Escherichia coli* O157:H7 in the Swedish pig population. Vet. Rec. 152:712–717.

Feder, I. E., M. Wallace, J. T. Gray, P. Fratamico, P. J. Fedorka-Cray, R. Pearce, J. E. Call, R. Perrine, and J. B. Luchansky. 2003. Isolation of *Escherichia coli* O157:H7 from intact colon fecal samples of swine. Emerg. Infect. Dis. 9:380–383.

Gyles, C. L., R. Friendship, K. Ziebell, S. Johnson, I. Yong, and R. Amezcua. 2002. *Escherichia coli* O157:H7 in pigs. Proc. Inter. Pig Vet. Soc. Abstract 191.

Hancock, D. D., T. E. Besser, M. L. Kinsel, P. I. Tarr, D. H. Rice, and M. G. Paros. 1994. The prevalence of *Escherichia coli* O157:H7 in dairy and beef cattle in Washington state. Epidemiol. Infect. 113:199–207.

Hancock, D. D., T. E. Besser, D. H. Rice, E. D. Ebel, D. E. Herriot, and L. V. Carpender. 1998. Multiple sources of *Escherichia coli* O157 in feedlots and dairy farms in the northwestern USA. Prev. Vet. Med. 35:11–19.

Heuvelink, A. E., J. T. M. Zwartkruis-Nahuis, F. L. A. M. Van den Biggelaar, W. J. Leeuwen, and E. de Boer. 1999. Isolation and characterization of verocytotoxin-producing *Escherichia coli* O157 from slaughter pigs and poultry. Int. J. Food Microbiol. 52:67–75.

Jay, M. T., M. Cooley, D. Carychao, G. W. Wiscomb, R. A. Sweitzer, L. Crawford-Miksza, J. A. Farrar, D. K. Lau, J. O. O'Connell, A. Millington, R. V. Asmundson, E. R. Atwill, and R. E. Mandrell. 2007. *Escherichia coli* O157:H7 in feral swine near spinach fields and cattle, central California coast. Emerg. Infect. Dis. 13:1908–1911.

Johnsen, G., Y. Wasteson, E. Heir, O. I. Berget, and H. Herikstad. 2001. *Escherichia coli* O157:H7 in faeces from cattle, sheep and pigs in the southwest part of Norway during 1998 and 1999. Int. J. Food Microbiol. 65:193–200.

Jordan, D. M., S. L. Booher, and H. W. Moon. 2005. *Escherichia coli* O157:H7 does not require intimin to persist in pigs. Infect. Immun. 73:1865–1867.

Jordan, D. M., N. A. Cornick, A. G. Torres, E. A. Dean-Nystrom, J. B. Kaper, and H. W. Moon. 2004. Long polar fimbriae contribute to colonization by *Escherichia coli* O157:H7 in vivo. Infect. Immun. 72:6168–6171.

Keen, J. E., T. E. Wittum, J. R. Dunn, J. L. Bono, and L. M. Durso. 2006. Shiga-toxigenic *Escherichia coli* O157 in agricultural fair livestock, United States. Emerg. Infect. Dis. 12:780–786.

Keene, W. E., E. Sazie, J. Kok, D. H. Rice, D. D. Hancock, V. K. Balan, T. Zhao, and M. P. Doyle. 1997. An outbreak of *Escherichia coli* O157:H7 infections traced to jerky made from deer meat. JAMA 277:1229–1231.

Kudva, I. T., P. G. Hatfield, and C. J. Hovde. 1996. *Escherichia coli* O157:H7 in microbial flora of sheep. J. Clin. Microbiol. 34:431–433.

Kudva, I. T., C. W. Hunt, C. J. Williams, U. M. Nance, and C. J. Hovde. 1997. Evaluation of dietary influences on *Escherichia coli* O157:H7 shedding by sheep. Appl. Environ. Microbiol. 63:3878–3886.

Mead, P. S., L. Slutsker, V. Dietz, L. F. McCraig, J. S. Bresee, C. Shapiro, P. M. Griffin, and R. V. Tauxe. 1999. Food-related illness and death in the United States. Emerg. Infect. Dis. 5: Sept.–Oct.: [on line] http://www.cdc.org.

Nakazawa, M., M. Akiba, and T. Sameshima. 1999. Swine as a potential reservoir of Shiga toxin-producing *Escherichia coli* O157:H7 in Japan. Emerg. Infect. Dis. Nov.–Dec.: [on line] http://www.cdc.org.

Sargeant, J. M., D. J. Hafer, J. R. Gillespie, R. D. Oberst, and S. J. A. Flood. 1999. Prevalence of *Escherichia coli* O157:H7 in white-tailed deer sharing rangeland with cattle. J. Am. Vet. Med. Assoc. 215:792–794.

Sheng, H., J. Y. Lim, H. J. Knecht, J. Li, and C. J. Hovde. 2006. Role of *Escherichia coli* O157:H7 virulence factors in colonization at the bovine terminal rectal mucosa. Infect. Immun. 74:4685–4693.

Vlisidou, I., F. Dziva, R. M. LaRagione, A. Best, J. Garmendia, P. Hawes, P. Monaghan, S. A. Cawthraw, G. Frankel, M. J. Woodward, and M. P. Stevens. 2006. Role of intimin-tir and the tir-cytoskelton coupling protein in the colonization of calves and lambs by *Escherichia coli* O157:H7. Infect. Immun. 74:758–764.

Wallace, J. S., T. Cheasty, and K. Jones. 1997. Isolation of Vero cytotoxin-producing *Escherichia coli* O157 from wild birds. J. Appl. Microbiol. 82:399–404.

Woodward, M. J., A. Best, K. A. Sprigings, G. R. Pearson, A. M. Skuse, A. Wales, C. M. Hayes, J. M. Roe, C. Low, and R. M. La Ragione. 2003. Non-toxigenic *Escherichia coli* O157:H7 strain NCTC12900 causes attaching-effacing lesions and eae-dependent persistence in weaned sheep. Int. J. Med. Microbiol. 293:299–308.

| 4 |

Traversing the Swine Gastrointestinal Tract: *Salmonella* Survival and Pathogenesis

Shawn M. D. Bearson and Bradley L. Bearson

Introduction

As one of the most consumed meats in the world, the presence of the foodborne pathogen *Salmonella* in pork is a food-safety concern. Furthermore, *Salmonella* in swine is an animal health issue, costing pork producers over $100 million annually. Based on the National Animal Health Monitoring System 1995 report (NAHMS 1997) and the Collaboration in Animal Health and Food Safety Epidemiology 2005 report (CAHFSE 2005), the estimated prevalence of *Salmonella* in swine oper-ations is 38% and 58%, respectively. Of the 10 most frequently identified *Salmonella* serovars from shedding hogs in the NAHMS study, 4 were also on the CDC's top-ten list of *Salmonella* serovars isolated from humans (Typhimurium, Enteritidis, Agona, Heidelberg). Both the national Hazard Analysis Critical Control Point pro-gram for slaughter plants as well as consumer education on proper handling and cooking of raw meat are important to reduce the incidence of *Salmonella* exposure in pork. However, consumption of contaminated pork is not the only possible cul-prit for foodborne outbreaks; *Salmonella*-tainted pig manure used as crop fertilizer can contaminate fruits and vegetables eaten raw by consumers or pollute human water supplies, emphasizing the importance of preharvest food safety (Guan and Holley 2003).

With over 2,500 serovars, *Salmonella enterica* isolated from the pig can be a broad host range serovar that causes enterocolitis (e.g., Typhimurium) or a narrow host range serovar resulting in systemic disease (Choleraesuis) (Schwartz 1999). Recent literature comparing the genome sequences of serovar Choleraesuis and serovar Typhimurium (Chiu et al. 2005) as well as the porcine response to infection with the 2 serovars (Skjolaas et al. 2006; Uthe et al. 2007) suggests that both the pathogen and the host perform vital roles in *Salmonella* host-specificity.

Both clinically and subclinically (carrier) infected pigs can shed *Salmonella*. As a result of their rooting behavior, the fecal-oral route is the most common route of *Salmonella* infection. Thus, a major source of *Salmonella* infection to a pig is another pig. Following ingestion of *Salmonella* by the pig, the pathogen must survive the volatile conditions of the stomach, compete with the resident gut microbiota, invade the intestinal epithelial lining, and evade as well as manipulate the host's immune system to achieve colonization. Hence, many attributes contribute to the pathogenesis of *Salmonella* in swine. Although *Salmonella* has been extensively studied over several decades in the mouse model, investigations at the molecular level of the virulence mechanisms employed by *Salmonella* in the pig and the porcine response to infection are limited. This chapter will specifically discuss *Salmonella* infections in swine, focusing on the molecular mechanisms known to date that are involved in *Salmonella* virulence and the porcine response to the pathogen.

Salmonella Pathogenesis

Salmonella Stress Survival

Microorganisms that colonize the gastrointestinal tract, including *Salmonella*, must endure the gastric acidity of the stomach. In addition to the low pH of the stomach, organic acids including lactic, acetic, propionic, and butyric acids may be present. Most investigations of *Salmonella* survival during exposure to acidic pH have focused on the Acid Tolerance Response (ATR) in *S.* Typhimurium. The ATR is an adaptive stress response that allows *Salmonella*, when exposed to a mild acidic pH, to induce acid shock proteins for protection against extreme acid stress. In an *ex vivo* swine stomach contents assay, strains containing mutations in previously identified ATR genes (*rpoS, fur,* and *phoP*) were shown to have reduced survival (Bearson et al. 2006). These three genes are global regulators with *rpoS* encoding an alternate sigma factor involved in stationary phase physiology and stress responses, *fur* encoding the Ferric Uptake Regulator (Fur) that regulates the uptake and utilization of iron by the bacterial cell, and *phoP* encoding a response regulator involved in macrophage survival and defense against antimicrobial peptides. The screening of transposon mutants in the *ex vivo* swine stomach contents assay identified strains with reduced survival following challenge during these hostile conditions (Bearson et al. 2006). The following genes were shown to be inactivated by transposon insertion in the strains sensitive to the swine stomach contents: *dnaK, dgt, pnp, usg, poxR, barA, sopB, traL, pefA, pefC, ynaI, rfaL,* and *asmA.* A multitude of cellular functions are represented by these gene products, including molecular chaperone, energy metabolism, transcription, translation, cellular regulation, and components of the cell envelope and fimbriae synthesis. Interestingly, not all of the genes sensitive to the swine stomach contents were sensitive to low pH alone. Thus, low pH is an important component of the porcine gastric environment, but additional lethal factors that appear to be low pH-dependent also exist. These factors could be antimicrobial peptides, microbial byproducts, antimicrobial peptides, bacteriophage, ammonia, alpha-amylase, lipase, surfactant, pepsin, bile, weak acids, etc.

Quorum Sensing in the Gastrointestinal Tract

Quorum sensing is a bacterial cell-to-cell communication mechanism whereby both commensal and pathogenic microorganisms (such as *Escherichia coli, Shigella* spp, *Salmonella* spp, *Klebsiella pneumoniae,* and *Enterobacter cloacae*) sense and respond to hormone-like chemical molecules called autoinducers in their environment. In the gastrointestinal tract, at least two chemical signals are present, autoinducer-3 (AI-3) and norepinephrine (NE) (Walters and Sperandio 2006). Bacteria produce AI-3 whereas the animal host produces NE. The QseBC two-component system senses (QseC) and responds (QseB) to AI-3 and NE to modulate gene expression, including bacterial flagellar and motility genes. Phentolamine, an α-adrenergic antagonist, eliminated the NE-altered motility of the wild-type strain, thereby identifying a potential target for disrupting colonization via inhibition of quorum sensing (Clarke et al. 2006; Bearson and Bearson 2007). Furthermore, an *in vivo* competition assay revealed decreased colonization of the gastrointestinal tract for a *qseC* mutant compared to wild-type *S.* Typhimurium, thereby indicating a role for quorum sensing in swine colonization (Bearson and Bearson 2007).

Salmonella Invasion

At least two types of disease pathology occur due to *Salmonella* infection in swine, enteritis and systemic disease (Schwartz 1999). Enteritis caused by most *Salmonella* serovars including Typhimurium is most commonly manifested as an elevation in body temperature, diarrhea, and lethargy. An invasion study involving ligated ileal gut-loops inoculated with *S.* Typhimurium in 4- to 5-week-old pigs indicated initial adherence of bacteria to microfold cells (M cells) within 5 minutes and invasion at the apical membrane of M cells, goblet cells and enterocytes within 10 minutes (Meyerholz et al. 2002). During the 60-minute infection time course, cellular invasion was rapid and multicellular. Loynachan and Harris (2005) found that following intranasal inoculation of 10- to 14-day old pigs with *S.* Typhim - urium, the ID_{50} for various tissues was approximately 1×10^5 colony forming units (cfu) for both tonsil and ileum tissue and 5 x 10^7 cfu for both cecum and colon contents.

Colonization of the swine intestine was attenuated in a *fimA* mutant of *S.* Typhimurium that is nonadhesive to porcine enterocytes (Althouse et al. 2003). The *fimA* gene encoding an adhesion, FimA, is the major subunit of type 1 fimbriae. The *hilA* gene (*h*yper-*i*nvasive *l*ocus A) encodes the HilA transcriptional regulator for invasion gene transcription. When compared to the wild-type *S.* Choleraesuis strain, a *hilA* mutant was at a competitive disadvantage following oral inoculation in pigs (Lichtensteiger and Vimr 2003). HilA is required for activation of gene expression from *Salmonella* pathogenicity island 1 (SPI-1) and invasion of intestinal cells (Jones 2005).

Bacterial invasion of epithelial cells requires a type III secretion system 1 (TTSS-1) encoded by SPI-1 for "injection" of effector proteins into the host cell cytoplasm (Figure 4.1). These *Salmonella* effector proteins are responsible for altering various host systems, including cytoskeletal rearrangements, to promote their uptake into the host cell. Multiple genes in SPI-1 (*sprB, sipC, spaP, invA,* and *invH*) when mutated have been shown to have a competitive disadvantage in swine compared to wild-type *S.* Choleraesuis following intranasal inoculation (Ku et al. 2005). Additionally, the SPI-1 gene *sipB* is important for *S.* Typhimurium intestinal invasion and colonization of swine (Boyen et al. 2006). Following invasion, intracellular survival and replication within membrane-bound *Salmonella*-containing vacuoles require genes located in *Salmonella* pathogenicity island 2 (SPI-2). Ku et al. (2005) also described four genes in SPI-2 (*spiA, spiR, ssaJ,* and *ssaV*) that when mutated resulted in less colonization of the porcine host than the wild-type *S.* Choleraesuis.

Attenuated Mutants of *Salmonella* as Vaccine Strains

Various *Salmonella* strains (including several strains discussed in the previous section) are attenuated in swine when specific genes or combinations of genes are mutated. Furthermore, some of these strains provide protective immunity during subsequent challenge with a virulent *Salmonella* strain. An *aroA* mutant of *S.* Typhimurium is attenuated for virulence; furthermore, diarrhea in *aroA*-vaccinated pigs was significantly reduced following subsequent challenge with a wild-type strain (Lumsden and Wilkie 1992). Coe and Wood (1992) exposed six-week-old, caesarean-derived, colostrum-deprived pigs to a Δ*cya* Δ*crp* mutant of *S.* Typhim - urium. The double deletion *cya crp* mutant was mildly virulent and induced a transient fever. Colonization of the ileum, cecum, liver, spleen, tonsils, and mandibular and ileocolic lymph nodes of pigs was noted for the double mutant; however, recovery of the mutant compared to the wild-type isogenic strain was

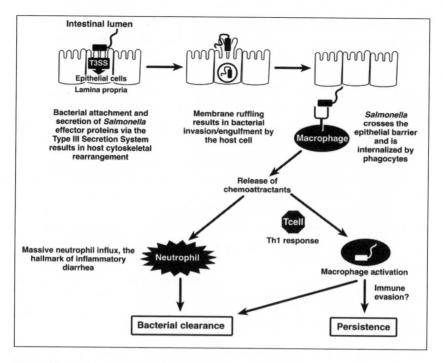

Figure 4.1. Model of *Salmonella* interaction with host cells.

reduced 100- to 1000-fold in the ileum. Prior vaccination of pigs with the Δ*cya* Δ*crp* mutant reduced clinical severity and colonization of the wild-type strain 21 days later.

In another study, four strains of *S.* Choleraesuis, Δ*cya* Δ*crp* and Δ*cya* Δ*crp-cdt* with and without the *Salmonella* virulence plasmid, were evaluated as vaccines for protection against wild-type *S.* Choleraesuis infection (Kennedy et al. 1999). No significant adverse effects were observed except a short-term elevation in body temperature for the two strains harboring the virulence plasmid. Oral challenge of pigs with a virulent *S.* Choleraesuis indicated that pigs vaccinated with any of these four strains had significantly better morbidity scores than nonvaccinated control pigs.

A neutrophil-adapted *S.* Choleraesuis strain (Scs 54) also lacks the *Salmonella* virulence plasmid but is otherwise genetically undefined. Vaccination of pigs with Scs 54 was protective against challenge with the virulent parental strain Scs 38 (Kramer et al. 1992). Three of five control pigs that received only Scs 38 died during the experiment whereas all of the vaccinated pigs survived.

Oral immunization of four-week-old pigs with a *gyrA cpxA rpoB* mutant followed by an oral challenge using a virulent *S.* Typhimurium DT104 strain prevented clinical symptoms of salmonellosis compared to nonimmunized pigs (Roesler et al. 2004). Isolation of *Salmonella* from vaccinated pigs showed significantly reduced rate of organ colonization and fecal shedding compared to control animals. In addition, vaccinated pigs developed higher specific immunoglobulin (Ig)A antibody activity compared to nonimmunized pigs.

In their signature-tagged mutagenesis screen for attenuated *S.* Choleraesuis strains, Ku et al. (2005) analyzed five mutants for protection of pigs against a virulent *S.* Choleraesuis challenge. When inoculated orally, two strains (*gifsy-1* and *ssaV*) were found to provide superior protection to pigs orally challenged 21-days later with a virulent *S.* Choleraesuis strain.

Several genes previously identified to be important for *Salmonella* pathogenesis in other animal models have also been described in pigs, especially gene products required for systemic disease. However, additional research is needed to fill gaps in our knowledge and to identify similarities and differences between *Salmonella* infections in swine and other animals. Furthermore, determining specific genes that promote *Salmonella* colonization of the swine gastrointestinal tract will reveal targets for future intervention strategies.

Porcine Response to *Salmonella*

Upon entry into the porcine gastrointestinal tract, *Salmonella* will encounter physical, biochemical, and cellular barriers. Although the pathogen provokes the development of adaptive immunity during the course of infection, it is the innate immune response that is vital during the initial stages of infection.

Antimicrobial Peptides

Mammals, amphibians, and insects produce antimicrobial peptides, small molecules with broad microbicidal activity against various microorganisms (Oswald 2006; Zhang et al. 2000a). Blecha and colleagues have shown enhanced expression of the porcine antimicrobial peptides PR-39, protegrin, LEAP-2, and hepcidin in response to infection with *Salmonella* (Sang et al. 2006; Wu et al. 2000). Originally identified in the porcine small intestine, the concentration of PR-39 in porcine serum increases 10–14 days postinfection with *S.* Choleraesuis and parallels neutrophil counts during the infection (Zhang et al. 1997). The porcine ß-defensin, PBD-1, is present at 20–100 µg ml^{-1} on the dorsal tongue surface and has synergistic activity against *Salmonella* with either porcine neutrophil peptide PG-3 or PR-39 (Shi et al. 1999). The PhoPQ two-component regulatory system of *Salmonella* is required for protection against several antimicrobial peptides including pig cecropin P1 (Groisman et al. 1992). In addition to their direct microbicidal function, antimicrobial peptides can also affect cytokine release and modulate host immunity (Bals and Wilson 2003), thereby making them intriguing models for pharmaceutical and agricultural therapy as well as potential targets for extrinsic modulation of host defenses.

Phagocytic Cells

Macrophages, monocytes, and neutrophils play a unique role in *Salmonella* infections, performing a vital function in clearance of the pathogen but also serving as targets for *Salmonella* invasion and potential vehicles for its dissemination (Donne et al. 2005; Riber and Lind 1999). Following invasion of the intestinal epithelium, macrophages in the lamina propria internalize *Salmonella* (Figure 4.1). The macrophage is activated to kill the microorganism and to release chemoattractants, recruiting additional immune cells to the site of invasion and initiating a T helper 1 (Th1) response. *In vitro* analysis of *Salmonella* serovars Typhimurium, Choleraesuis, and Dublin with porcine alveolar macrophages identified Typhimurium with the highest persistence in the macrophage, which may correlate with its virulence in a broad range of hosts as well as its ability to establish carrier status in the host (Watson et al. 2000). Once phagocytosed, elimination of *Salmonella* is dependent upon the killing capacity of the phagocytic cell; however, *Salmonella* can suppress these killing mechanisms; for example, by preventing the fusion of lysosomes with *Salmonella*-containing phagosomes. Donne et al. (2005) illustrated the importance of respiratory burst in the porcine monocyte for *Salmonella* killing and the ability of the organism to suppress the activity.

Neutrophils

The hallmark of *Salmonella* infection and the development of inflammatory diarrhea in the host is neutrophil influx (Tukel et al. 2006). Foster et al. (2003, 2005) demonstrated the rapid protective effect of neutrophil migration to the intestinal villi: piglets preinoculated with an attenuated *Salmonella enterica* strain were protected from salmonellosis during a subsequent challenge with a virulent strain 24 hours later. The production of reactive oxygen species by the neutrophil was required, although the authors suggest that neutrophil induction is not responsible for clinical symptoms or intestinal pathology in the pig. Further evidence of the importance of neutrophils in response to *Salmonella* was presented by van Diemen et al. (2002) in a reference population of swine bred for resistance to disease with Choleraesuis: pigs resistant to salmonellosis had a higher number of circulating neutrophils and better polymorphonuclear neutrophil (PMN) function than *Salmonella*-susceptible pigs. Stabel et al. (2002) suggested that the slow rate of

Choleraesuis uptake by the neutrophil during the initial stages of colonization may provide an opportunity for the serovar to establish systemic infection and/or carrier status in swine.

Cytokines and Chemokines

The host response to *Salmonella* infection is communicated through the language of cytokines and chemokines. Multiple cell types including epithelial and phagocytic cells produce, secrete, and respond to these small molecules with each having a distinct role in the immune response (Tzianabos and Wetzler 2004). In the last few years, several researchers have discussed the expression of immune-related porcine genes during infection with *Salmonella in vivo* (blood and various tissues), *ex vivo* (tissue explants), and *in vitro* (cell culture). These expression studies have revealed both similarities and differences in the pig compared to human and murine hosts. For example, interleukin (IL)12 is a stimulator of interferon-γ (IFNG) levels and is transcriptionally induced in mice and humans during *Salmonella* infection (Jouanguy et al. 1999). However, IL12 is down-regulated in swine (Uthe et al. 2007), suggesting that another IFNG-inducing cytokine regulates intracellular infections in swine (Domeika et al. 2002). Another example, the neutrophil chemoattractant, IL8 is greatly expressed in swine during *Salmonella* infection (Hyland et al. 2006a; Skjolaas et al. 2006), but is lacking in mice. Furthermore, the expression of tumor necrosis factor-α (TNF) in murine infections is important for *Salmonella* clearance, but its role in swine infections has been inconsistent and may be tissue dependent (Cho and Chae 2003; Hyland, Brown et al., 2006; Splichal et al. 2005; Trebichavsky et al. 2003; Uthe et al. 2007). Similar to the immune responses of other hosts, pro-inflammatory cytokines IFNG and IL1β have been shown by multiple investigators to be highly up-regulated (Hyland, Kohrt et al. 2006; Splichal et al. 2002; Trebichavsky et al. 2003; Zhao et al. 2006; Uthe et al. 2007), as well as SLC11A1 (NRAMP1), a divalent cation efflux pump of macrophages implicated in resistance to *Salmonella* (Blackwell et al. 2001; Lalmanach et al. 2001; Zaharik et al. 2002; Zhang et al. 2000b). In addition, a role has been suggested for heat-shock response proteins, involved in molecular chaperoning and protein folding, during Choleraesuis infection (Uthe et al. 2006).

Comparative studies indicate differences in the transcriptional response of pigs and porcine cell lines infected with Typhimurium and Choleraesuis (Burkey et al. 2006; Burkey et al. 2007; Skjolaas et al. 2006, 2007; Uthe et al. 2007). Transcriptional alterations occurred more rapidly in Typhimurium-infected pigs but were transient, whereas Choleraesuis induced a delayed but extended transcriptional induction in response to *Salmonella* infection (Uthe et al. 2007). Furthermore, a key immune response system (NFκB) is suppressed during Typhimurium infection, potentially providing a strategy for *Salmonella* Typhimurium to evade a strong immune response and contributing to the establishment of a subclinical (carrier) infection in the pig (Wang et al. 2007). Niewold et al. (2007) also suggested that Typhimurium evades a strong porcine response by down-regulating the local inflammatory response. Differences in transcriptional response during the initial stages of infection with a narrow (Choleraesuis) or broad (Typhimurium) host range *Salmonella* serovar may be a pivotal factor determining the establishment of a systemic or localized infection, respectively. Genetic mapping (Kim et al. 2005, 2006) and single nucleotide polymorphism (SNP) analysis on genes differentially regulated during *Salmonella* infection may identify potential targets for disease control.

Continued characterization of the pig's response to *Salmonella* infection at the molecular level will identify porcine genetic targets for improved resistance to *Salmonella* through industry breeding programs as well as immune response mechanisms to enhance through immunomodulation. Dvorak et al. (2006) recently determined through gene expression profiling that more than 40% of mRNA enriched from the porcine Peyer's patch region (the preferred site of *Salmonella* invasion) represented genes of unknown function, indicating that many unique immunological and physiological properties of the pig remain to be characterized (and potentially exploited to control *Salmonella*).

Summary

Controlling *Salmonella* infections in swine is important for food safety and animal health. Various measures to reduce colonization and disease have included the following: feed to enhance protective conditions of the gastrointestinal tract such as lower stomach pH and competitive gut microbial flora (Mikkelsen and Jensen 2003; Mikkelsen et al. 2004); feed additives such as chlorate and subtherapeutic levels of antibiotics to inhibit bacterial growth (Anderson et al. 2005; Edrington et al. 2001); probiotics such as lactic acid bacteria that inhibit *Salmonella* (Casey et al. 2004; Tsai et al. 2005); and vaccination to prevent colonization and disease (Haesebrouck et al. 2004). Regardless of the control strategy, molecular systems of either the pathogen or the pig are targeted for disease prevention.

The presence of *Salmonella* in U.S. swine herds is generally considered ubiquitous. *Salmonella* in swine has at least three ramifications: clearly *Salmonella* can be pathogenic to swine; contaminated pork is a food-safety risk to consumers; and *Salmonella* shed in feces poses an environmental risk during disposal. Therefore, the greatest impact to protect swine producers and the public from dissemination of *Salmonella* would be to prevent the colonization of swine by *Salmonella*. In medicine, an effective method to prevent infectious disease is through vaccination of the host to stimulate an immune response against the pathogen. Although multiple vaccines that target *Salmonella* have been constructed, the complexity of the

Salmonella genus and the epidemiology of the pathogen prevent complete protection against *Salmonella* colonization of swine. This is due to at least two factors: first, the >2,500 serovars have subtle differences in outer membrane structure that may afford protection from immune surveillance even when the host has previously been exposed to *Salmonella*. Secondly, in swine, at least two types of disease pathology occur, enteritis and systemic disease due to *Salmonella* infection. The immune response that protects against enteritis may not be optimal for protection against systemic infection and vice versa. Thus, characterizing the mechanisms on both sides of the infection to understand host-specificity and the host-pathogen interaction is vital in achieving disease control, both clinically and subclinically (carrier).

References

Althouse C., S. Patterson, P. Fedorka-Cray, and R. E. Isaacson. 2003. Type 1 fimbriae of *Salmonella enterica* serovar Typhimurium bind to enterocytes and contribute to colonization of swine in vivo. Infect. Immun. 71:6446–6452.

Anderson, R. C., R. B. Harvey, J. A. Byrd, T. R. Callaway, K. J. Genovese, T. S. Edrington, Y. S. Jung, J. L. McReynolds, and D. J. Nisbet. 2005. Novel preharvest strategies involving the use of experimental chlorate preparations and nitro-based compounds to prevent colonization of food-producing animals by foodborne pathogens. Poult. Sci. 84:649–654.

Bals, R., and J. M. Wilson. 2003. Cathelicidins—a family of multifunctional antimicrobial peptides. Cell. Mol. Life Sci. 60:711–720.

Bearson, B. L., and S. M. D. Bearson. 2008. The role of the QseC quorum-sensing sensor kinase in colonization and norepinephrine-enhanced motility of *Salmonella enterica* serovar Typhimurium. Microb. Pathog. 44:271–278.

Bearson, S. M., B. L. Bearson, and M. A. Rasmussen. 2006. Identification of *Salmonella enterica* serovar Typhimurium genes important for survival in the swine gastric environment. Appl. Environ. Microbiol. 72:2829–2836.

Blackwell J. M., T. Goswami, C. A. Evans, D. Sibthorpe, N. Papo, J. K. White, S. Searle, E. N. Miller, C. S. Peacock, H. Mohammed, and M. Ibrahim. 2001. SLC11A1 (formerly NRAMP1) and disease resistance. Cell. Microbiol. 3:773–784.

Boyen, F., F. Pasmans, F. Van Immerseel, E. Morgan, C. Adriaensen, J. P. Hernalsteens, A. Decostere, R. Ducatelle, and F. Haesebrouck. 2006. *Salmonella* Typhimurium SPI-1 genes promote intestinal but not tonsillar colonization in pigs. Microbes Infect. 8:2899–2907.

Burkey, T. E., K. A. Skjolaas, S. S. Dritz, and J. E. Minton. 2007. Expression of Toll-like receptors, interleukin 8, macrophage migration inhibitory factor, and osteopontin in tissues from pigs challenged with *Salmonella enterica* serovar Typhimurium or serovar Choleraesuis. Vet. Immunol. Immunopathol. 115:309–319.

CAHFSE. 2005. Collaboration in Animal Health and Food Safety Epidemiology annual report. APHIS. At http://www.aphis.usda.gov/cahfse/results/index.htm. Accessed December 2006.

Casey, P. G., G. D. Casey, G. E. Gardiner, M. Tangney, C. Stanton, R. P. Ross, C. Hil, and

G. F. Fitzgerald. 2004. Isolation and characterization of anti-*Salmonella* lactic acid bacteria from the porcine gastrointestinal tract. Lett. Appl. Microbiol. 39:431–438.

Chiu, C. H., P. Tang, C. Chu, S. Hu, Q. Bao, J. Yu, Y. Y. Chou, H. S. Wang, and Y. S. Lee. 2005. The genome sequence of *Salmonella enterica* serovar Choleraesuis, a highly invasive and resistant zoonotic pathogen. Nucleic Acids Res. 33:16901698.

Cho, W. S., and C. Chae. 2003. Expression of inflammatory cytokines (TNF-a, IL-1, IL-6 and IL-8) in colon of pigs naturally infected with *Salmonella typhimurium* and *S. choleraesuis*. J. Vet. Med. Physiol. Pathol. Clin. Med. 50:484–487.

Clarke, M. B., D. T. Hughes, C. Zhu, E. C. Boedeker, and V. Sperandio. 2006. The QseC sensor kinase: a bacterial adrenergic receptor. Proc. Natl. Acad. Sci. USA 103:10420–10425.

Coe, N. E., and R. L. Wood. 1992. The effect of exposure to a delta cya/delta crp mutant of *Salmonella typhimurium* on the subsequent colonization of swine by the wild-type parent strain. Vet. Microbiol. 31:207–220.

Domeika, K., M. Berg, M. L. Eloranta, and G. V. Alm. 2002. Porcine interleukin-12 fusion protein and interleukin-18 in combination induce interferon-g production in porcine natural killer and T cells. Vet. Immunol. Immunopathol. 86:11–21.

Donne, E., F. Pasmans, F. Boyen, F. Van Immerseel, C. Adriaensen, J. P. Hernalsteens, R. Ducatelle, and F. Haesebrouck. 2005. Survival of *Salmonella* serovar Typhimurium inside porcine monocytes is associated with complement binding and suppression of the production of reactive oxygen species. Vet. Microbiol. 107:205–214.

Dvorak, C. M., G. N. Hirsch, K. A. Hyland, J. A. Hendrickson, B. S. Thompson, M. S. Rutherford, and M. P. Murtaugh. 2006. Genomic dissection of mucosal immunobiology in the porcine small intestine. Physiol. Genomics 28:5–14.

Edrington, T. S., R. B. Harvey, L. A. Farrington, and D. J. Nisbet. 2001. Evaluation of subtherapeutic use of the antibiotics apramycin and carbadox on the prevalence of antimicrobial-resistant *Salmonella* infection in swine. J. Food Prot. 64:2067–2070.

Foster, N., S. Hulme, M. Lovell, K. Reed, and P. Barrow. 2005. Stimulation of gp91 phagocytic oxidase and reactive oxygen species in neutrophils by an avirulent *Salmonella enterica* serovar *infantis* strain protects gnotobiotic piglets from lethal challenge with serovar Typhimurium strain F98 without inducing intestinal pathology. Infect. Immun. 73:4539–4547.

Foster, N., M. A. Lovell, K. L. Marston, S. D. Hulme, A. J. Frost, P. Bland, and P. A. Barrow. 2003. Rapid protection of gnotobiotic pigs against experimental salmonellosis following induction of polymorphonuclear leukocytes by avirulent *Salmonella enterica*. Infect. Immun. 71:2182–2191.

Groisman, E. A., C. Parra-Lopez, M. Salcedo, C. J. Lipps, and F. Heffron. 1992. Resistance to host antimicrobial peptides is necessary for *Salmonella* virulence. Proc. Natl. Acad. Sci. USA 89:11939–11943.

Guan, T. Y., and R. A. Holley. 2003. Pathogen survival in swine manure environments and transmission of human enteric illness—a review. J. Environ. Qual. 32:383–392.

Haesebrouck, F., F. Pasmans, K. Chiers, D. Maes, R. Ducatelle, and A. Decostere. 2004. Efficacy of vaccines against bacterial diseases in swine: what can we expect? Vet. Microbiol. 100:255–268.

Hyland, K. A., D. R. Brown, and M. P. Murtaugh. 2006. *Salmonella enterica* serovar Choleraesuis infection of the porcine jejunal Peyer's patch rapidly induces IL-1b and IL-8 expression. Vet. Immunol. Immunopathol. 109:1–11.

Hyland, K. A., L. Kohrt, L. Vulchanova, and M. P. Murtaugh. 2006. Mucosal innate immune response to intragastric infection by *Salmonella enterica* serovar Choleraesuis. Mol. Immunol. 43:1890–1899.

Jones, B. D. 2005. *Salmonella* invasion gene regulation: a story of environmental awareness. J. Microbiol. 43:110–117.

Jouanguy, E., R. Doffinger, S. Dupuis, A. Pallier, F. Altare, and J. L. Casanova. 1999. IL-12 and IFN-g in host defense against *mycobacteria* and *salmonella* in mice and men. Curr. Opin. Immunol. 11:346–351.

Kennedy, M. J., R. J. Yancey Jr., M. S. Sanchez, R. A. Rzepkowski, S. M. Kelly, and R. Curtiss III. 1999. Attenuation and immunogenicity of Dcya Dcrp derivatives of *Salmonella choleraesuis* in pigs. Infect. Immun. 67:4628–4636.

Kim, J. W., S. H. Zhao, J. J. Uthe, S. M. Bearson, and C. K. Tuggle. 2005. Physical mapping of eight pig genes whose expression level is acutely affected by *Salmonella* challenge. Anim. Genet. 36:359–362.

Kim, J. W., S. H. Zhao, J. J. Uthe, S. M. Bearson, and C. K. Tuggle. 2006. Assignment of the scavenger receptor class B, member 2 gene (SCARB2) to porcine chromosome 8q11—>q12 by somatic cell and radiation hybrid panel mapping. Cytogenet. Genome Res. 112:342H.

Kramer, T. T., M. B. Roof, and R. R. Matheson. 1992. Safety and efficacy of an attenuated strain of *Salmonella choleraesuis* for vaccination of swine. Am. J. Vet. Res. 53:444–448.

Ku, Y. W., S. P. McDonough, R. U. Palaniappan, C. F. Chang, and Y. F. Chang. 2005. Novel attenuated *Salmonella enterica* serovar Choleraesuis strains as live vaccine candidates generated by signature-tagged mutagenesis. Infect. Immun. 73:8194–8203.

Lalmanach, A. C., A. Montagne, P. Menanteau, and F. Lantier. 2001. Effect of the mouse *Nramp1* genotype on the expression of IFN-gamma gene in early response to *Salmonella* infection. Microbes Infect. 3:639–644.

Lichtensteiger, C. A., and E. R. Vimr. 2003. Systemic and enteric colonization of pigs by a *hilA* signature-tagged mutant of *Salmonella choleraesuis*. Microb. Pathog. 34:149–154.

Loynachan, A. T., and D. L. Harris. 2005. Dose determination for acute *Salmonella* infection in pigs. Appl. Environ. Microbiol. 71:2753–2755.

Lumsden, J. S., and B. N. Wilkie. 1992. Immune response of pigs to parenteral vaccination with an aromatic-dependent mutant of *Salmonella typhimurium*. Can. J. Vet. Res. 56:296–302.

Meyerholz, D. K., T. J. Stabel, M. R. Ackermann, S. A. Carlson, B. D. Jones, and J. Pohlenz. 2002. Early epithelial invasion by *Salmonella enterica* serovar Typhimurium DT104 in the swine ileum. Vet. Pathol. 39:712–720.

Mikkelsen, L. L., and B. B. Jensen. 2003. The stomach as a barrier that reduces the occurrence of pathogenic bacteria in pigs. Page 66 in Ninth International Symposium on Digestive Physiology in Pigs. Vol. 2. Banff, AB, Canada.

Mikkelsen, L. L., P. J. Naughton, M. S. Hedemann, and B. B. Jensen. 2004. Effects of physical properties of feed on microbial ecology and survival of *Salmonella enterica* serovar Typhimurium in the pig gastrointestinal tract. Appl. Environ. Microbiol. 70:3485–3492.

NAHMS. 1997. Shedding of *Salmonella* by finisher hogs in the US. In National animal health monitoring system, swine 1995: grower/finishers. Animal and Plant Health

Inspection Service. http://www.aphis.usda.gov/vs/ceah/ncahs/nahms/swine/swine95/sw95salm. pdf. Accessed December 2006.

Niewold, T. A., E. J. Veldhuizen, J. van der Meulen, H. P. Haagsman, A. A. de Wit, M. A. Smits, M. H. Tersteeg, and M. M. Hulst. 2007. The early transcriptional response of pig small intestinal mucosa to invasion by *Salmonella enterica* serovar *typhimurium* DT104. Mol. Immunol. 44:1316–1322.

Oswald, I. P. 2006. Role of intestinal epithelial cells in the innate immune defence of the pig intestine. Vet. Res. 37:359–368.

Riber, U., and P. Lind. 1999. Interaction between *Salmonella typhimurium* and phagocytic cells in pigs. Phagocytosis, oxidative burst and killing in polymorphonuclear leukocytes and monocytes. Vet. Immunol. Immunopathol. 67:259–270.

Roesler, U., H. Marg, I. Schroder, S. Mauer, T. Arnold, J. Lehmann, U. Truyen, and A. Hensel. 2004. Oral vaccination of pigs with an invasive *gyrA-cpxA-rpoB Salmonella* Typhimurium mutant. Vaccine 23:595–603.

Sang, Y., B. Ramanathan, J. E. Minton, C. R. Ross, and F. Blecha. 2006. Porcine liver-expressed antimicrobial peptides, hepcidin and LEAP-2: cloning and induction by bacterial infection. Dev. Comp. Immunol. 30:357–366.

Schwartz, K. J. 1999. *Salmonellosis.* Pages 535–551 in Disease of swine. B. E. Straw, S. D'Allaire, W. L. Mengeling, and D. J. Taylor, eds. 8th ed. Iowa State University Press, Ames, IA.

Shi, J., G. Zhang, H. Wu, C. Ross, F. Blecha, and T. Ganz. 1999. Porcine epithelial beta-defensin 1 is expressed in the dorsal tongue at antimicrobial concentrations. Infect. Immun. 67:3121–3127.

Skjolaas, K. A., T. E. Burkey, S. S. Dritz, and J. E. Minton. 2006. Effects of *Salmonella enterica* serovars Typhimurium (ST) and Choleraesuis (SC) on chemokine and cytokine expression in swine ileum and jejunal epithelial cells. Vet. Immunol. Immunopathol. 111:199–209.

Skjolaas, K. A., T. E. Burkey, S. S. Dritz, and J. E. Minton. 2007. Effects of *Salmonella enterica* serovar Typhimurium, or serovar Choleraesuis, *Lactobacillus reuteri* and *Bacillus licheniformis* on chemokine and cytokine expression in the swine jejunal epithelial cell line, IPEC-J2. Vet. Immunol. Immunopathol. 115:299–308.

Splichal, I., I. Trebichavsky, Y. Muneta, and Y. Mori. 2002. Early cytokine response of gnotobiotic piglets to *Salmonella enterica* serotype *typhimurium*. Vet. Res. 33:291–297.

Splichal, I., I. Trebichavsky, A. Splichalova, and P. A. Barrow. 2005. Protection of gnotobiotic pigs against *Salmonella enterica* serotype Typhimurium by rough mutant of the same serotype is accompanied by the change of local and systemic cytokine response. Vet. Immunol. Immunopathol. 103:155–161.

Stabel, T. J., P. J. Fedorka-Cray, and J. T. Gray. 2002. Neutrophil phagocytosis following inoculation of *Salmonella choleraesuis* into swine. Vet. Res. Commun. 26:103–109.

Trebichavsky, I., I. Splichal, A. Splichalova, Y. Muneta, and Y. Mori. 2003. Systemic and local cytokine response of young piglets to oral infection with *Salmonella enterica* serotype Typhimurium. Folia Microbiol. 48:403–407.

Tsai, C. C., H. Y. Hsih, H. H. Chiu, Y. Y. Lai, J. H. Liu, B. Yu, and H. Y. Tsen. 2005. Antagonistic activity against *Salmonella* infection *in vitro* and *in vivo* for two *Lactobacillus* strains from swine and poultry. Int. J. Food Microbiol. 102:185–194.

Tukel, C., M. Raffatellu, D. Chessa, R. P. Wilson, M. Akcelik, and A. J. Bäumler. 2006. Neutrophil influx during non-typhoidal salmonellosis: who is in the driver's seat? FEMS Immunol. Med. Microbiol. 46:320–329.

Tzianabos, A. O., and L. M. Wetzler. 2004. Cellular communication. ASM Press, Washington, DC.

Uthe, J. J., T. J. Stabel, S. H. Zhao, C. K. Tuggle, and S. M. Bearson. 2006. Analysis of porcine differential gene expression following challenge with *Salmonella enterica* serovar Choleraesuis using suppression subtractive hybridization. Vet. Microbiol. 114:60–71.

Uthe, J. J., A. Royaee, J. K. Lunney, T. J. Stabel, S. H. Zhao, C. K. Tuggle, and S. M. D. Bearson. 2007. Porcine differential gene expression in response to *Salmonella enterica* serovars Choleraesuis and Typhimurium. Mol. Immunol. 44:2900–2914.

van Diemen, P. M., M. B. Kreukniet, L. Galina, N. Bumstead, and T. S. Wallis. 2002. Characterisation of a resource population of pigs screened for resistance to salmonellosis. Vet. Immunol. Immunopathol. 88:183–196.

Walters, M., and V. Sperandio. 2006. Quorum sensing in *Escherichia coli* and *Salmonella*. Int. J. Med. Microbiol. 296:125–131.

Wang, Y., L. Qu, J. J. Uthe, S. M. D. Bearson, D. Kuhar, J. K. Lunney, O. P. Couture, D. Nettleton, J. C. Dekkers, and C. K. Tuggle. 2007. Global transcriptional response of porcine mesenteric lymph nodes to *Salmonella enterica* serovar Typhimurium. Genomics 90:72–84.

Watson, P. R., S. M. Paulin, P. W. Jones, and T. S. Wallis. 2000. Interaction of *Salmonella* serotypes with porcine macrophages in vitro does not correlate with virulence. Microbiology 146:1639–1649.

Wu, H., G. Zhang, J. E. Minton, C. R. Ross, and F. Blecha. 2000. Regulation of cathelicidin gene expression: induction by lipopolysaccharide, interleukin-6, retinoic acid, and *Salmonella enterica* serovar Typhimurium infection. Infect. Immun. 68:5552–5558.

Zaharik, M. L., B. A. Vallance, J. L. Puente, P. Gros, and B. B. Finlay. 2002. Host-pathogen interactions: Host resistance factor *Nramp1* up-regulates the expression of *Salmonella* pathogenicity island-2 virulence genes. Proc. Natl. Acad. Sci. USA 99:15705–15710.

Zhang, G., C. R. Ross, S. S. Dritz, J. C. Nietfeld, and F. Blecha. 1997. *Salmonella* infection increases porcine antibacterial peptide concentrations in serum. Clin. Diagn. Lab. Immunol. 4:774–777.

Zhang, G., C. R. Ross, and F. Blecha. 2000a. Porcine antimicrobial peptides: new prospects for ancient molecules of host defense. Vet. Res. 31:277–296.

Zhang, G., H. Wu, C. R. Ross, J. E. Minton, and F. Blecha. 2000b. Cloning of porcine *NRAMP1* and its induction by lipopolysaccharide, tumor necrosis factor alpha, and interleukin-1b: role of CD14 and mitogen-activated protein kinases. Infect. Immun. 68:1086–1093.

Zhao, S. H., D. Kuhar, J. K. Lunney, H. Dawson, G. Guidry, J. J. Uthe, S. M. Bearson, J. Recknor, D. Nettleton, and C. K. Tuggle. 2006. Gene expression profiling in *Salmonella* Choleraesuis-infected porcine lung using a long oligonucleotide microarray. Mamm. Genome. 17:777–789.

∎ 5 ∎

On-Farm Interventions to Reduce Epizootic Bacteria in Food-Producing Animals and the Environment

*Robin C. Anderson, Aleksandar Božic, Todd R. Callaway,
Yong S. Jung, Kenneth J. Genovese, Thomas S. Edrington,
Roger B. Harvey, Jackson L. McReynolds,
J. Allen Byrd, and David J. Nisbet*

Introduction

Foodborne disease infections are estimated to afflict more than 76 million people in the United States each year (Mead et al. 1999). *Campylobacter* and *Salmonella* are responsible for most cases of the bacterial foodborne disease, causing more than 1.9 and 1.3 million infections, respectively (Mead et al. 1999). *Escherichia coli* O157 and non-O157 shiga toxin-producing *E. coli* are estimated to cause more than 62,000 and 31,000 cases of foodborne disease annually (Mead et al. 1999). Costs associated with foodborne illness caused by the bacteria are considerable, exceeding $2.4 billion for *Salmonella* and $445 million for *E. coli* O157 (Crutchfield and Roberts 2000; ERS 2007). Cost for foodborne *Campylobacter* infections are estimated at $1.2 billion annually (Crutchfield and Roberts 2000). Although *Listeria monocytogenes* is estimated to cause fewer than 2,500 cases of disease, its associated annual cost exceeds $6.9 billion, due largely to the long-term impact of congential and newborn infections (Crutchfield and Roberts 2000).

Considerable effort has been and continues to be expended by government agencies and all involved in the production of food to ensure that the food produced is safe and wholesome. Contemporary concepts now recognize that applying preharvest interventions with existing or newly develped postharvest technologies (chemical dehairing, steam pasteurization, steam vacuuming, and hot water and organic acid rinses and irradiation) in a "multi-hurdle" approach could result in the most effective way to minimize contamination of meat products by foodborne pathogens (Acuff et al. 1987; Castell-Perez and Moreira 2004; Cherrington et al. 1991; Dorsa 1997; Farkas 1998; Hardin et al. 1995; Keeton and Eddy 2004; Koohmaraie et al. 2005; Micheals et al. 2004; Ricke 2003). In 2003, for example, the United States Department of Agriculture's Food Safety Inspection Service listed among their research priorities "Development of improved on-farm, feedlot, and antemortem interventions for reducing the incidence and levels of pathogens in raw products" (FSIS 2003). Risk assessments provide support for the concept that preharvest interventions would reduce human exposure to pathogens (Hynes and Wachsmuth 2000; Vugia et al. 2003). Presently, we review some of the research conducted within our laboratory aimed at developing safe and efficacious feed or water supplements to reduce the incidence and survivability of foodborne pathogens in the gut of food animals.

Experimental Chlorate Product

All major species of food-producing animals can be carriers of *Salmonella* spp. and ruminants, especially cattle, are recognized as important reservoirs of enterohemorrhagic *E. coli*. These organisms can reside within the gastrointestinal tracts of their animal hosts, often without causing apparent symptoms of disease or gastrointestinal upset. Like most members of the family *Enterobacteriaceae,* these bacteria possess the ability to respire anaerobically using nitrate as an anaerobic electron acceptor (Brenner 1984; Stewart 1988). Respiratory nitrate reductase activity, conferred by membrane-bound nitrate reductases, has long been known to catalyze the intracellular reduction of the relatively innocuous chlorate ion to cytotoxic chlorite (Pichinoty and Piéchaud 1968; Stewart 1988). Considering that most strict anaerobic gut bacteria lack respiratory nitrate reductase activity, also referred to as dissimilatory nitrate reductase activity, we hypothesized that an experimental product containing chlorate (ECP) could be applied within the last days before slaughter to selectively kill *Salmonella* and other enterobacteria, while not harming the beneficial population of gut bacteria (Anderson et al. 2000a). Support for our hypothesis was first generated during an *in vitro* experiment where concentrations of *Salmonella* serovar Typhimurium DT104 and *E. coli* O157:H7 were reduced in a dose-dependent manner. Approximately 10^6 colony forming units (CFU)/ml of these pathogens were reduced to below our limit of detection (10 CFU/ml) during *in vitro* incubation of mixed ruminal bacteria supplemented with 1.25 and 5 mM sodium chlorate ion (Anderson, Buckley, Kubena et al. 2000). Concentrations of total culturable anaerobes were not reduced by the chlorate treatments thus supporting the concept that most beneficial anaerobes are unaffected by chlorate. Notable exceptions include certain *Clostridium, Desulfovibrio, Propionibacterium,* and *Selenomonas* spp. as well as *Denitrobacterium detoxificans, Viellonella alcalescens,* and *Wolinella succinogenes,* bacteria reported to possess some form of respiratory nitrate reductase activity (Alaboudi 1982; Allison and Reddy 1984; Anderson, Rasmussen et al. 2000; Inderlied and Delwiche 1973; Richardson et al. 2001; Wolin et al. 1961). It is arguable if these bacteria would be essential for optimal gut function of the host within the last few days of the animal's life. While it has been shown that chlorate can kill *Viellonella alcalescens* and *Clostridium perfringens* (Anderson, Harvey et al. 2006; McReynolds 2004), the latter case suggests a possible application of the technology to treat or prevent necrotic enteritis, more research may be warranted to fully characterize the effect of chlorate against other nitrate-respiring anaerobes.

Conversion of chlorate to chlorite is likely catalyzed by membrane-bound NarG nitrate reductase rather than the constitutively expressed NarZ enzyme, which contributes only a fraction of the activity attributable to NarG (Moreno-Vivián et al. 1999; Wang et al. 1999). Involvement of the periplasmic (Nap) respiratory nitrate reductase enzyme is not likely as it does not use chlorate as a substrate (Wang et al. 1999), and this may explain the lack of *in vitro* bactericidal activity of chlorate against *Campylobacter* spp., which exclusively possess a periplasmic enzyme (unpublished results).

Subsequent studies have since proven that oral administration of ECP significantly reduced *Salmonella* colonization in the gut of pigs, broilers, and turkeys (Anderson, Buckley, Callaway et al. 2001; Anderson, Callaway, Buckley et al. 2001; Anderson, Hume et al. 2004; Byrd et al. 2003; Jung, Anderson, Byrd et al. 2003; Moore et al. 2002). Intraruminal, drinking water, or feed administration of ECP to cattle and sheep significantly reduced gut concentrations of generic *E. coli* and/or *E. coli* O157:H7 (Anderson et al. 2002, 2005; Callaway et al. 2002, 2003; Edrington et al. 2003) and showed that a chlorate product designed to bypass the rumen enhanced the bactericidal effect in the lower gastrointestinal tract (Anderson et al. 2005; Edrington et al. 2003). For instance, feeding an ECP with bypass characteristics to cattle effectively reduced fecal *E. coli* concentrations whereas administration of a nonprotected ECP in drinking water reduced ruminal but not fecal *E. coli* concentrations (Anderson et al. 2005).

Whereas both *E. coli* and *Salmonella* possess respiratory nitrate reductase, evidence from *in vitro* studies reveal that co-supplementation with small amounts of nitrate markedly enhanced the bactericidal activity of chlorate against *Salmonella,* but not necessarily against *E. coli* (Anderson, Buckley, Kubena et al. 2000; Anderson et al. 2007). Conversely, even in the absence of added nitrate the bactericidal activity of chlorate against *E. coli* was equivalent to that of nitrate-supplemented *Salmonella*

(Anderson et al. 2000, 2007). These observations implicate that, in the case of *Salmonella*, induction and subsequent expression of the nitrate reductase activity is enhanced by nitrate but that chlorate itself is sufficient to induce expression of the enzyme in *E. coli*. This observation likely may have practical implications for the design of anti-*Salmonella* products for application in animals as these products may require small amounts of nitrate to achieve optimal efficacy. Results from subsequent studies have supported the concept that nitrate may be required for optimal *Salmonella* control, demonstrating that the bactericidal effect of chlorate treatment was enhanced nearly 100-fold or more during *in vitro* incubations of porcine fecal suspensions with added nitrate as well as in pigs and broilers if preceded by a short nitrate preadaptation period (Anderson, Jung, Genovese et al. 2006; 2007; Jung, Anderson, Byrd et al. 2003). Preadaptation of pigs with small amounts of nitrate enhanced the efficacy of a chlorate treatment against endogenous *E. coli* even at half the minimum efficacious dose (Anderson, Jung, Genovese et al. 2006). A similar nitrate preadaptation period did not enhance the bactericidal activity of chlorate against *E. coli* in the lower gut of cattle (Fox et al. 2005). It is possible, however, that rapid ruminal reduction of nitrate in the cattle may have precluded its delivery, and therefore its beneficial effect, in the lower gut (Fox et al. 2005). While the potential benefits of prior or coadministration of nitrate on chlorate's bactericidal efficacy against *E. coli* remain unresolved, it is clear that nitrate treatment is needed for optimal efficacy against *Salmonella*. Nitrate is used as a terminal electron acceptor by both *E. coli* and *Salmonella* and in the absence of added chlorate may support a small and temporary enrichment of *Salmonella* in fecal incubations. However, *in vitro* and *in vivo* studies have demonstrated that the potential for an undesired enrichment can be safely minimized or eliminated by supplying nitrate at < 2.5 mM nitrate (Anderson et al. 2007; Jung and Anderson 2004). Nitrate preadaptation may serve another useful purpose as well as it may select for nitrate respiring bacteria *in situ* and thus minimize the growth of chlorate-resistant populations. By definition, nitrate respiring enterobacteria are chlorate sensitive (Stewart 1988). Addition of nitrate for this purpose is probably unnecessary; however, evidence indicates that chlorate resistance is unlikely to arise in mixed populations of gut bacteria. When resistant cultures generated in pure culture were added to a mixed population, the resistant bacteria did not persist in the competitive environment (Callaway et al. 2001).

The chlorate technology has now been licensed to an industry partner and the process of obtaining approval as a feed additive has been initiated. Research examining pharmocokinetic aspects of gastrointestinal chlorate metabolism and the metabolic fate of chlorate in the animal indicate that residual chlorate in edible tissue is within established provisional safety limits (Oliver et al. 2007; Smith, Anderson et al. 2005; Smith, Oliver et al. 2005; Smith et al. 2006).

Short-Chain Nitroalkanes

We have long realized that ideal preharvest food-safety strategies must be economically acceptable for livestock producers because they are being expected to absorb the costs of implementing these interventions. In that regard, research has been initiated at our laboratory to develop a cost-recoverable preharvest food-safety strategy that combines the bactericidal activity of certain nitrocompounds (2-nitropropanol, nitroethanol, and nitroethane) with their ability to reduce ruminal methane production by up to 90% *in vitro* and up to 40% *in vivo* (Anderson et al. 2003, 2006; Gutierrez-Bañuelos et al. 2007). Energy losses from ruminal methanogenesis are not inconsequential, costing U.S. cattle feeders up to, if not more than, $900,000 per day (or $3.6 billion over a 10-year period) (Van Kessel, personal communication) for finishing cattle losing 2 to 4% of their gross energy intake as methane (Johnson and Johnson 1995).

We first recognized the inhibitory activity of 2-nitro-1-propanol, nitroethanol, and nitroethane in chlorate co-supplementation studies where we observed that their substitution in place of nitrate into fecal suspensions synergistically enhanced the bactericidal activity of chlorate against *Salmonella* Typhimurium (Anderson et al. 2007). Additionally, 2-nitro-1-propanol and nitroethanol were observed to inhibit *Salmonella* even in the absence of chlorate (Anderson et al. 2007) and have shown potential to be developed into feed additives to reduce *Salmonella* colonization in the gut of broilers and pigs (Jung, Anderson, Genovese et al. 2003; Jung, Anderson, Edrington et al. 2004). These nitrocompounds also inhibit the *in vitro* growth of *Listeria, Campylobacter, Yersinia, Enterococcus faecalis,* vancomycin resistant *Enterococcus faecium, Staphylococcus aureus, Streptococccus agalactiae* and, to a lesser extent, *E. coli* O157:H7 (Božic et al. 2007; Dimitrijevic et al. 2005; Horrocks et al. 2007; Jung and Anderson 2004; Jung, Anderson, Callaway et al. 2004). Short-chain nitrocompounds have also been shown to inhibit growth of uric-acid-degrading bacteria and thus may have application for decreasing ammonia emissions from poultry operations (Kim et al. 2005). At present, the only known bacterium possessing appreciable nitroalkane-reducing activity is the ruminal anaerobe *Denitrobacterium detoxificans.* This obligate nonfermentative anaerobe that conserves energy for growth exclusively via anaerobic respiration by oxiding hydrogen, formate, or lactate for the reduction of nitrate or select short-chain nitroalkanes (Anderson et al. 2000).

Lauric Acid and Lauricidin

Medium-chain fatty acids such as lauric acid and its glycerol monoester, monolaurin (Lauricidin), are also known to be potent inhibitors of ruminal methane production (Dohme et al. 2001; Machmüller et al. 2001; Soliva et al. 2003) and could also have application as cost-recoverable preharvest feed additives. In the case of these compounds, however, their application would be against Gram-positive bacteria as their activity appears directed to this cell wall type. Perhaps the most relevant use of medium-chain fatty acids would be as an antibiotic alternative directed toward control of *Clostridium perfringens* infections. Skrivanova et al. (2005) showed that lauric acid and a variety of other medium-chain fatty acids inhibited antimicrobial activity against this pathogen. Similarly, lauric acid and Lauricidin have been shown to markedly inhibit other important Gram-postive bacteria such as vancomycin-resistant *Enterococcus faecium*, mastitis causing *Streptococcus agalactiae*, and *Staphylococcus aureus*, the efficient starch utilizer *Streptococcus bovis, Listeria*, and *Pseudomonas* spp. (Božic et al. 2007; Vasseur et al. 2001). With respect to *Listeria monocytogenes*, we observed > 4 $\log_{10}$ unit reductions during *in vitro* incubation with mixed populations of ruminal fluid supplemented with 5 mg/ml lauric acid or Lauricidin (Božic et al. 2005). However, despite being widely disseminated among animals and the farm environment (Nightingale et al. 2004), *Listeria monocytogenes* is not presently recognized as a preharvest problem but rather is considered to primarily be a problem within the processing environment. Moreover, many Gram-positive ruminal bacteria contribute to fiber digestion, which may limit the use of lauric acid and Lauricidin as feed additives throughout the feeding period. Assimilation of lauric acid into edible tissue and subsequent consumption could be detrimental to human health as excessive saturated fatty acid intake is a contributor to coronary heart disease (Hu et al. 2001). Thus, the use of lauric acid as feed additive throughout long periods of the fattening stage may not be beneficial from a human health perspective. If, however, on-farm control of *Listeria* should become warranted it is not unreasonable to suspect that a more restrictive use of medium-chain fatty acids only during the last day or several days before slaughter may significantly reduce gut carriage of *Listeria* with minimal effects on production efficiency or the composition of fat accretion.

Competitive Exclusion

Competitive exclusion is another strategy having potential to substitute as an alternative to antibiotics currently used to treat enteric disease. Conceptually, competitive exclusion is the application of beneficial and commensal gut bacteria to the neonatal gastrointestinal tract to facilitate colonization with healthy gut flora with the capacity to out-compete and thus exclude pathogens (Mead 2000). The rationale for using competitive exclusion is that modern intensive agricultural production practices may delay or perturb the natural establishment of a healthy gut flora. For instance, young poultry are typically hatched in hygienic hatcheries devoid of influences that could provide exposure to a maternal inoculum source. Additionally, young swine experience a dietary change upon weaning which can upset the maturing gastrointestinal ecosystem, which may predispose the pig to enteric infections.

Nurmi and Rantala reported in 1973 that young chicks inoculated with gut bacteria collected from healthy *Salmonella*-free adult chickens had enhanced resistance against colonization by *Salmonella*. A number of subsequent studies have demonstrated protection of newly hatched chicks following administration with a variety of defined or undefined mixtures of healthy gut bacteria (Barnes et al. 1979; Corrier et al. 1995; Nisbet et al. 1993, 1994; Reid and Barnum 1984; Stavric and D'Aoust 1993). More recently, protective effects of competitive exclusion treatments against colonization of cattle and pigs by pathogens such as *Salmonella*, enterotoxigenic *E. coli*, and *E. coli* O157:H7 have been published (Anderson et al. 1999; Fedorka-Cray et al. 1999; Genovese et al. 2000, 2001, 2003; Zhao et al. 1998). Likewise, protective effects of probiotic bacterial preparations against *E. coli* O157:H7 in ruminants have been reported (Brashears, Galyean et al. 2003; Brashears, Jaroni et al. 2004; Lema et al. 2001; Ohya et al. 2000), although some of these probiotic bacteria did not necessarily originate from a ruminant host (Lema et al. 2001).

In the United States, competitive exclusion cultures are regulated as animal drugs and as such are currently under the jurisdiction of the United States Food and Drug Administration. At present, only one competitive eclusion product, trade named PREEMPT, has been approved as a drug to reduce *Salmonella* colonization in poultry (Federal Register 1998). PREEMPT demand was not enough to sustain production (Muirhead 2003), purportedly because poultry processors have been able to maintain *Salmonella* levels below allowable regulatory levels. As pressure to reduce antibiotic usage in food animals increases, however, demand for environmentally compatible alternatives to contemporary antibiotic technologies will likely increase as well. The National Pork Board, for instance, has been a supporter of research aimed at developing such technologies. In a series of field trials conducted in five states from August 2002 to February 2004, approximately 37,000 neonatal pigs orally treated with a recombined porcine-derived continuous flow culture-maintained competitive exclusion culture (RPCF) had significantly reduced *E. coli*-associated mortality and medication costs than nontreated pigs (Harvey et al. 2005). Moreover, evidence now suggests that this competitive exclusion culture exerts its protective effect against enteropathogen colonization by stimulating the immune system. Germ-free piglets administered RPCF at birth had serum immunoglobulin levels 20–100-fold higher than those of noncolonized control piglets when examined at six weeks of age (Butler et al. 2000). Ongoing studies further suggest that RPCF stimulates the innate immune system of neonatal pigs by upregulating specific cytokines associated with immune function development (unpublished data). Practical implications of this later finding indicate that competitive exclusion cultures may more aptly be considered biological agents (immunostimulator/adjuvant) rather than as drugs which would warrant their regulation by the Center for Veterinary Biologics, APHIS, USDA, rather than by the Food and Drug Administration, a jurisdictional change which could expedite approval and implementation of this technology for pork producers.

Conclusion

In summary, results from these *in vitro* and *in vivo* experiments indicate that novel experimental preparations containing chlorate, select nitroalkanes, the medium-chain fatty acid lauric acid (and its glycerol monoester, monoluarin), and competitive exclusion cultures may have application in the preharvest control of enteric pathogens. Regulatory issues need to be resolved, however, before any of these technologies can be implemented in the field and none are as yet commercially available.

References

Acuff, G. R., C. Vanderzant, J. W. Savell, D. K. Jones, D. B. Griffin, and J. G. Ehlers. 1987. Effect of acid decontamination of beef subprimal cuts on the microbiological and sensory characteristics of steaks. Meat Sci. 19:217–226.

Alaboudi, A. R. 1982. Microbiological studies of nitrate and nitrite reduction in the ovine rumen. PhD Diss. University of Saskatchewan, Saskatoon.

Allison, M. J., and C. A. Reddy. 1984. Adaptations of gastrointestinal bacteria in response to changes in dietary oxalate and nitrate. Pages 248–256 in Current perspectives in microbial ecology. Proc. 3rd Intl. Symp. on Microbial Ecology. M. J. Klug and C. A. Reddy, eds. American Society for Microbiology, Washington, DC.

Anderson, R. C., S. A. Buckley, T. R. Callaway, K. J. Genovese, L. F. Kubena, R. B. Harvey, and D. J. Nisbet. 2001. Effect of sodium chlorate on *Salmonella* Typhimurium concentrations in the weaned pig gut. J. Food Prot. 64:255–258.

Anderson, R. C., S. A. Buckley, L. F. Kubena, L. H. Stanker, R. B. Harvey, and D. J. Nisbet. 2000. Bactericidal effect of sodium chlorate on *Escherichia coli* O157:H7 and *Salmonella* Typhimurium DT104 in rumen contents *in vitro*. J. Food Prot. 63:1038–1042.

Anderson, R. C., T. R. Callaway, T. J. Anderson, L. F. Kubena, N. K. Keith, and D. J. Nisbet. 2002. Bactericidal effect of sodium chlorate on *Escherichia coli* concentrations in bovine ruminal and fecal concentrations *in vivo*. Microb. Ecol. Health Dis. 14:24–29.

Anderson, R. C., T. R. Callaway, S. A. Buckley, T. J. Anderson, K. J. Genovese, C. L. Sheffield, and D. J. Nisbet. 2001. Effect of oral sodium chlorate administration on

Escherichia coli O157:H7 in the gut of experimentally infected pigs. Int. J. Food Microbiol. 71:125–130.

Anderson, R. C., T. R. Callaway, J. S. Van Kessel, Y. S. Jung, T. S. Edrington, and D. J. Nisbet. 2003. Effect of select nitrocompounds on ruminal fermentation: an initial look at their potential to reduce economic and environmental costs associated with ruminal methanogenesis. Bioresource Technol. 90:59–63.

Anderson, R. C., M. A. Carr, R. K. Miller, D. A. King, G. E. Carstens, K. J. Genovese, T. R. Callaway, T. S. Edrington, Y. S. Jung, J. L. McReynolds, M. E. Hume, R. C. Beier, R. O. Elder, and D. J. Nisbet. 2005. Effects of experimental chlorate preparations as feed and water supplements on *Escherichia coli* colonization and contamination of beef cattle and carcasses. Food Microbiol. 22:439–447.

Anderson, R. C., G. E. Carstens, R. K. Miller, T. R. Callaway, C. L. Schultz, T. S. Edrington, R. B. Harvey, and D. J. Nisbet. 2006. Effect of oral nitroethane and 2-nitropropanol administration on methane-producing activity and volatile fatty acid production in the ovine rumen. Bioresourc. Technol. 97:2421–2426.

Anderson, R. C., R. B. Harvey, T. L. Poole, N. Ramlachan, T. R. Callaway, J. A. Byrd, T. S. Edrington, J. L. McReynolds, and D. J. Nisbet. 2006. Comparative susceptibilities of representative enterobacteriaceae and gut commensal bacteria to the active agent of an experimental chlorate product. Page 463 in Proceedings of the 19th International Pig Veterinary Society Congress, July 16–19. J. P. Nielsen and S. E. Jorsal, eds. Copenhagen, Denmark.

Anderson, R. C., M. E. Hume, K. J. Genovese, T. R. Callaway, Y. S. Jung, T. S. Edrington, T. L. Poole, R. B. Harvey, K. M. Bischoff, and D. J. Nisbet. 2004. Effect of drinking water administration of experimental chlorate ion preparations on *Salmonella enterica* serovar Typhimurium colonization in weaned and finished pigs. Vet. Res. Comm. 28:174–184.

Anderson, R. C., Y. S. Jung, K. J. Genovese, J. L. McReynolds, T. R. Callaway, T. S. Edrington, R. B. Harvey, and D. J. Nisbet. 2006. Low level nitrate or nitroethane preconditioning enhances the bactericidal effect of suboptimal experimental chlorate treatment against *Escherichia coli* and *Salmonella* Typhimurium but not *Campylobacter* in swine. Foodborne Path. Dis. 3:461–465.

Anderson, R. C., Y. S. Jung, C. E. Oliver, S. M. Horrocks, K. J. Genovese, R. B. Harvey, T. R. Callaway, T. S. Edrington, and D. J. Nisbet. 2007. Effects of nitrate or nitro-supplementation, with or without added chlorate, on *Salmonella enterica* serovar Typhimurium and *Escherichia coli* in swine feces. J. Food Prot. 70:308–315.

Anderson, R. C., M. A. Rasmussen, N. S. Jensen, and M. J. Allison. 2000. *Denitrobacterium detoxificans* gen. Nov., sp. nov., a ruminal bacterium that respires on nitrocom-pounds. Int. J. Syst. Evol. Microbiol. 50:633–638.

Anderson, R. C., L. H. Stanker, C. R. Young, S. A. Buckley, K. J. Genovese, R. B. Harvey, J. R. DeLoach, and D. J. Nisbet. 1999. Effect of competitive exclusion treatment on colonization of early weaned pigs by *Salmonella* serovar Choleraesuis. Swine Health Product. 7:155–160.

Barnes, E. M., C. S. Impey, and B. J. H. Stevens. 1979. Factors affecting the incidence and anti-*Salmonella* activity of the anaerobic cecal flora of the young chick. J. Hyg. 82:263–283.

Božic, A. K., R. C. Anderson, G. E. Carstens, T. R. Callaway, and D. J. Nisbet. 2007. In vitro effects of the methane-inhibitors nitroethane, 2-nitro-1-propanol, lauric acid, and lauricidin on select populations of Gram-positive bacteria. Page 153 in Abstracts of the Symposium on Veterinary medicine, animal husbandry and economy in animal health and food safety production. Herceg Novi, Montenegro.

Brashears, M. M., M. L. Galyean, G. H. Loneragan, J. E. Mann, and K. Killinger-Mann. 2003. Prevalence of *Escherichia coli* O157:H7 and performance by beef feedlot cattle given lactobacillus direct-fed microbials. J. Food Prot. 66, 748–754.

Brashears, M. M., D. Jaroni, and J. Trimble. 2003. Isolation, selection and characterization of lactic acid bacteria for a competitive exclusion product to reduce shedding of *Escherichia coli* O157:H7 in cattle. J. Food Prot. 66, 355–363.

Brenner, D. J. 1984. *Enterobacteriaceae*. Pages 408–420 in Bergey's manual of systematic bacteriology, Vol. 1. N. R. Krieg and J. G. Holt, eds. Williams & Wilkins, Baltimore, MD.

Butler, J. E., J. Sun, P. Weber, P. Navarro, and D. Francis. 2000. Antibody repertoire development in fetal and newborn piglets, III. Colonization of the gastrointestinal tract selectively diversifies the preimmune repertoire in mucosal lymphoid tissues. Immunology 100:119–130.

Byrd, J. A., R. C. Anderson, T. R. Callaway, R. W. Moore, K. D. Knape, L. F. Kubena, R. L. Ziprin, and D. J. Nisbet. 2003. Effect of experimental chlorate product administration in the drinking water on *Salmonella* Typhimurium contamination of broilers. Poult. Sci. 82:1403–1406.

Callaway, T. R., R. C. Anderson, S. A. Buckley, T. L. Poole, K. J. Genovese, L. F. Kubena, and D. J. Nisbet. 2001. *Escherichia coli* O157:H7 becomes resistant to sodium chlorate in pure, but not mixed culture *in vitro*. J. Appl. Microbiol. 91:427–434.

Callaway, T. R., R. C. Anderson, K. J. Genovese, T. L. Poole, T. J. Anderson, J. A. Byrd, L. F. Kubena, and D. J. Nisbet. 2002. Sodium chlorate supplementation reduces *E. coli* O157:H7 populations in cattle. J. Anim. Sci. 80:1683–1689.

Callaway, T. R., T. S. Edrington, R. C. Anderson, K. J. Genovese, T. L. Poole, R. O. Elder, J. A. Byrd, K. M. Bischoff, and D. J. Nisbet. 2003. *Escherichia coli* O157:H7 populations in sheep can be reduced by chlorate supplementation. J. Food Prot. 66:194–199.

Castell-Perez, M., and R. G. Moreira. 2004. Decontamination systems. Pages 337–347 in Preharvest and post harvest food safety. Contemporary issues and future directions. R. C. Beier, S. D. Pillai, and T. D. Phillips, eds. Blackwell Publishing, Ames IA.

Cherrington, C. A., M. Hinton, G. C. Mead, and I. Chopra. 1991. Organic acids: chemistry, antibacterial activity and practical applications. Adv. Microbial Physiol. 32:87–108.

Corrier, D. E., D. J. Nisbet, C. M. Scanlan, A. G. Hollister, and J. R. DeLoach. 1995. Control of *Salmonella* Typhimurium colonization in broiler chicks with a continuous-flow characterized mixed culture of cecal bacteria. Poult. Sci. 74:916–924.

Crutchfield, S. R., and T. Roberts. 2000. Food safety efforts accelerate in the 1990's. Food Review 23:44–49.

Dimitrijevic, M., R. C. Anderson, T. R. Callaway, Y. S. Jung, R. B. Harvey, S. C. Ricke, and D. J. Nisbet. 2006. Inhibitory effect of select nitrocompounds on growth and survivability of *Listeria monocytogenes* in vitro. J. Food Protect. 69:1061–1065.

Dohme, F., A. Machmüller, and M. Kreuzer. 2001. Ruminal methanogenesis as influenced by individual fatty acids supplemented to complete ruminant diets. Lett. Appl. Microbiol. 32:47–51.

Dorsa, W. J. 1997. New and established carcass decontamination procedures commonly used in the beef-processing industry. J. Food Protect. 60:1146–1151.

Edrington, T. S., T. R. Callaway, R. C. Anderson, K. J. Genovese, Y. S. Jung, J. L. McReynolds, K. M. Bischoff, and D. J. Nisbet. 2003. Reduction of *E. coli* O157:H7 populations in sheep by supplementation of an experimental sodium chlorate product. Small Rumin. Res. 49:173–181.

ERS. 2007. Foodborne illness cost calculator. United States Department of Agriculture, Economic Research Service. Available at: http://www.ers.usda.gov/Data/FoodborneIllness/. Accessed April 23, 2008.

Farkas, J. 1998. Irradiation as a method for decontaminating food, a review. Int. J. Food Microbiol. 44:189–204.

Federal Register. 1998. Volume 63(88):25163.

Fedorka-Cray, P. J., J. S. Bailey, N. J. Stern, N. A. Cox, S. R. Ladely, and M. Musgrove. 1999. Mucosal competitive exclusion to reduce *Salmonella* in swine. J. Food Protect. 62:1376–1380.

Fox, J. T., R. C. Anderson, G. E. Carstens, R. K. Miller, Y. S. Jung, J. L. McReynolds, T. R. Callaway, T. S. Edrington, and D. J. Nisbet. 2005. Effect of nitrate adaptation on the bactericidal activity of an experimental chlorate product against *Escherichia coli* in cattle. Int. J. Appl. Res. Vet. Med. 3:76–80.

FSIS. 2003. Food Safety and Inspection Service Research Priorities, United States Department of Agriculture, Washington DC, Available at: http://www.fsis.usda.gov/OA/programs/research_priorities.htm.

Genovese, K. J., R. C. Anderson, R. B. Harvey, T. R. Callaway, T. L. Poole, T. S. Edrington, P. J. F. Cray, and D. J. Nisbet. 2003. Competitive exclusion of *Salmonella* from the gut of pigs. J. Food Prot. 66:1353–1359.

Genovese, K. J., R. C. Anderson, R. B. Harvey, and D. J. Nisbet. 2000. Competitive exclusion treatment reduces the mortality and fecal shedding associated with enterotoxigenic *Escherichia coli* infection in nursery-raised neonatal pigs. Can. J. Vet. Res. 64:204–207.

Genovese, K. J., R. B. Harvey, R. C. Anderson, and D. J. Nisbet. 2001. Protection of suckling neonatal pigs against an infection with an enterotoxigenic *Escherichia coli* expressing 987P fimbriae infection by the administration of a bacterial competitive exclusion culture. Microbial Ecol. Health Dis. 13:223–228.

Gutierrez-Bañuelos, H., R. C. Anderson, G. E. Carstens, L. J. Slay, N. Ramlachan, S. M. Horrocks, T. R. Callaway, T. S. Edrington, and D. J. Nisbet. 2007. Zoonotic bacterial populations, gut fermentation characteristics and methane production in feedlot steers during oral nitroethane treatment and after the feeding of an experimental chlorate product. Anaerobe 13:21–31.

Hardin, M. D., G. R. Acuff, L. M. Lucia, J. S. Oman, and J. W. Savell. 1995. Comparison of methods for decontamination from beef carcass surfaces. J. Food Protect. 58:368–374.

Harvey, R. B., R. C. Anderson, K. J. Genovese, T. R. Callaway, and D. J. Nisbet. 2005. Use of competitive exclusion to control enterotoxigenic strains of *E. coli* in weaned pigs. J. Anim. Sci. 83(Suppl. E):E44–E47.

Horrocks, S. M., Y. S. Jung, J. K. Huwe, R. B. Harvey, S. C. Ricke, G. E. Carstens, T. R. Callaway, R. C. Anderson, N. Ramlachan, and D. J. Nisbet. 2007. Effects of short-chain nitrocompounds against *Campylobacter jejuni* and *Campylobacter coli* in vitro. J. Food Sci. 72:M50–M55.

Hu, F. B., J. E. Manson, and W. C. Willett. 2001. Types of dietary fat and risk of coronary heart disease: a critical review. J. Amer. Coll. Nutr. 20:5–19.

Hynes, N. A., and I. K. Wachsmuth. 2000. *Escherichia coli* O157:H7 risk assessment in ground beef: a public health tool. Page 46 in Abstracts of the 4th International Symposium and Workshop on Shiga Toxin (Verocytotoxin)-Producing *Escherichia coli* Infections. Kyoto, Japan.

Inderlied, C. B., and E. A. Delwiche. 1973. Nitrate reduction and the growth of *Viellonella alcalescens*. J. Bacteriol. 114:1206–1212.

Johnson, K. A., and D. E. Johnson. 1995. Methane emissions from cattle. J. Anim. Sci. 73:2483–2492.

Jung, Y. S., and R. C. Anderson. 2004. Pre-nitrate adaptation and chlorate supplementation to reduce *Salmonella, Escherichia coli*, and *Yersinia enterocolitica* in swine. Final report to the National Pork Board, NPB Project # 03–136.

Jung, Y. S., R. C. Anderson, J. A. Byrd, T. S. Edrington, R. W. Moore, T. R. Callaway, J. L. McReynolds, and D. J. Nisbet. 2003. Reduction of *Salmonella* Typhimurium in experimentally challenged broilers by nitrate adaptation and chlorate supplementation in drinking water. J. Food Prot. 66:660–663.

Jung, Y. S., R. C. Anderson, T. R. Callaway, T. S. Edrington, K. J. Genovese, R. B. Harvey, T. L. Poole, and D. J. Nisbet. 2004. Inhibitory activity of 2-nitropropanol against select food-borne pathogens *in vitro*. Lett. Appl. Microbiol. 39:471–476.

Jung, Y. S., R. C. Anderson, T. S. Edrington, K. J. Genovese, J. A. Byrd, T. R. Callaway, and D. J. Nisbet. 2004. Experimental use of 2-nitropropanol for reduction of Salmonella Typhimurium in the ceca of broiler chicks. J. Food Prot. 67:1045–1047.

Jung, Y. S., R. C. Anderson, K. J. Genovese, T. S. Edrington, T. R. Callaway, J. A. Byrd, K. M. Bischoff, R. B. Harvey, J. McReynolds, and D. J. Nisbet. 2003. Reduction of *Campylobacter* and *Salmonella* in pigs treated with a select nitrocompound. Pages 205–207 in Proceedings of the 5th International Symposium on the Epidemiology and Control of Foodborne Pathogens in Pork. Hersonissos, Crete.

Keeton, J. T., and S. M. Eddy. 2004. Chemical methods for decontamination of meat and poultry. In: Beier R. C., Pillai S. D., and Phillips T. D. (eds) Pages 319–333 in Preharvest and post harvest food safety: contemporary issues and future directions. R. C. Beier, S. D. Pillai, and T. d. Phillips, eds. Blackwell Publishing, Ames, IA.

Kim, W. K., R. C. Anderson, A. L. Ratliff, D. J. Nisbet, and S. C. Ricke. 2005. Growth inhibition by nitrocompounds of selected uric-acid utilizing microorganisms isolated from poultry manure. J. Environ. Sci. Health Part B. 41:97–107.

Koohmaraie, M., T. M. Arthur, J. M. Bosilevac, M. Guerini, S. D. Shackelford, and T. L. Wheeler. 2005. Post-harvest interventions to reduce/eliminate pathogens in beef. Meat Sci. 71:79–91.

Lema, M., L. Williams, and D. R. Rao. 2001. Reduction of fecal shedding of enterohemorrhagic *Escherichia coli* O157:H7 in lambs by feeding microbial feed supplement. Small Rum. Res. 39:31–39.

Machmüller, A., F. Dohme, C. R. Soliva, M. Wanner, and M. Kreuzer. 2001. Diet composition affects the level of ruminal methane suppression by medium-chain fatty acids. *Aust. J. Agr. Res.* 52:713–722.

McReynolds, J. L. 2004. The effects of an experimental chlorate product on the microbial ecology in *Gallus gallus* var. *domesticus*. PhD Diss., Texas A&M University, College Station, TX.

Mead, G. C. 2000. Prospects for "competitive exclusion" treatment to control salmonellas and other food-borne pathogens. Vet. J. 159:111–123.

Mead, P. S., L. Slutsker, V. Dietz, L. F. McCaig, J. S. Bresee, C. Shapiro, P. M. Griffin, and R. V. Tauxe. 1999. Food-related illness and death in the United States. Emerg. Infect. Dis. 5:607–625.

Micheals, B., C. Keller, M. Blevins, G. Paoli, T. Ruthman, E. Todd, and C. J. Grittith. 2004. Prevention of food worker transmission of foodborne pathogens: risk assessment and evaluation of effective hygiene intervention strategies. Food Ser. Technol. 4:31–49.

Moore, R. W., J. A. Byrd, K. D. Knape, R. C. Anderson, T. R. Callaway, T. S. Edrington, L. F. Kubena, and D. J. Nisbet. 2006. The effect of an experimental chlorate product on *Salmonella* recovery of turkeys when administered prior to feed and water withdrawal. Poult. Sci. 85:2101–2105.

Moreno-Vivián C, P., Cabello, M. Martínez-Luque, R. Blasco, and F. Castillo. 1999. Prokaryotic nitrate reduction: molecular properties and functional distinction among bacterial nitrate reducases. J. Bacteriol. 181:6573–6584.

Muirhead, S. 2003. MS Bioscience to halt production of *Salmonella* product. Feedstuffs 74:27.

Nightingale, K. K., Y. H. Schukken, C. R. Nightingale, E. D. Fortes, A. J. Ho, Z. Her, Y. T. Grohn, P. L. McDonough, and M. Wiedmann. 2004. Ecology and transmission of *Listeria monocytogenes* infecting ruminants and in the farm environment. Appl. Environ. Microbiol. 70:4458–4467.

Nisbet, D. J., D. E. Corrier, and J. R. DeLoach. 1993. Effect of mixed cecal microflora maintained in continuous culture and of dietary lactose on *Salmonella* Typhimurium colonization in broiler chicks. Avian Dis. 37:528–535.

Nisbet, D. J., S. C. Ricke, C. M. Scanlan, D. E. Corrier, A. G. Hollister, and J. R. DeLoach. 1994. Inoculation of broiler chicks with a continuous-flow derived bacterial culture facilitates early cecal bacterial colonization and increases resistance to *Salmonella* Typhimurium. J. Food Protect. 57:12–15.

Ohya, T., T. Marubashi, and H. Ito. 2000. Significance of fecal volatile fatty acids in shedding of *Escherichia coli* O157 from calves: experimental infection and preliminary use of a probiotic product. J. Vet. Med. Sci. 62:1151–1155.

Oliver, C. E., A. L. Craigmill, J. S. Caton, R. C. Anderson, and D. J. Smith. 2007. Pharmacokinetics of ruminally-dosed sodium [36Cl]chlorate in beef cattle. J. Vet. Pharmacol. Therapeut. 30:358–365.

Pichinoty, F., and M. Piéchaud. 1968. Recherche des nitrate-réductases bactérériennes A et B: méthodes. Ann. Inst. Pasteur (Paris) 114:77–98.

Reid, C. R., and D. A. Barnum. 1984. The effects of treatments of cecal contents on the protective properties against *Salmonella* in poults. Avian Dis. 29:1–11.

Richardson, D. J., B. C. Berks, D. A. Russell, S. Spiro, and C. J. Taylor. 2001. Functional, biochemical and genetic diversity of prokaryotic nitrate reductases. Cell. Mol. Life Sci. 58:165–178.

Ricke, S. C. 2003. Perspectives on the use of organic acids and short chain fatty acids as antimicrobials. Poult. Sci. 82:632–639.

Skrivanova, M., G. Marounek, G. Dlouha, and J. Ka ka. 2005. Susceptibility of *Clostridium perfringens* to C2-C18 fatty acids. Lett. Appl. Microbiol. 41:77–81.

Smith, D. J., R. C. Anderson, D. A. Ellig, and G. L. Larsen. 2005. Tissue distribution, elimination, and metabolism of dietary sodium [^{36}Cl]chlorate in beef cattle. J. Agric. Food Chem. 53:4272–4280.

Smith, D. J., R. C. Anderson, and J. K. Huwe. 2006. Effect of sodium [^{36}Cl] chlorate dose on total radioactive residues and residues of parent chlorate in growing swine. J. Agric. Food Chem. 54:8648–8653.

Smith, D. J., C. E. Oliver, J. S. Caton, and R. C. Anderson. 2005. Effect of sodium [^{36}Cl] chlorate dose on total radioactive residues and residues of parent chlorate in beef cattle. J. Agric. Food Chem. 53:7352–7360.

Soliva, C. R., I. K. Hindrichsen, L. Melle, M. Kreuzer, and A. Machmüller. 2003. Effects of mixtures of lauric acid and myristic acid on rumen methanogens and methanogenesis *in vitro*. Lett. Appl. Microbiol. 37:35–39.

Stavric, S., and J. Y. D'Aoust. 1993. Undefined and defined bacterial preparations for competitive exclusion of *Salmonella* in poultry. J. Food Protect. 56:173–180.

Stewart, V. 1988. Nitrate respiration in relation to facultative metabolism in enterobacteria. Microbiol. Rev. 52:190–232.

Vasseur, C., N. Rigaud, M. Hebraud, and J. Labadie. 2001. Combined effects of NaCl, NaOH, and biocides (monolaurin or lauric acid) on inactivation of *Listeria monocytogenes* and *Pseudomonas* spp. J. Food Prot. 64:1442–1445.

Vugia, D., J. Hadler, S. Chaves, D. Blythe, K. Smith, D. Morse, P. Cieslak, T. Jones, A. Cronquist, D. Goldman, J. Guzewich, F. Angulo, P. Griffin, and R. Tauxe. 2003. Preliminary FoodNet data on the incidence of foodborne illnesses: selected sites, United States, 2002. MMWR 52:340–343.

Wang, H., C. Tseng, and R. P. Gunsalus. 1999. The *napF* and *narG* nitrate reductase operons in *Escherichia coli* are differentially expressed in response to submicromolar concentrations of nitrate but not nitrite. J. Bacteriol. 181:5303–5308.

Wolin, M. J., E. A. Wolin, and N. J. Jacobs. 1961. Cytochrome-producing anaerobic vibrio, *Vibrio succinogenes* Sp.n. J. Bacteriol. 81:911–917.

Zhao, T., M. P. Doyle, B. G. Harmon, C. A. Brown, P. O. E. Mueller, and A. H. Parks. 1998. Reduction of carriage of enterohemorrhagic *Escherichia coli* O157:H7 in cattle by inoculation with probiotic bacteria. J. Clin. Microbiol. 36:641–647.

▌ 6 ▌

Colonization and Pathogenesis of Foodborne *Salmonella* in Egg-Laying Hens

Lisa M. Norberg, Jackson L. McReynolds, Woo-Kyun Kim,
Vesela I. Chalova, David J. Nisbet, and Steven C. Ricke

Introduction

Salmonellosis is a foodborne disease that affects over 1.4 million people each year in the United States alone, of which more than 500 cases are fatal (CDC 2004). Frenzen et al. (1999) has estimated the annual cost of foodborne *Salmonella* infection to be nearly $2.3 billion in the United States. The majority of this cost is due to loss of productivity in the workforce and medical bills (Frenzen et al. 1999). While human *Salmonella* cases are at their lowest levels since 1987, they are not on the decline (Cogan and Humphrey 2003). The CDC estimates that for every one case that is reported, 37 go unrecognized (CDC 2004), thus the total number of outbreaks is much greater and the cost estimates are quite conservative. While there are estimated to be nearly 2,400 different serovars of *Salmonella* believed to cause foodborne illness, two are considered to be the most dominant.

The two serotypes that cause the majority of the cases are *Salmonella* subspecies *enterica* serovar Enteritidis (SE) and serovar *Salmonella* Typhimurium (ST). SE cases are generally believed to be derived from shell eggs from chickens. These eggs come from hens that appear asymptomatic but carry SE in their gastrointestinal and reproductive tracts, which can potentially be transmitted to the interior of the egg prior to shell formation; in addition, these contaminated eggs are indistinguishable from noncontaminated, normal eggs (Cogan and Humphrey 2003; Wegener et al. 2003). This fact along with undercooking contaminated eggs leads to SE infection. Patrick et al. (2004) estimates that of all the outbreaks of SE from 1985 through 1999, 80% were egg associated. Among this 80%, 28% of the outbreaks were from foods that contained raw eggs such as ice cream, egg nog, and Caesar salad dressing. Of the outbreaks, 27% were attributed to traditional egg dishes such as omelets, French toast, and other foods that use egg batter (Patrick et al. 2004).

While the incidence of SE in egg contents is estimated to be 0.005%, it is still a prominent food-safety issue, as approximately 3.2 million eggs are contaminated annually in the United States alone (CDC 2004). Physiological stresses, such as molting, increase the susceptibility of SE infection in the hen (Poppe 1999). Modified molting diets and prebiotics incorporation appear to be alternative approaches to control preharvest SE contamination of layer flocks. Understanding the physiology of the hen and the route of transmission of *Salmonella* will facilitate the formulation of effective alternative molt diets that benefit the hen while inhibiting *Salmonella* colonization.

Molting in Laying Hens

Concepts and Management Programs

Molting in avian species is defined as periodic shedding and replacement of feathers. During this time, most birds also undergo reproductive rejuvenation in which egg production ceases and the reproductive tract regresses (Berry 2003). Domestic laying hens will naturally undergo a molt after an extensive egg-laying period; however, this process generally takes approximately four months (North and Bell 1990), which raises economic concerns as the hens continue to be fed during nonproductive times (McDaniel and Aske 2000).

The molting process can be accelerated by a management practice commonly called an induced or forced molt (other terms include pause, forced rest, and recycling; Berry 2003). The induced molt method, which was developed in the 1960s, uniformly rests all hens and returns them to a more consistent high rate of lay for an extended period (McDaniel and Aske 2000). By the mid-1970s induced molting had gained popularity throughout the United States (Bell 2003; Park et al. 2004b). The U.S. commercial egg industry commonly uses induced molt procedures to rejuvenate flocks for a second or third laying cycle and to increase profits. Accord-ing to Bell (2003), approximately 75% of commercial laying facilities in the United States used an induced molt program in order to rejuvenate flocks for increased productivity.

Implementing an induced molt program can result in a 30% higher profit margin for producers when compared to an all-pullet operation (Bell 2003). Induced molt management practices increase profits by optimizing the use of replacement pullets, considering a nonmolted program would require 47% more hens to keep houses at maximum capacity (Bell 2003). In addition to increased profit margins, an induced molt rejuvenates the hens' reproductive tract to produce higher-quality eggs that are more marketable (Keshavarz and Quimby 2002). The main purpose of molting is to cease egg production in order for the hens to enter a nonreproductive state, which increases egg production and egg quality post molt (Webster 2003). On the average, laying hens undergo an induced molt at 65 to 70 weeks of age and commonly return to egg production, mortality, and egg quality values seen in hens aged 40 to 50 weeks of age (Bell 2003). An induced molt is seen as advantageous because hens are uniformly molted and can return to 50% production in less than 6 weeks (Parkhurst and Mountney 1988).

Physiology of a Molt

An induced molt is usually initiated by decreasing the photoperiod from 16 hours light : 8 hours dark to 8 hours light : 16 hours dark (Andrews et al. 1987) a week prior to removing feed, which allows for continued production while the birds are photosensitized in order for a more complete and rapid molt after fasting (Andrews et al. 1987). Changing day length by either increasing or decreasing it causes changes in circadian and circannual rhythms. The reduction in photoperiod is known to initiate molt and is related to a more complete molt. Reducing photoperiod acts on the hypothalamic-hypophyseal axis (Andrews et al. 1987) and initiates gonadal regression (Berry 2003). Upon the loss of gonadotrophin, ovaries regress and the follicles become atretic while the yolk is resorbed. During this time, ovary weight decreases thus decreasing overall body weight. In addition, a week after the photoperiod is reduced, feed is removed or the diet is changed to a low-energy molt diet, decreasing adipose deposits and overall body weight (Brake 1993).

Optimal body weight loss during a molt is between 27 to 32%, and at this time the shell gland lipid decreases (Brake 1993). Shell gland lipid naturally increases as a hen ages, causing an increase in shell-less eggs; however, once 25% body weight loss is achieved, lipid accumulation in the uterus is decreased (Brake 1993; Berry 2003), consequently decreasing the incidence of shell-less eggs and increasing postmolt production. In addition, molting increases the concentration of shell gland calcitriol receptor and calbindin, which are responsible for increased shell strength postmolt (Brake 1993). Eggshell formation is also closely related to bone metabolism in laying hens with tibia bone-breaking strength, weight, and percentage ash all being negatively correlated with eggshell weight and shell thickness (Kim et al. 2005). Using dual energy X-ray absorptiometry along with conventional bone density assays, Kim et al. (2006) suggested that hens lose considerable quantities of bone mineral during molting. In a followup study, Kim et al. (2007) showed that the medullary and cancellous bones are labile during this process.

During ovary regression, ovarian steroid synthesis of estradiol and proges-terone is decreased thus decreasing the synthesis of yolk precursors in the liver, which are dependent on estrogen (Berry 2003). The reduction of precursors such as phospholipoprotein and energy stores (glycogen) has been shown to influence liver weight loss (Berry and Brake 1985) and ovary involution (Berry 2003). Pro-lactin, the hormone that is responsible for broodiness and rises in concentration as egg laying proceeds, also influences ovary regression and overall body weight loss (Berry 2003). Eventually, the level of prolactin reaches high enough levels to inhibit the hypothalamic release of gonadotropin-releasing hormone (GnRH) and luteinizing hormone (LH) from the pituitary. Prolactin has also been implicated in the reduction of ovarian steroidogenesis (Berry 2003), which, as mentioned pre-viously, induces ovary regression.

Corticosterone is an adrenal glucocorticoid stress hormone affected by the hypothalamic-pituitary-adrenal (HPA) axis, which is activated by the need to mobilize body energy stores (stress). It serves a number of roles during molting such as regulation of behavioral patterns, coping cycles, and general well-being (Cheng et al. 2001). In addition, corticosterone can inhibit growth, reduce the size of gonads, increase heterophil/lymphocyte ratio, and reduce antibody response to a specific antigen, as well as increase fear (Littin and Cockrem 2001). The increases in corticosterone levels are the first significant endocrine changes seen in molted hens and are initiated by stress (Nasir et al. 1999). During a molt, corticosterone levels increase initially and levels decrease as the molt continues. Upon refeeding layer ration at the end of a molt, the levels again increase (Berry 2003) and then revert to normal levels as the energy level of feed is increased (Webster 2003). The increased release in corticosterone during molting can cause a decrease in passage rate in molted hens, which in turn decreases feed intake (Nasir et al. 1999) and forces the hen to save energy by maintaining metabolic activity at a low level, which contributes to body weight loss (DeJong et al. 2002). Furthermore, increased corticosterone levels retard spleen weights as immunological organs such as the spleen are sensitive to corticosterone (Post et al. 2003), also contributing to body weight loss.

Heterophil and lymphocyte ratios are frequently used as stress indicators in laying hens during molting (DeJong et al. 2002). Studies have shown that during molting there is a positive relationship between heterophil and lymphocytes in the blood and corticosterone levels (DeJong et al. 2002). The increase in heterophil and lymphocyte ratios can be explained in the same manner as the increase in cor-ticosterone levels as corticosterone causes changes in circulating populations of leukocytes (Webster 2003). Limited feeding has also been shown to increase the ratio (DeJong et al. 2002). The change in circulating populations of leukocytes has been proven to affect the immunological defenses against infection and disease (Webster 2003).

In addition to ovary regression, another goal of a successful molt is feather loss and replacement, which is believed to be controlled by the thyroid hormones. An increase in thyroid hormones results in an increase in feather loss. While ovary regression and feather loss function under separate control, considerable interest has arisen regarding the relationship between ovarian and thyroid steroids (Berry 2003).

Feed Withdrawal Molting

While there are several molting methods, feed withdrawal (FW) historically was the most popular due to ease of application, low cost, and low mortality rates (Keshavarz and Quimby 2002; Bell 2003). Some FW periods were relatively short, completed in as little as 4 days (Webster 2003), while some were as long as 14 days (Bell 2003). Following the FW period, the lighting scheme was returned to normal (16 hours light : 8 hours dark) and the hens were fed layer ration and returned to production on a low-energy corn-based maintenance diet that allows for a resting period (North and Bell 1990). A short resting period of 0 to 7 days resulted in flocks returning to peak production in as little as 4 weeks; whereas, a longer resting period could last as long as 21 days and results in peak production at 10 to 11 weeks postmolt (North and Bell 1990).

Feed withdrawal molting methods were seen as logical because wild birds exhibited similar behavior when they undergo a natural molt; they lose as much as 40% of their body weight, half of which is attributed to ovarian regression (Berry 2003) while refusing food until the later stages of the molt (Mrosovsky and Sherry 1980). Berry (2003) stated that survival with little or no food for relatively long periods of time is a normal feature of a chicken's physiology.

Feed withdrawal molting methods were shown to alter behavior in force-molted hens (Webster 2003). Some behavioral activities that were studied include gakel calls, non-nutritive pecking, preening, head shaking, aggression, and general activity (sitting, standing, walking, among other behaviors). Gakel calls are vocalizations thought to be signs of frustration in molted hens. Non-nutritive pecking is pecking at objects such as cage wires, the floor, or any other object placed into the bird's view. This was a typical response of birds when feed is withdrawn and occurred in response to hunger. Birds are natural foragers and as a result of feed removal, they continue this behavior and search for food in their environment. Preening may be performed as a displacement action in situations of conflict or frustration; however, some researchers have suggested that preening in molted hens is related to integument stimulation as feathers are being pushed out (Webster

2000). Aggression has been shown to increase during periods of FW in molted hens. When hens were deprived of feed they tended to not only peck at non-nutritive objects but other hens as well. If one hen is more dominant in the group, she will continue to peck at subordinate hens, possibly until death (Webster 2000). Birds stressed by FW molting show signs of hyperactivity, increased drinking, and increased non-nutritive pecking (pecking at nonfood substances). These behaviors are characteristic of hunger and frustration (DeJong et al. 2002).

Animal welfare concerns affected the means by which commercial producers molt their hens. Efforts were made to reduce or even eliminate the use of programs that require complete removal of feed from hens. For this reason, alternative methods that do not require complete removal of feed were extensively considered. As a result, a variety of feedstuffs were developed as dietary alternatives to FW to alleviate concerns about increased *Salmonella* infection (Holt 2003), effects of fasting (Berry 2003), and stress (Post et al. 2003).

Salmonella Invasion in Molted Hens

Intestinal Colonization

In addition to stress associated with FW, hens in controlled experiments exhibited an increased susceptibility to SE infection, marked by increased intestinal shedding and dissemination of SE to internal organs such as the liver, spleen, and ovary (Holt and Porter 1992). Consequently, it was concluded that practices such as FW could increase the susceptibility of SE infection in hens and increase the risk of human salmonellosis from SE-contaminated eggs (Holt and Porter 1992). Feed withdrawal molting increased stress in experimental hens, which led to a compromised immune system (Holt 2003). By compromising the immune system, hens are more susceptible to infection by a number of organisms, in particular SE (Holt 1993; Holt 2003). The indigenous microflora in the alimentary canal naturally provides a hostile environment for infectious agents such as SE; however, during FW molting, the environment is altered (Holt and Porter 1992). The first line of defense in the chicken's alimentary canal is the crop, which is a nonsecretory organ that is populated by lactobacilli with a low pH of 4 to 5 (Holt 2003). When feed enters the crop, it is fermented by lactic acid bacteria, which in turn decreases the pH and inhibits the growth of enteric pathogens (Hinton et al. 2000). However, during FW, microbiological changes occur in the crop due to the absence of feed, thus allowing pH to increase and increasing susceptibility to pathogen infection (Hinton et al. 2000; Durant et al. 1999; Ricke 2003a). When feed is present in the intestines, particularly the ceca, volatile fatty acids (VFA) are produced, which act on enteric pathogens much in the same way lactic acid does in the crop (Hentges 1983; Russell and Diez-Gonzalez 1998; Van Der Wielen et al. 2000; Ricke 2003b).

Again, without feed in the alimentary canal, enteric bacteria are more able to colonize the intestines. Upon colonization of the intestinal epithelium, SE is able to invade a variety of organs including the spleen, liver, ceca, ovary, and oviduct (Berry and Brake 1985; Gast 1994). Once organ invasion is achieved, eggs from infected hens can also become contaminated with SE.

Egg Contamination

SE egg contamination has been reported to occur in two ways: either in-egg by which the eggs are contaminated during the formation in the ovary and oviduct or by on-egg contamination caused by feces (Barua and Yoshimura 2004) and other environmental factors (Davies and Breslin 2001). In both modes of transmission, the hen shows little to no signs of being infected with SE (Guard-Petter 2001). While in-egg contamination is rarely seen without being associated with shell contamination, it is likely due to ovarian infection (Gast 1994). The ovary is the site of egg yolk formation (Burley and Vadehra 1989); therefore, when the ovary is contaminated, SE is deposited in the yolk (Gast et al. 2004). The yolk does contain antibodies that can inhibit bacterial growth (Gast 1994); however, at storage temperatures above 20°C, bacteria flourish (Gast and Holt 2001). *Salmonella* contamination may also occur during the passage of the egg through the oviduct in association with albumen (Keller et al. 1995). Colonization of the oviduct by SE has been established experimentally by using different infection routes including intravaginal and intracloacal (Miyamoto et al. 1997), oral (Kinde et al. 2000), and intravenous (Gast et al. 2002). Once in the lumen of the oviduct, SE can develop a persisting colonization of the oviduct tissue and cause contamination of the forming eggs (De Buck et al. 2004).

Shell contamination occurs after the formation of the egg shell either in the oviduct or by fecal contents as the cloaca is the common orifice for the reproductive and digestive tracts (Takata et al. 2003). While the occurrence of on-egg contamination is also low, it is increased by relatively high humidity, low temperatures during storage (Humphrey 1994), and cracked shells (Guard-Petter 2001). A variety of chemical and physical approaches have been explored in our laboratory for limiting external eggshell contamination by *Salmonella* and other bacteria (Kuo et al. 1996; Kuo, Carey, and Ricke 1997; Kuo, Ricke, and Carey, 1997; Kuo, Kwon et al. 1997; Kwon et al. 1997; McKee et al. 1998; Knape et al. 1999, 2001; Ricke et al. 2001).

Detection and Control of SE in Molting Laying Hens

Detection in Layer Houses and Feed

SE reduction in the egg-laying industry will no doubt require more comprehensive control programs that anticipate sources of SE in the egg layer house and potential for infection of susceptible hens. Vaccination along with better surveillance methods that would allow for detection of low levels of SE and other *Salmonella* spp. in the layer house before large numbers became prevalent would be an important component for any control program. Although several environmental sampling approaches have been implemented over the years, most require some manual input which can limit effectiveness for continuous monitoring. Detection of airborne *Salmonella* spp. and other microorganisms in chicken house environments has been examined as one means for more continuous monitoring of *Salmonella* spp. that has the potential to minimize hands-on maintenance (Peña et al. 1999; Kwon et al. 1999; Kwon, Woodward, Pillai et al. 2000; Kwon, Woodward, Corrier et al. 2000; Endley et al. 2001; Pillai and Ricke 2002; Woodward et al. 2004). Air samples have the potential to be fairly autonomous but expense and reliability issues remain to be worked out.

In addition to environmental sources of SE, other components of layer house management will need to be considered. Among these sources, contamination of layer-hen feeds and feed ingredients remains an ongoing potential issue (Maciorowski et al. 2004, 2006, 2007; Ricke 2005; Park et al. 2008). *Salmonella* can survive at detectable levels for several weeks and even months in dry feeds (Ha, Jones et al. 1997; Ha, Maxciorowski et al. 1998). Although a number of feed amendments have been examined for limiting *Salmonella* in poultry feeds the effectiveness remains variable (Ha, Jones et al. 1997; Ha, Maciorowski et al.1998; Park et al. 2003; Maciorowski et al. 2004). However, it has been recently shown that some feed antimicrobial treatments may in fact inhibit recovery of *Salmonella* from the feed samples and mask the actual levels of survivors in culture media (Carrique-Mas et al. 2007). Since this may compromise evaluation of effectiveness of the antimicrobial compounds, it becomes imperative to develop detection methods that overcome this masking effect. Consequently, development of better detection methods that are not only more rapid but more specific for distinguishing the physiological status of *Salmonella* in feeds is needed (Maciorowski et al. 2005, 2006). Molecular methods that could be adapted to quantify *Salmonella* in feed matrix backgrounds and potentially assess virulence status would enable more realistic assessment of risk to the poultry flock. Real-time PCR methods have been successfully developed to quantify *Salmonella* virulence expression in the chicken ceca (Dunkley et al. 2007a; Kundinger et al. 2007), and it is possible these approaches would work in feed matrices as well.

Alternative Molting Diets

Historically researchers have examined alternative diets to FW that provide similar benefits while not altering the health of the animals. General dietary modification strategies have involved either constructing diets that are deficient in some nutrients such as sodium or contain an excess of a particular compound such as zinc. In the past, studies have been conducted using diets mixed with low and high zinc concentrations (Bell 2003; Park, Birkhold et al. 2004; Park et al 2004a, b, c; Ricke, Woodward et al. 2004; Ricke, Park et al. 2004; Ricke, Hume et al. 2004), thyroxine (Keshavarz and Quimby 2002), and low sodium concentrations (Berry and Brake 1985) to induce molt. Such diets have yielded inconsistent results, are costly, and can cause negative behavior such as cannibalistic pecking (Biggs et al. 2004; Webster 2003). Low calcium diets have also been used; however, ovary and oviducts did not regress to a nonproductive state, production did not cease completely, and the diets have been shown to cause osteoporosis and temporary paralysis (Webster 2003). A second general approach has incorporated the use of insoluble plant fibers such as grape pomace (Keshavarz and Quimby 2002), cotton meal (Davis et al. 2002), jojoba meal (Arnouts et al. 1993; Vermaut et al. 1997), wheat middlings (Seo et al. 2001), and alfalfa (Donalson et al. 2005; Dunkley et al. 2007a, b, c; Dunkley, McReynolds, Dunkley, Kubena et al. 2007; Dunkley, McReynolds, Dunkley, Njongmeta et al. 2007; Dunkley, Friend et al. 2008; Dunkley, Kim et al. 2008; Landers, Howard et al. 2005; Landers, Woodward et al. 2005; Landers, Moore, Dunkley et al. 2008; Landers, Moore, Herrera et al. 2008; Woodward et al. 2005; McReynolds et al. 2005, 2006). Of these high-fiber diets alfalfa has been one of the most extensively studied for potential to retain an effective molt and limit SE colonization. Specific issues associated with alfalfa as a molt induction diet are discussed in the following section.

Alfalfa as an Alternative Molting Diet

Alfalfa is high in protein (17%), low in energy (1,200 kcal/kg), and relatively high in calcium (1.44%; NRC 1994). Alfalfa has been used for many years to add pigment to egg yolks as hens do not produce yolk pigment in their bodies (Madiedo and Sunde 1964). Xanthophyll, which is present at 220 mg/kg (Madiedo et al. 1964; NRC 1994) in alfalfa, is estimated to contribute to 70% of the total yolk color while zeaxanthin contributes the remaining 30% (Madiedo and Sunde 1964). Pro-Xan is an alfalfa leaf concentrate made from the wet fraction of fresh alfalfa (Kuzmicky et al. 1977) that includes protein, amino acids, and other nutrients that has approximately 1.7 times the xanthophyll availability when compared to alfalfa meal (Kuzmicky and Kohler 1977). Alfalfa increases egg quality by adding pigment to the yolk; however, with increased amounts of alfalfa, the occurrence of blood spots also increased, causing eggs to be unmarketable, thus causing losses to producers (Sauter et al. 1965).

Several characteristics of alfalfa make it an optimal molting diet for laying hens. According to Bell (2003) an ideal molting diet should be inexpensive, result in low mortalities, be easy to apply, and lead to postmolt production comparable to that of FW molt. In addition, a molt diet should cause a cease in egg production to allow for reproductive rejuvenation. Eggs laid by alfalfa-molted hens have been shown to be overall heavier, longer, and exhibit higher albumen heights (Landers et al. 2005a). In addition, no differences were found between taste/texture and color of eggs laid by feed-deprived and alfalfa-molted hens. According to these studies, alfalfa induces molt just as well as FW and has the ability to reduce SE colonization (Woodward et al. 2005; Dunkley et al. 2007a). No differences in consumer preference in eggs from alfalfa-molted hens were detected (Landers, Howard et al. 2005).

Alternative molting methods must also effectively induce molt in hens while maintaining the indigenous microflora in order to reduce colonization or prevent SE infection (Ricke 2003a; Park, Birkhold et al. 2004). Within the intestinal tract is a diverse, complex microbial ecosystem with the majority of bacteria residing in the cecum (Salanitro et al. 1974; Van Der Wielen et al. 2000; Apajalahti et al. 2004; Ricke and Pillai 1999; Ricke, Woodward et al. 2004; Saengkerdsub et al. 2007). The cecum in birds is much different as compared to mammalian ceca, due to the increased surface area, which is helpful in hydrolysis, absorption, and fermentation (Vispo and Karasov 1997). Most of the bacteria in the cecum are considered anaerobic and include species such as *Lactobacilli, Bifidobacterium,* and *Propionio-bacterium* (Salanitro et al. 1974). The microflora in the ceca work together to maintain a stable ecosystem in order to form a natural resistance to infections produced by enteric pathogens (Hentges 1983; Ricke, Woodward et al. 2004); this is accomplished by forming a physical barrier to keep intestinal bacteria in check and protect against enteric pathogens by discriminating between enteric and resident microflora (Lu and Walker 2001; Ricke, Woodward et al. 2004). Enteric pathogens possess specialized processes, which allow them to penetrate the intestinal epithelium. Inside the intestinal epithelium the pathogen can adhere to the surface, colonize, and establish permanent residence, which can cause disease if not prevented by the natural microflora (Lu and Walker 2001).

Alfalfa is fermentable by cecal microflora *in vitro* and when fed helps to retain the indigenous microflora (Hume et al. 2003; Dunkley, Dunkley et al. 2007; Dunkley et al. 2007b ; Saengkerdsub et al. 2006). A study was conducted by Woodward et al. (2005) where an alfalfa diet was also shown to successfully induce molt as compared to the traditional FW molt. Hens that underwent the alfalfa molt showed greater lactic acid and VFA concentrations than FW hens, indicating that alfalfa could serve as an inhibitory diet for SE colonization (Woodward et al. 2005). Combining layer ration with alfalfa also induced molt similarly to alfalfa alone as well as contributed to greater fermentation than feed withdrawal and increase feed intake as compared to 100% alfalfa diets (Donalson et al. 2005).

Prebiotics

Prebiotics have been defined by Gibson and Roberfroid (1995) as indigestible food ingredients which stimulate the growth and/or activity of a select number of bacteria in the colon and improve the host's health. In order for a food ingredient to be considered a prebiotic, it must have certain characteristics. Prebiotics have been shown to alter gastrointestinal microflora, alter the immune system, prevent colonic cancer, reduce pathogen invasion including pathogens such as SE and *Escherichia coli,* and reduce cholesterol and odor compounds (Cummings at al. 2001; Cummings and Macfarlane 2002; Patterson and Burkholder 2003).

Prebiotics are short-chain carbohydrates that are indigestible by human, animal, and poultry digestive systems. The major effects of prebiotics have been reviewed by Cummings and Macfarlane (2002) and include production of short-chain fatty acids and lactate, selective increases in *Bifidobacteria* and *Lactobacilli,* increase in pathogen resistance, and improved calcium and magnesium absorption. The use of prebiotics in human as well as animal diets is a generally new concept in the United States. In Japan, prebiotics are a normal ingredient in many diets, especially weaning piglets, and their use is ever increasing in Europe (Flickinger et al. 2003). Common prebiotics currently available for human and animal consumption include isomaltooligosaccharides, oligomates, palatinoses, polydextrose, raftilines, soybean oligosaccharides, xylooligosaccharides, and the most popular being fructooligosaccharides.

In addition to using alternative diets to induce molt and potentially reduce SE colonization, researchers have used feed additives to limit colonization and infection in chickens. SE has been shown to be reduced by the presence of mannose and lactose in the diet (Oyofo et al. 1989); however, the results are variable (Corrier et al. 1990; Hinton et al. 2000). Other carbohydrates such as dextrose, maltose, and sucrose had no effect on SE colonization (Oyofo et al. 1989). The addition of prebiotics such as oligosaccharides to poultry diets have also been shown to inhibit SE colonization while beneficially affecting the indigenous microflora (Bailey et al. 1991; Orban et al. 1997; Fernandez et al. 2002).

While the indigenous microflora flourish in the presence of prebiotics, enteric pathogens such as *E. coli, Clostridium perfringens* (Cummings and Macfarlane, 2002) and *Salmonella* (Bailey et al. 1991; Donalson et al. 2007) are limited in their ability to colonize. These pathogens are hindered due to the fact that they are unable to use prebiotic sources as sole carbon energy sources, and when fermentation by indigenous microflora increases in the presence of prebiotics so do volatile fatty acid concentrations, which decrease the pH to levels antagonistic to many pathogenic bacteria (Cummings and Macfarlane 2002). In addition, increased activity by the microflora limits attachment of pathogenic bacteria, which is essential for infection (Flickinger et al. 2003).

Fructooligosaccharides

The most commonly used prebiotic in both human as well as animal diets is fructooligosaccharide (FOS), which is a naturally occurring oligosaccharide usually of plant origin and is the only product recognized and used as a colonic food ingredient and prebiotic (Bomba et al. 2002; Gibson and Roberfroid 1995). FOS is composed of one molecule of glucose and one to three molecules of fructose (Bengmark 1998) and can be marketed commercially as Raftilose or Nutraflora or can be synthesized from food sources (Kaplan and Hutkins 2000). Common foods that contain fructooligosaccharides are garlic, onion, artichoke, and asparagus (Gibson and Roberfroid 1995), and common animal feed ingredients that contain FOS include alfalfa meal, barley, peanut hulls, wheat middlings, and wheat bran (Flickinger et al. 2003).

Due to the β-linkages possessed by FOS, they are able to resist enzymatic degradation and absorption in the upper gastrointestinal tract to reach the cecum where the majority of fermentation occurs in chickens (Gibson and Roberfroid 1995; Júskiewicz et al. 2004; Xu et al. 2003). Once in the ceca, FOS are selectively fermented by strains of bifidobacteria subsequently decreasing the pH. The decrease in pH is attributed to the production of VFAs by bifidobacteria, mainly acetate, lactate (Gibson and Roberfroid 1995), and butyrate (LeBlay et al. 1999). While this is the primary method of pathogen control, Gibson and Roberfroid (1995) report this may not be the only method. When compared to glucose as a growth promotant for bifidobacteria as well as lactobacillus, FOS proved to be a comparable substrate (Kaplan and Hutkins 2000).

The effective level of FOS to include in diets has been difficult to determine. Bailey et al. (1991) reported that 0.375% FOS was not enough to inhibit SE colonization; however, 0.75% was sufficient to inhibit SE colonization. The results from Waldroup et al. (1993) showed similar results, concluding that there were no effects of 0.375% FOS on broiler body weight, feed efficiency, dressing percentage, or abdominal fat. Xu et al. (2003) reported that 4% FOS improved growth, while higher levels of 8% produced poorer results. Feeding higher levels of FOS (greater than 20%) has been shown to cause flatulence and loose stools (Flickinger et al. 2003). In addition to adding FOS to feed, it has also been added to water at a level of 2% (Janssens et al. 2004). This study showed that at 2% there were no effects on *S.* Typhimurium excretion and water intake increased markedly.

Donalson, Kim et al. (2008) observed that *in vitro* incubation of laying-hen cecal bacteria in the presence of FOS combined with alfalfa or layer ration increased formation of propionate, butyrate, total volatile fatty acids, and lactic acids. Addition of FOS in this *in vitro* system also synergistically decreased *S.* Typhimurium populations, but only when the fermentation had been allowed to proceed for 24 hours prior to *S*. Typhimurium inoculation (Donalson et al. 2007). However when FOS was incorporated into alfalfa-layer ration molt diets, minimal differences were seen in SE reductions and fermentation products when compared to hens fed these same diets without FOS (Donalson, McReynolds et al. 2008). Based on these *in vitro* studies, it would appear that FOS may influence the fermentation of cecal bacteria, but sufficient time needs to be allowed for adaptation of the microflora. However, the *in vivo* results suggest that FOS may have limited ability to alter the fermentative microbial population and/or their fermentation capacity when other fermentable dietary components are present. Detailed profiling of the microbial population in future studies would help to answer this question.

Conclusions and Future Considerations

Molting practices such as feed withdrawal have been shown to increase the susceptibility of laying hens to SE thus increasing the risk of human salmonellosis. Alternative molt diets have been developed to alleviate these food-safety concerns as well as animal-welfare concerns, as the primary method of molting is feed withdrawal. Of these diets, high fiber diets such as alfalfa have proven to be acceptable according to the experiments presented here and by other researchers. Alfalfa combined with layer ration at different ratios has been shown to induce molt, increase postmolt egg quality and postmolt egg production as well as the conventional feed withdrawal method. This is greatly beneficial to the egg industry, as alfalfa is readily available and inexpensive.

The positive effect of prebiotics, in particular of FOS, on *Salmonella* reduction beyond that of alfalfa was more apparent *in vitro* rather than *in vivo*. However, more experiments and research is needed to optimize the delivery of FOS to the host and to determine in more details the response of the hen gastrointestinal microbial population to this additive. In conclusion, foodborne *Salmonella* spp. contamination remains a problem for the poultry industry and will require further efforts to design more interventions at the different stages of production. The layer industry in particular has opportunities to more extensively reduce SE occurrence in the flock. Combinations of increased surveillance, preventative approaches, and dietary modifications during molting should enhance the control of this pathogen.

Acknowledgments

This review was supported by the United States Department of Agriculture (USDA-NRI grant number 2002–02614) and U.S. Poultry and Egg Association grant #485. L.M.N. was partially supported by a Maurice Stein Fellowship awarded by the Poultry Science Association, Savoy, IL.

References

Andrews, D. K., W. D. Berry, and J. Brake. 1987. Effect of lighting program and nutrition on reproductive performance of molted single comb white leghorn hens. Poult. Sci. 66:1298–1305.

Apajalahti, J., A. Kettunen, and H. Graham. 2004. Characteristics of the gastrointestinal microbial communities, with special reference to the chicken. World's Poult. Sci. 60:223–232.

Arnouts, S., J. Buyse, M. M. Cokelaere, and E. Decuypere. 1993. Jojoba meal (*Simmondsia chinensis*) in the diet of broiler breeder pullets: physiological and endocrinological effects. Poult. Sci. 72:1714–1721.

Bailey, J. S., L. C. Blankenship, and N. A. Cox. 1991. Effect of fructooligosaccharide on *Salmonella* colonization of the chicken intestine. Poult. Sci. 70:2433–2438.

Barua, A., and Y. Yoshimura. 2004. Ovarian cell-mediated immune response to *Salmonella enteritidis* infection in laying hens (*Gallus domesticus*). Poult. Sci. 83:997–1002.

Bell, D. D. 2003. Historical and current molting practices in the U.S. table egg industry. Poult. Sci. 82:965–970.

Bengmark, S. 1998. Immunonutrition: Role of biosurfactants, fiber, and probiotic bacteria. Nutrition 14:585–594.

Berry, W. D. 2003. The physiology of induced molting. Poult. Sci. 82:971–980.

Berry, W. D., and J. Brake. 1985. Comparison of parameters associated with molt induced by fasting, zinc and low dietary sodium in caged layers. Poult. Sci. 64:2027–2036.

Biggs, P. E., M. E. Persia, K. W. Koelkebeck, and C. M. Parsons. 2004. Further evaluation of nonfeed removal methods for molting programs. Poult. Sci. 83:745–752.

Bomba, A., R. Nemcová, S. Gancar íková, R. Herich, P. Guba, and D. Mudro ová. 2002. Improvement of the probiotic effect of micro-organisms by their combination with maltodextrins, fructo-oligosaccharides and polyunsaturated fatty acids. Br. J. Nutr. 88:S95–S99.

Brake, J. 1993. Recent advances in induced molting. Poult. Sci. 72:929–931.

Burley, R. W., and D. V. Vadehra. 1989. An outline of the physiology of avian egg formation. Pages 17–23 in The avian egg: chemistry and biology. Wiley, New York, NY.

Carrique-Mas, J. J., S. Bedford, and R. H. Davies. 2007. Organic acid and formaldehyde treatment of animal feeds to control *Salmonella*: efficacy and masking during culture. J. Appl. Microbiol. 103:88–96.

Center for Disease Control (CDC). 2004. Disease information: *Salmonellosis*. Online ed. At http://www.cdc.gov/ncidod/dbmd/diseaseinfo/salmonellosis_g.htm. Accessed September 28, 2004.

Cheng, H. W., G. Dillworth, P. Singleton, Y. Chen, and W. M. Muir. 2001. Effects of group selection for productivity and longevity on blood concentrations of serotonin, catecholamines, and corticosterone of laying hens. Poult. Sci. 80:1278–1285.

Cogan, T. A., and T. J. Humphrey. 2003. The rise and fall of *Salmonella* Enteritidis in the UK. J. Appl. Microbiol. 94:114S–119S.

Corrier, D. E., A. Hinton Jr., R. L. Ziprin, R. C. Beier, and J. R. DeLoach. 1990. Effect of dietary lactose on cecal pH, bacteriostatic volatile fatty acids and Salmonella typhimurium colonization of broiler chicks. Avian Dis. 34:617–625.

Cummings, J. H., and G. T. Macfarlane. 2002. Gastrointestinal effects of prebiotics. Br. J. Nutr. 87:S145–S151.

Cummings, J. H., G. T. Macfarlane, and H. N. Englyst. 2001. Prebiotic digestion and fermentation. Am. J. Clin. Nutr. 73:415S-420S.

Davies, R., and M. Breslin. 2001. Environmental contamination and detection of *Salmonella enterica* serovar *enteritidis* in laying flocks. Vet. Rec. 149:699–704.

Davis, A. J., M. M. Lordelo, and N. Dale. 2002. The use of cottonseed meal with or without added soapstock in laying hen diets. J. Appl. Poult. Res. 11:127–133.

De Buck, J., F. Van Immerseel, F. Haesebrouck, and R. Ducatelle. 2004. Colonization of the chicken reproductive tract and egg contamination by *Salmonella*. J. Appl. Microbiol. 97:233–245.

De Jong, I. C., S. Van Voorst, D. A. Ehlhardt, and H. J. Blokhuis. 2002. Effects of restricted feeding on physiological stress parameters in growing broiler breeders. Br. Poult. Sci. 43:157–168.

Donalson, L. M., W. K. Kim, C. L. Woodward, P. Hererra, L. F. Kubena, D. J. Nisbet, and S. C. Ricke. 2005. Utilizing different ratios of alfalfa and layer ration for molt induction and performance in commercial laying hens. Poult. Sci. 84:362–369.

Donalson, L. M., W.- K. Kim, V. I. Chalova, P. Herrera, C. L. Woodward, J. L. McReynolds, L. F. Kubena, D. J. Nisbet, and S. C. Ricke. 2007. *In vitro* anaerobic incubation of *Salmonella enterica* serotype Typhimurium and laying hen cecal bacteria in poultry feed substrates and a fructooligosaccharide prebiotic. Anaerobe 13:208–214.

Donalson, L. M., W. K. Kim, V. I. Chalova, P. Herrera, J. L. McReynolds, V. G. Gotcheva, D. Vidanovi , C. L. Woodward, L. F. Kubena, D. J. Nisbet, and S. C. Ricke. 2008. In vitro fermentation response of laying hen cecal bacteria to combinations of fructooligosacharide prebiotics with alfalfa or layer ration. Poult. Sci. 87:1263–1275.

Donalson, L. M., J. L. McReynolds, W. K. Kim, V. I. Chalova, C. L. Woodward, L. F. Kubena, D. J. Nisbet, and S. C. Ricke. 2008. The influence of a fructooligosaccharide (FOS) prebiotic combined with alfalfa molt diets on the gastrointestinal tract fermentation, *Salmonella* Enteriditis infection and intestinal shedding in laying hens. Poult. Sci. 87:1253–1262.

Dunkley, C. S., J. L. McReynolds, K. D. Dunkley, L. F. Kubena, D. J. Nisbet, and S. C. Ricke. 2007. Molting in *Salmonella* Enteritidis-challenged laying hens fed alfalfa crumbles. III. Blood plasma metabolite response. Poult. Sci. 86:2492–2501.

Dunkley, C. S., J. L. McReynolds, K. D. Dunkley, L. N. Njongmeta, L. R. Berghman, L. F. Kubena, D. J. Nisbet, and S. C. Ricke. 2007. Molting in *Salmonella* Enteritidis-challenged laying hens fed alfalfa crumbles. IV. Immune and stress protein response. Poult. Sci. 86:2502–2508.

Dunkley, C. S., T. H. Friend, J. L. McReynolds, W. K. Kim, K. D. Dunkley, L. F. Kubena, D. J. Nisbet, and S. C. Ricke. 2008. Behavior of laying hens on alfalfa crumble molt diet. Poult. Sci. 87:815–822.

Dunkley, C. S., T. H. Friend, J. L. McReynolds, C. L. Woodward, W. K. Kim, K. D. Dunkley, L. F. Kubena, D. J. Nisbet, and S. C. Ricke. 2008. Behavioral responses to different alfalfa-layer ration combinations fed during molting. Poult. Sci. 87:1005–1011.

Dunkley, K. D., C. S. Dunkley, N. L. Njongmeta, T. R. Callaway, M. E. Hume, L. F. Kubena, D. J. Nisbet, and S. C. Ricke. 2007. Comparison of in vitro fermentation and molecular microbial profiles of high-fiber feed substrates incubated with chicken cecal inocula. Poult. Sci. 86:801–810.

Dunkley, K. D., J. L. McReynolds, M. E. Hume, C. S. Dunkley, T. R. Callaway, L. F. Kubena, D. J. Nisbet, and S. C. Ricke. 2007a. Molting in *Salmonella* Enteriditis-challenged laying hens fed alfalfa crumbles. I. *Salmonella* Enteriditis colonization and virulence gene *hilA* response. Poult. Sci. 86:1633–1639.

Dunkley, K. D., J. L. McReynolds, M. E. Hume, C. S. Dunkley, T. R. Callaway, L. F. Kubena, D. J. Nisbet, and S. C. Ricke. 2007b. Molting in *Salmonella* Enteritidis challenged laying hens fed alfalfa crumbles II. Fermentation and microbial ecology response. Poult. Sci. 86:2101–2109.

Durant, J. A., D. E. Corrier, J. A. Byrd, L. H. Stanker, and S. C. Ricke. 1999. Feed deprivation affects crop environment and modulates *Salmonella enteritidis* colonization and invasion of leghorn hens. Appl. Environ. Micro. 65:1919–1923.

Endley, S., J. Peña, S. C. Ricke, and S. D. Pillai. 2001. The applicability of *hns* and *fimA* primers for detecting *Salmonella* in bioaerosols associated with animal and municipal wastes. World J. Microbiol. & Biotechnol. 17:363–369.

Fernandez, R., M. Hinton, and B. Van Gils. 2002. Dietary mannan-oligosaccharides and their effect on chicken caecal microflora in relation to *Salmonella* Enteritidis colonization. Avian Path. 31:49–58.

Flickinger, E. A., J. Van Loo, and G. C. Fahey Jr. 2003. Nutritional responses to the presence of inulin and oligofructose in the diets of domesticated animals: a review. Crit. Rev. in Food Sci. and Nutr. 43:19–60.

Frenzen, P. D., T. L. Riggs, J. C. Buzby, T. Breuer, T. Roberts, D. Voetsch, S. Reddy, and the FoodNet Working Group. 1999. *Salmonella* cost estimate update using FoodNet data. Food Review 22:10–15.

Gast, R. K. 1994. Understanding *Salmonella enteritidis* in laying chickens: the contributions of experimental infections. Int. J. Food Micro. 21:107–116.

Gast, R. K., J. Guard-Bouldin, and P. S. Holt. 2004. Colonization of reproductive organs and internal contamination of eggs after experimental infection of laying hens with *Salmonella heidelberg* and *Salmonella enteritidis*. Avian Dis. 48:863–869.

Gast, R. K., J. Guard-Petter, and P. S. Holt. 2002. Characteristics of Salmonella enteritidis contamination in eggs after oral, aerosol, and intravenous inoculation of laying hens. Avian Dis. 46:629–635.

Gast, R. K., and P. S. Holt. 2001. Assessing the frequency and consequences of *Salmonella enteritidis* deposition on the egg yolk membrane. Poult. Sci. 80:997–1002.

Gibson, G. R., and M. B. Roberfroid. 1995. Dietary modulation of the human colonic microbiota: introducing the concept of prebiotics. J. Nutr. 125:1401–1412.

Guard-Petter, J. 2001. The chicken, the egg and *Salmonella enteritidis*. Environ. Micro. 3:421–430.

Ha, S. D., F. T. Jones, Y. M. Kwon, and S. C. Ricke. 1997. Survival of an unirradiated *Salmonella typhimurium* marker strain inoculated in poultry feeds after irradiation. J. Rapid Methods Automation Microbiol. 5:47–59.

Ha, S. D., K. G. Maciorowski, Y. M. Kwon, F. T. Jones, and S. C. Ricke. 1998a. Survivability of indigenous feed microflora and a *Salmonella typhimurium* marker strain in poultry mash treated with buffered propionic acid. Anim. Feed Sci. Technol. 75:145–155.

Ha, S. D., K. G. Maciorowski, Y. M. Kwon, and S. C. Ricke. 1997. Indigenous poultry feed microflora response to ethyl alcohol and buffered propionic acid addition. J. Rapid Methods Automation Microbiol. 5:309–319.

Ha, S. D., K. G. Maciorowski, Y. M. Kwon, F. T. Jones, and S. C. Ricke. 1998b. Indigenous feed microflora and *Salmonella typhimurium* marker strain survival in poultry mash diets containing varying levels of protein. Anim. Feed Sci. Technol. 76:23–33.

Hentges, D. J. 1983. Role of the intestinal microflora in host defense against infection. Pages 311–331 in Human intestinal microflora in health and disease. Academic Press, New York, NY.

Hinton, A., Jr., R. J. Buhr, and K. D. Ingram. 2000. Reduction of *Salmonella* in the crop of broiler chickens subjected to feed withdrawal. Poult. Sci. 79:1566–1570.

Holt, P. S. 1993. Effect of induced molting on the susceptibility of White Leghorn hens to a Salmonella enteritidis infection. Avian Dis. 37:412–417.

Holt, P. S. 2003. Molting and *Salmonella enterica* serovar Enteritidis infection: the problem and some solutions. Poult. Sci. 82:1008–1010.

Holt, P. S., and R. E. Porter Jr. 1992. Microbiological and histopathological effects of an induced-molt fasting procedure on a Salmonella enteritidis infection in chickens. Avian Dis. 36:610–618.

Hume, M. E., L. F. Kubena, T. S. Edrington, C. J. Donskey, R. W. Moore, S. C. Ricke, and D. J. Nisbet. 2003. Poultry digestive microflora biodiversity as indicated by denaturing gradient gel electrophoresis. Poult. Sci. 82:1100–1107.

Humphrey, T. J. 1994. Contamination of egg shell and contents with *Salmonella enteritidis:* a review. Intl. J. Food Micro. 21:31–40.

Janssens, G. P. J., S. Millet, F. Van Immerseel, J. DeBuck, and M. Hesta. 2004 The impact of prebiotics and salmonellosis on apparent nutrient digestibility and *Salmonella* Typhimurium var. *Copenhagen* excretion in adult pigeons (*Columba Livia Domestica*). Poult. Sci. 83:1884–1890.

Jukiewicz, J., Z. Zdu czyk, and J. Jankowski. 2004. Selected parameters of gastrointestinal tract metabolism of turkeys fed diets with flavomycin and different inulin content. World's Poult. Sci. J. 60:177–185.

Kaplan, H., and R. W. Hutkins. 2000. Fermentation of fructooligosaccharides by lactic acid bacteria and bifidobacteria. Appl. Environ. Micro. 66:2682–2684.

Keller, L. H., C. E. Benson, K. Krotec, and R. J. Eckroade. 1995. *Salmonella enteritidis* colonization of the reproductive tract and forming and freshly laid eggs of chickens. Infect. Immun. 63:2443–2449.

Keshavarz, K., and F. W. Quimby. 2002. An investigation of different molting techniques with an emphasis on animal welfare. J. Appl. Poult. Res. 11:54–67.

Kim, W. K., L. M. Donalson, S. A. Bloomfield, H. A. Hogan, L. F. Kubena, D. J. Nisbet, and S. C. Ricke. 2007. Molt performance and bone density of cortical, medullary, and cancellous bone in laying hens during feed restriction or alfalfa-based feed molt. Poult. Sci. 86:1821–1830.

Kim, W. K., L. M. Donalson, P. Herrera, L. F. Kubena, D. J. Nisbet, and S. C. Ricke. 2005. Comparisons of molting diets on skeletal quality and eggshell parameters in hens at the end of the second egg-laying cycle. Poult. Sci. 84:522–527.

Kim, W. K., L. M. Donalson, A. D. Mitchell, L. F. Kubena, D. J. Nisbet, and S. C. Ricke. 2006. Effects of alfalfa and fructooligosaccharide on molting parameters and bone qualities using dual energy X-ray absorptiometry and conventional bone assays. Poult. Sci. 85:15–20.

Kinde, H., H. L. Shivaprasad, B. M. Daft, D. H. Read, A. Ardans, R. Breitmeyer,

G. Rajashekara, K. V. Nagaraja, and I. A. Gardner. 2000. Pathologic and bacteriologic findings in 27-week-old commercial laying hens experimentally infected with Salmonella enteritidis, phage type 4. Avian Dis. 44:239–248.

Knape, K. D., J. B. Carey, R. P. Burgess, Y. M. Kwon, and S. C. Ricke. 1999. Comparison of chlorine with an iodine-based compound on eggshell surface microbial populations in a commercial egg washer. J. Food Safety 19:185–194.

Knape, K. D., J. B. Carey, and S. C. Ricke. 2001. Response of foodborne Salmonella spp. marker strains inoculated on egg shell surfaces to disinfectants in a commercial egg washer. J. Environ. Sci. Health B36:219–227.

Kundinger, M. M., I. B. Zabala-Díaz, V. I. Chalova, W.- K. Kim, R. W. Moore, and S. C. Ricke. 2007. Characterization of rsmC as a potential reference gene for Salmonella Typhimurium gene expression during growth in spent media. Sens. Instrumen. Food Qual. 1:99–103.

Kuo, F.-L., J. B. Carey, and S. C. Ricke. 1997. UV irradiation of shell eggs: Effect on populations of aerobes, molds, and inoculated Salmonella typhimurium. J. Food Prot. 60:639–643.

Kuo, F.-L., J. B. Carey, S. C. Ricke, and S. D. Ha. 1996. Peroxidase catalyzed chemical dip, egg shell surface contamination, and hatching. J. Appl. Poult. Res. 5:6–13.

Kuo, F.-L., Y. M. Kwon, J. B. Carey, B. M. Hargis, D. P. Krieg, and S. C. Ricke. 1997. Reduction of Salmonella contamination on chicken egg shells by a peroxidase-catalyzed sanitizer. J. Food Sci. 62:873–874, 884.

Kuo, F.-L., S. C. Ricke, and J. B. Carey. 1997. Shell egg sanitation: UV radiation and egg rotation to effectively reduce populations of aerobes, yeasts, and molds. J. Food Prot. 60:694–697.

Kuzmicky, D. D., and G. O. Kohler. 1977. Nutritional value of alfalfa leaf protein concentrate (Pro-Xan) for broilers. Poult. Sci. 56:1510–1516.

Kuzmicky, D. D., A. L. Livingston, R. E. Knowles, G. O. Kohler, E. Guenthner, O. E. Olson, and C. W. Carlson. 1977. Xanthophyll availability of alfalfa leaf protein concentrate (Pro-Xan) for broilers and laying hens. Poult. Sci. 56:1504–1509.

Kwon, Y. M., D. P. Kreig, F.-L. Kuo, J. B. Carey, and S. C. Ricke. 1997. Biocidal activity of a peroxidase-catalyzed sanitizer against selected bacteria on inert carriers and egg shells. J. Food Safety 16:243–254.

Kwon, Y. M., C. L. Woodward, D. E. Corrier, J. A. Byrd, S. D. Pillai, and S. C. Ricke. 2000. Recovery of a marker strain of Salmonella typhimurium in litter and aerosols from isolation rooms containing infected chickens. J. Environ. Sci. Health B35:517–525.

Kwon, Y. M., C. L. Woodward, J. Peña, D. E. Corrier, S. D. Pillai, and S. C. Ricke. 1999. Comparison of methods for processing litter and air filter matrices from poultry houses to optimize polymerase chain reaction detection of Salmonella typhimurium. J. Rapid Methods Automation Microbiol. 7:103–111.

Kwon, Y. M., C. L. Woodward, S. D. Pillai, J. Peña, D. E. Corrier, J. A. Byrd, and S. C. Ricke. 2000. Litter and aerosol sampling of chicken houses for rapid detection of Salmonella typhimurium contamination using gene amplification. J. Industrial Microbiol. Biotech. 24:379–382.

Landers, K. L., Z. R. Howard, C. L. Woodward, S. G. Birkhold, and S. C. Ricke. 2005. Potential of alfalfa as an alternative molt induction diet for laying hens: egg quality and consumer acceptability. Bioresource Technol. 96:907–911.

Landers, K. L., C. L. Woodward, X. Li, L. F. Kubena, D. J. Nisbet, and S. C. Ricke. 2005. Alfalfa as a single dietary source for molt induction in laying hens. Bioresource Technol. 96:565–570.

Landers, K. L., R. W. Moore, C. S. Dunkley, P. Herrera, W. K. Kim, D. A. Landers, Z. R. Howard, J. L. McReynolds, J. A. Bryd, L. F. Kubena, D. J. Nisbet, and S. C. Ricke. 2008. Immunological cell and serum metabolite response of 60-week-old commercial laying hens to an alfalfa meal molt diet. Bioresource Technol. 99:604–608.

Landers, K. L., R. W. Moore, P. Herrera, D. A. Landers, Z. R. Howard, J. L. McReynolds, J. A. Bryd, L. F. Kubena, D. J. Nisbet, and S. C. Ricke. 2008. Organ weight and serum triglyceride responses of older (80 week) commercial laying hens fed an alfalfa meal molt diet. Bioresource Technol. 99:6692–6696.

Le Blay, G., C. Michel, H. M. Blottière, and C. Cherbut. 1999. Prolonged intake of fructo-oligosaccharides induces a short-term elevation of lactic acid-producing bacteria and a persistent increase in cecal butyrate in rats. J. Nutr. 129:2231–2235.

Littin, K. E., and J. F. Cockrem. 2001. Individual variation in corticosterone secretion in laying hens. Br. Poult. Sci. 42:536–546.

Lu, L., and W. A. Walker. 2001. Pathologic and physiologic interactions of bacteria with the gastrointestinal epithelium. Am. J. Clin. Nutr. 73:1124S–1130S.

Maciorowski, K. G., P. Herrera, F. T. Jones, S. D. Pillai, and S. C. Ricke. 2007. Effects on poultry and livestock of feed contamination with bacteria and fungi. Animal Feed Sci. Technol. 133:109–136.

Maciorowski, K. G., P. Herrera, M. M. Kundinger, and S. C. Ricke. 2006. Animal feed production and contamination by foodborne *Salmonella*. J. Consumer Prot. Food Safety 1:197–209.

Maciorowski, K. G., F. T. Jones, S. D. Pillai, and S. C. Ricke. 2004. Incidence, sources, and control of food-borne *Salmonella* spp. in poultry feeds. World's Poultry Sci. J. 60:446–457.

Maciorowski, K. G., S. D. Pillai, F. T. Jones, and S. C. Ricke. 2005. Polymerase chain reaction detection of foodborne *Salmonella* spp. in animal feeds. Crit. Rev. Microbiol. 31:45–53.

Madiedo, G., E. F. Richter, and M. I. Sunde. 1964. A comparison between chemical determination for xanthophylls and yolk pigmentation scores for yellow corn, alfalfa, algae, lake weed and marigold petals. Poult. Sci. 43:990–994.

Madiedo, G., and M. L. Sunde. 1964. The effect of algae, dried lake weed, alfalfa and ethoxyquin on yolk color. Poult. Sci. 43:1056–1061.

McDaniel, B. A., and D. R. Aske. 2000. Egg prices, feed costs, and the decision to molt. Poult. Sci. 79:1242–1245.

McKee, S. R., Y. M. Kwon, J. B. Carey, A. R. Sams, and S. C. Ricke. 1998. Comparison of a peroxidase-catalyzed sanitizer with other egg sanitizers using a laboratory-scale sprayer. J. Food Safety 18:173–183.

McReynolds, J., L. Kubena, J. Byrd, R. Anderson, S. Ricke, and D. Nisbet. 2005. Evaluation of *Salmonella enteriditis* in molting hens after administration of an experimental chlorate product (for nine days) in the drinking water and feeding an alfalfa molt diet. Poult. Sci. 84:1186–1190.

McReynolds, J. L., R. W. Moore, L. F. Kubena, J. A. Byrd, C. L. Woodward, D. J. Nisbet, and S. C. Ricke. 2006. Effect of various combinations of alfalfa and standard layer diet on susceptibility of laying hens to *Salmonella* Enteriditis during forced molt. Poult. Sci. 85:1123–1128.

Miyamoto, T., E. Baba, T. Tanaka, K. Sasai, T. Fukata, and A. Arakawa. 1997. Salmonella enteritidis contamination of eggs from hens inoculated by vaginal, cloacal, and intravenous routes. Avian Dis. 41:296–303.

Mrosovsky, N., and D. F. Sherry. 1980. Animal anorexias. Science 207:837–842.

National Research Council (NRC). 1994. Nutrient requirements of poultry. 9th rev. ed. National Academy Press, Washington, DC.

Nasir, A., R. P. Moudgal, and N. B. Singh. 1999. Involvement of corticosterone in food intake, passage time and *in vivo* uptake of nutrients in the chicken (*Gallus domesticus*). Br. Poult. Sci. 40:517–522.

North, M. O., and D. D. Bell. 1990. Commercial chicken production manual. 4th ed. Chapman & Hall, New York.

Orban, J. I., J. A. Patterson, A. L. Sutton, and G. N. Richards. 1997. Effect of sucrose thermal oligosaccharide caramel, dietary vitamin-mineral level, and brooding temperature on growth and intestinal bacterial populations of broiler chickens. Poult. Sci. 76:482–490.

Oyofo, B. A., R. E. Droleskey, J. O. Norman, H. H. Mollenhauer, R. L. Ziprin, D. E. Corrier, and J. R. DeLoach. 1989. Inhibition by mannose of *in vitro* colonization of chicken small intestine by *Salmonella typhimurium*. Poult. Sci. 68:1351–1356.

Park, S. Y., S. G. Birkhold, L. F. Kubena, D. J. Nisbet, and S. C. Ricke. 2003. Survival of a *Salmonella typhimurium* poultry marker strain added as a dry inoculum to zinc and sodium organic acid amended feeds. J. Food Safety 23:263–274.

Park, S. Y., S. G. Birkhold, L. F. Kubena, D. J. Nisbet, and S. C. Ricke. 2004a. Effects of high zinc diets using zinc propionate on molt induction, organs, and postmolt egg production and quality in laying hens. Poult. Sci. 83:24–33.

Park, S. Y., S. G. Birkhold, L. F. Kubena, D. J. Nisbet, and S. C. Ricke. 2004b. Review on the role of dietary zinc in poultry nutrition, immunity, and reproduction. Biological Trace Element Res. 101:147–163.

Park, S. Y., W. K. Kim, S. G. Birkhold, L. F. Kubena, D. J. Nisbet, and S. C. Ricke. 2004a. Using a feed-grade zinc propionate to achieve molt induction in laying hens and retain postmolt egg production and quality. Biol. Trace Element Res. 101:165–179.

Park, S. Y., W. K. Kim, S. G. Birkhold, L. F. Kubena, D. J. Nisbet, and S. C. Ricke. 2004b. Induced moulting issues and alternative dietary strategies for the egg industry in the United States. World's Poult. Sci. J. 60:196–209.

Park, S. Y., C. L. Woodward, L. F. Kubena, D. J. Nisbet, S. G. Birkhold, and S. C. Ricke. 2008. Environmental dissemination of foodborne *Salmonella* in preharvest poultry production: Reservoirs, critical factors and research strategies. Critical Rev. Environmental Sci. Technol. 38:73–111.

Parkhurst, C. R., and G. J. Mountney. 1988. Poultry meat and egg production. Chapman & Hall, New York, NY.

Patrick, M. E., P. M. Adcock, T. M. Gomez, S. F. Altekruse, B. H. Holland, R. V. Tauxe, and D. L. Swerdlow. 2004. *Salmonella* Enteritidis infections, United States, 1985–1999. Emerg. Infect. Dis. 10:1–7.

Patterson, J. A., and K. M. Burkholder. 2003. Application of prebiotics and prebiotics in poultry production. Poult. Sci. 82:627–631.

Peña, J., S. C. Ricke, C. L. Shermer, T. Gibbs, and S. D. Pillai. 1999. A gene amplification—hybridization sensor based methodology to rapidly screen aerosol samples for specific bacterial gene sequences. J. Environ. Sci. Health A34:529–556.

Pillai, S. D., and S. C. Ricke. 2002. Bioaerosols from municipal and animal wastes: background and contemporary issues. Can. J. Microbiol. 48:681–696.

Poppe, C. 1999. Epidemiology of *Salmonella enterica* serovar Enteritidis. Pages 3–18 in *Salmonella enterica* serovar Enteritidis in humans and animals—epidemiology, pathogenesis, and control. A. M. Saeed, R. K. Gast, M. E. Potter, and P. G. Wall, eds. Iowa State University Press, Ames, IA.

Post, J., J. M. J. Rebel, and A. A. H. M. ter Huurne. 2003. Physiological effects of elevated plasma corticosterone concentrations in broiler chickens: an alternative means by which to assess the physiological effects of stress. Poult. Sci. 82:1313–1318.

Ricke, S. C. 2003a. The gastrointestinal tract ecology of *Salmonella* Enteritidis colonization in molting hens. Poult. Sci. 82:1003–1007.

Ricke, S. C. 2003b. Perspectives on the use of organic acids and short chain fatty acids as antimicrobials. Poult. Sci. 82:632–639.

Ricke, S. C. 2005. Ensuring the safety of poultry feed. Pages 174–194 in Food safety control in the poultry industry. C. C. Mead, ed. Woodhead Publishing Ltd., Cambridge, UK.

Ricke, S. C., S. G. Birkhold, and R. K. Gast. 2001. Eggs and egg products. Pages 473–481 in Compendium of methods for the microbiological examinations of foods. 4th ed. F. P. Downes and K. Ito, eds. American Public Health Association, Washington, DC.

Ricke, S. C., M. E. Hume, S. Y. Park, R. W. Moore, S. G. Birkhold, L. F. Kubena, D. J. Nisbet. 2004. Denaturing gradient gel electrophoresis (DGGE) as a rapid method for assessing gastrointestinal tract microflora responses in laying hens fed similar zinc molt induction diets. J. Rapid Meth. Auto. Micro. 12:69–81.

Ricke, S. C., S. Y. Park, R. W. Moore, Y. W. Kwon, C. L. Woodward, J. A. Byrd, D. J. Nisbet, and L. F. Kubena. 2004. Feeding low calcium and zinc molt diets sustains gastro-intestinal fermentation and limits *Salmonella enterica* serovar Enteritidis colonization in laying hens. J. Food Safety 24:291–308.

Ricke, S. C., and S. D. Pillai. 1999. Conventional and molecular methods for understanding probiotic bacteria functionality in gastrointestinal tracts. Crit. Reviews Microbiol. 25:19–38.

Ricke, S. C., C. L. Woodward, Y. M. Kwon, L. F. Kubena, and D. J. Nisbet. 2004. Limiting avian gastrointestinal tract *Salmonella* colonization by cecal anaerobic bacteria and a potential role for methanogens. Pages 141–150 in Preharvest and postharvest food safety: contemporary issues and future directions. R. C. Beier, S. D. Pillai, T. D. Phillips, and R. L. Ziprin, eds. Blackwell Publishing Professional, Ames, IA.

Russell, J. B., and F. Diez-Gonzalez. 1998. The effects of fermentation acids on bacterial growth. Adv. in Microb. Phys. 39:205–234.

Saengkerdsub, S., R. C. Anderson, H. H. Wilkinson, W. K. Kim, D. J. Nisbet, and S. C. Ricke. 2007. Identification and quantification of methanogenic archaea in adult chicken ceca. Appl. Environ. Microbiol. 73:353–356.

Saengkerdsub, S., W. K. Kim, R. C. Anderson, D. J. Nisbet, and S. C. Ricke. 2006. Effects of nitrocompounds and feedstuffs on in vitro methane production in chicken cecal contents and rumen fluid. Anaerobe 12:85–92.

Salanitro, J. P., I. G. Blake, and P. A. Muirhead. 1974. Studies on the cecal microflora of commercial broiler chickens. Appl. Micro. 28:439–447.

Sauter, E. A., C. F. Petersen, C. E. Lampman, and A. C. Wiese. 1965. A study of the influence of dehydrated alfalfa meal on the production of blood spots in eggs. Poult. Sci. 44:52–62.

Seo, K.-H., P. S. Holt, and R. K. Gast. 2001. Comparison of *Salmonella* Enteritidis infection in hens molted via long-term feed withdrawal versus full-fed wheat middling. J. Food Prot. 64:1917–1921.

Takata, T., J. Liang, H. Nakano, Y. Yoshimura. 2003. Invasion of *Salmonella* Enteritidis in the tissues of reproductive organs in laying Japanese quail: an immunocytochemical study. Poult. Sci. 82:1170–1173.

Van Der Wielen, P. W. J. J., S. Biesterveld, S. Notermans, H. Hofstra, B. A. P. Urlings, and F. Van Knapen. 2000. Role of volatile fatty acids in development of the cecal microflora in broiler chickens during growth. Appl. Environ. Micro. 66:2536–2540.

Vermaut, S., K. De Coninck, G. Flo, M. Cokelaere, M. Onagbesan, and E. Decuypere. 1997. Effect of deoiled jojoba meal on feed intake in chickens: satiating or taste effect? J. Agric. Food Chem. 45:3158–3163.

Vispo, C., and W. H. Karasov. 1997. The interaction of avian gut microbes and their host: an elusive symbiosis. Pages 116–155 in Gastrointestinal Microbiology. R. I. Mackie and B. A. White, eds. Chapman and Hall, New York, NY.

Waldroup, A. L., J. T. Skinner, R. E. Hierholzer, and P. W. Waldroup. 1993. An evaluation of fructooligosaccharide in diets for broiler chickens and effects on salmonellae contamination of carcasses. Poult. Sci. 72:643–650.

Webster, A. B. 2000. Behavior of white leghorn laying hens after withdrawal of feed. Poult. Sci. 79:192–200.

Webster, A. B. 2003. Physiology and behavior of the hen during induced molt. Poult. Sci. 82:992–1002.

Wegener, H. C., T. Hald, D. L. F. Wong, M. Madsen, H. Korsgaard, F. Bager, P. Gerner-Smidt, and K. Mølbak. 2003. *Salmonella* control programs in Denmark. Emerg. Infect. Dis. 9:774–780.

Woodward, C. L., Y. M. Kwon, L. F. Kubena, J. A. Byrd, R. W. Moore, D. J. Nisbet, and S. C. Ricke. 2005. Reduction of *Salmonella enterica* serovar Enteritidis colonization and invasion by an alfalfa diet during molt in leghorn hens. Poult. Sci. 84:185–193.

Woodward, C. L., S. Y. Park, D. R. Jackson, X. Li, S. G. Birkhold, S. D. Pillai, and S. C. Ricke. 2004. Optimization and comparison of bacterial load and sampling time for bioaerosol detection in a poultry layer house. J. Appl. Poultry Res. 13:433–442.

Xu, Z. R., C. H. Hu, M. S. Xia, X. A. Zhan, and M. Q. Wang. 2003. Effects of dietary fructooligosaccharide on digestive enzyme activities, intestinal microflora and morphology of male broilers. Poult. Sci. 82:1030–1036.

Postharvest Foodborne Pathogen Ecology

∎ 7 ∎

Preharvest Food-Safety Issues That Carry Over into the Plant

Scott M. Russell

Introduction

Over the years, the USDA has instituted more strict regulations in the processing plant with regard to the presence of pathogenic bacteria, such as *Salmonella,* on poultry carcasses. The intention of these regulations is to create an opportunity for continuous improvement by setting a sampling protocol that 20% of the plants at the baseline level will fail and be forced to improve their process. Regulatory samples are comprised of only one carcass per day to preclude making any assessment on a given lot of carcasses and the evaluation is strictly on the evisceration process.

When a poultry company exceeds the *Salmonella* standard set by the USDA, the initial reaction is to place blame on the plant employees. Various companies have spent enormous amounts of time and money attempting to reduce *Salmonella* levels on finished carcasses by making changes in the plant. Unfortunately, this is not always successful. Although the processing-plant employees have the regulatory responsibility to ensure that the product meets or exceeds the *Salmonella* regulation, they have absolutely no control over production factors. Yet, the production people feel that "this is a processing problem" and do not want to get involved. In actuality, it is a "company" problem, and cooperation between processing and production is essential to identify and solve contamination situations. Management within a company must work together to identify and solve problems. Numerous factors during breeding, hatching, grow-out, and transportation (chick and broiler) can directly impact the level of *Salmonella* on the finished product.

Breeders

The breeder chickens have been a cause of concern for the poultry industry for many years with regard to *Salmonella*. *Salmonella* may be transferred on the surface of the egg shell due to fecal contamination during laying, or it may be encased within the egg. Cox et al. (2003) found that *Salmonella* could be found in the reproductive tracts of roosters and in hen sperm storage tubules. These findings indicate that roosters may be infecting hen's reproductive tracts during insemination. Thus, to prevent vertical transmission of *Salmonella*, many companies have instituted a vaccination program. These programs are showing positive benefits in decreasing *Salmonella* populations within the breeder chickens.

In some European countries, all breeder flocks are tested for *Salmonella*. If the flock is positive, it is slaughtered and the eggs are not used. This has dramatically reduced *Salmonella* populations in the broiler chickens. However, due to the sheer scale of the industry in the United States, this approach is seriously impractical.

Another approach that many European countries use is competitive exclusion. The idea is that by feeding the chickens populations of "good bacteria," these good bacteria will colonize the intestines of the chicken and occupy all of the attachment sites. Then, when *Salmonella* is ingested: (1) it does not have anywhere to attach, and (2) bacteriocins produced by the good bacteria will kill the *Salmonella*. The cultures used in these countries are generally undefined. The U.S. Food and Drug Administration will not allow these cultures to be used in the United States. In Europe, however, the undefined cultures have been demonstrated to be successful at reducing *Salmonella*.

Hatchery

To reduce the prevalence of *Salmonella* on poultry carcasses during processing, intervention strategies should be implemented during the hatching phase of poultry production. *Salmonella* spp. may be found in the nest box of breeder chickens, cold egg-storage rooms at the farm, on the hatchery truck, or in the hatchery environment (Cox et al. 2000). These bacteria may then be spread to fertilized hatching eggs on the shell, or in some cases, may penetrate the shell and reside just beneath the surface of the eggshell.

Cox et al. (1990, 1991) reported that broiler and breeder hatcheries were highly contaminated with *Salmonella* spp. Within the broiler hatchery, 71% of eggshell fragments, 80% of chick conveyor belts swabs, and 74% of pad samples placed under newly hatched chicks contained *Salmonella* spp. (Cox et al. 1990).

Cason et al. (1994) reported that, although fertile hatching eggs were contaminated with high levels of *Salmonella* Typhimurium, they were still able to hatch. The authors stated that paratyphoid salmonellae do not cause adverse health effects in the developing and hatching chick. During the hatching process, *Salmonella* spp. is readily spread throughout the hatching cabinet due to rapid air movement by circulation fans. When eggs were inoculated with a marker strain of *Salmonella* during hatching, greater than 80% of the chicks in the trays above and below the inoculated eggs were contaminated (Cason et al. 1994). In an earlier study, Cason et al. (1993) demonstrated that salmonellae on the exterior of eggs or in eggshell membranes could be transmitted to baby chicks during pipping.

Salmonella may persist in hatchery environments for long periods of time. When chick fluff contaminated with *Salmonella* was held for four years at room temperature, up to 1,000,000 *Salmonella* cells per gram could be recovered from these samples (Muira et al. 1964). Goren et al. (1988) isolated salmonellae from three different commercial hatcheries in Europe and reported that the same serotypes found in the hatcheries could be found on processed broiler chicken carcass skin. Thus, proper disinfection of the hatchery environment and fertile hatching eggs is essential for reducing *Salmonella* on ready-to-cook carcasses. Suggestions for elimination of *Salmonella* in the hatchery include

1. Install a disinfectant fogging system or electrostatic spraying system in the hatchery plenum, setters, and hatchers that are linked to a timer system.
2. Spray disinfectant every 30 minutes during setting and hatching to prevent cross-contamination.
3. Thoroughly clean and sanitize setters and hatchers regularly using documented sanitation standard operating procedures.
4. Regularly monitor eggshell fragments, chick paper pads, and chick dander from the bottom of the hatching cabinet for *Salmonella*.

Another popular method for reducing *Salmonella* in broiler chickens is to spray the chicks with a live vaccine in the hatchery. Using vaccines, companies have observed *Salmonella* reductions of 20 to 50%. As with breeders, European countries often use undefined competitive exclusion cultures. In Europe the growers have found that undefined competitive exclusion cultures have been demonstrated to be successful at reducing *Salmonella*.

Broilers

The modern broiler chicken has been bred over the years to be a veritable "eating machine." During growout, broiler chickens eat approximately every four hours. Frequent eating is advantageous because birds that eat this frequently gain weight and put on edible muscle rapidly. This attribute may be considered a disadvantage for maintaining the sanitary quality of the bird during processing. At the end of the growout period, prior to catching the birds and cooping them for transportation to the processing plant, the feed is removed from the birds for a period of approximately three to seven hours. During this time, birds become hungry and begin to search for food. Because there is no food available to them in the feeders, they begin to search for feed on the floor, which may be contaminated. Studies have shown that many birds entering the processing plant have high levels of *Salmonella* in their crops as a result of this litter pecking (Byrd et al. 2001).

Salmonella in the crops of chickens that have consumed litter may be spread from carcass to carcass during the crop removal process (Hargis et al. 1995; Barnhart et al. 1999). During cropping, the cropper piston is inserted into the vent area of the carcass and continues through the entire carcass, spinning as it goes. The piston has sharp grooves on the end of it that pick up the crop and wraps the crop around the end of the cropper piston. As the piston moves through the neck opening, the cropper piston comes in contact with a brush that removes the crop from the piston. Then, the piston, while spinning, goes back through the entire carcass. If the crop breaks during this removal process, the contents leak onto the cropper piston and are transferred to the interior and exterior of the carcass, possibly spreading *Salmonella*.

Studies have been conducted in which the crops of live birds were filled with fluorescein dye (Dr. Allen Byrd, unpublished data). After thirty minutes, the birds were processed. By examining the carcasses at different stages of processing under a black light, crop contents that were transferred to the inside or outside of the carcass could be clearly visualized. These studies have shown that commercial croppers result in a large amount of contamination of the inside and outside of the carcasses. Thus, efforts should be made to control *Salmonella* in the crop prior to the crop removal process.

Some companies have been successful at controlling *Salmonella* in the crop by acidifying the bird's drinking water during the feed withdrawal process. Acetic, citric, or lactic acids and poultry water treatment (PWT) have all been used to acidify the crop to the extent that *Salmonella* are unable to survive. Byrd et al. (2001) found that lactic acid was most effective and that 0.44% lactic acid in the waterers of broilers during the feed withdrawal period reduced *Salmonella* contaminated crops by 80%. This effect carried over to the pre-chill carcasses on which the prevalence of *Salmonella* was reduced by 52.4% (Byrd et al. 2001). Suggestions for elimination of *Salmonella* in the crop prior to processing are as follows:

1. Apply lactic acid to drinking water of the chickens before the feed withdrawal period.

2. Begin by applying small amounts and gradually increase levels until they reach 0.5% (0.64 oz. of lactic acid/gallon of water).
3. Occasionally have the QA employees check the pH of the crops of birds at the plant to ensure that they are being acidified.

Feed Withdrawal

Prior to evisceration, carcasses should be evaluated to determine if the birds have undergone proper feed withdrawal. By examining the abdominal cavity to see if it is concave (small amount of feces in the intestines) or convex (large amount of feces in the intestines), it is possible to determine if the birds have been withdrawn from feed long enough.

Moreover, following evisceration, the intestinal tracts hanging from the birds should be flat and not full of feces or bloated with gas (Northcutt et al. 1997). In the processing plant, birds held off feed for extended periods may have intestines that are distended with gas, which, if nicked during evisceration, may rupture and disperse contents onto the carcass, other carcasses, or processing machinery. Extended periods of feed withdrawal also cause the tensile strength of the intestines to become weak increasing the propensity for them to be torn during evisceration (Northcutt et al., 1997).

If birds are not held off feed long enough (< 8 hours), the intestines will be full of digesta (Northcutt et al. 1997). If pressure is applied to the outside of a bird with full intestines, the contents may come out of the vent and spread onto the carcass. Full colons can easily be nicked or cut during the venting process, especially if any line jerking or swinging occurs, causing intestinal contents to leak onto the carcass. Insufficient feed withdrawal time is perhaps the most important factor in meeting the zero tolerance standard for visible fecal contamination on carcasses entering the chiller. Reprocessing levels as high as 75% and line speeds as low as 20 birds/minute have been reported in plants due to excessive contamination as a result of insufficient feed withdrawal times.

Transportation

Conventional cage-dump systems used in the poultry industry are very difficult to clean and sanitize. Dry excreta should be removed before washing if possible. Implementing rinsing systems that do not thoroughly clean excreta off coops simply rehydrates the excreta, allowing *Salmonella* to proliferate. Thus, if disinfection systems are used, it is best to ensure that they are capable of thoroughly removing excreta prior to sanitizing the coops.

Scalder Sanitation

As the weather begins to warm, a number of poultry companies will begin to experience problems keeping the birds cool in the growout house. As a result, some growers will use foggers to keep them cool. In some cases, these fogging systems wet the birds, allowing litter to attach to feathers. This may result in a serious problem once the birds arrive at the processing plant, because the litter comes off of the birds, resulting in excessive organic material in the scald water. Scalder water containing high concentrations of fecal material comes in contact with the external surface of birds and, during picking, bacteria contained in this dirty water may be massaged into open feather follicles.

To reduce this problem, some companies have installed a bird brush and washer prior to scalding. Large brushes and chlorinated water physically remove the feces from the feathers and skin of the birds. One company using this technique decreased the amount of fecal material going into the scalder by approximately 90%. This decreases the amount of organic material on the surface of carcasses as they go into the chiller.

The next important step in removing organic material from carcasses is the scalder. The scalder is one of the most important areas in the processing plant in which cross-contamination with *Salmonella* can occur. Although most scalders are not set up to be truly counter-current, they should be. The water should move against the carcasses, going from the exit of the scalder toward the entrance. This opposing water flow is essential to wash the birds and remove contamination from the birds as they travel through the scalder. Counter-current flow may be accomplished by adding a steel barrier between the lines of chickens going in either direction. By separating these lines, bacteria that are washed off of the external surface of the chickens entering the scalder are not transferred to those exiting the scalder. The rate of water flow should be high, so as to dilute the concentration of foreign material and bacteria in the scalder. There is a common adage that goes "dilution is the solution to pollution" and it applies in this case. Plants that are not equipped with multistage scalders (scalders with successive, separate tanks) should attempt to make their scalders multistage.

By introducing plenty of fresh water into the scalder (at the exit end), a significant portion of the organic material can be removed from the surfaces of the carcass. If this material is allowed to remain on the surfaces of carcasses, they are likely to spread contamination as they progress down the line to the chiller. If the chiller contains high levels of organic material, then oxidative sanitizers, such as chlorine, will have little effect on bacterial concentrations. Thus, maintaining proper flow direction and water flow rate should increase the efficacy of chlorine as it is used later on in the process to kill bacteria.

Conclusion

Reducing *Salmonella* on fully processed, ready-to-cook carcasses requires a comprehensive approach that includes the entire integrated broiler operation. Efforts should be made during breeding, hatching, growout, and transportation to reduce *Salmonella* prior to processing. All emphasis should not be placed on the processing plant, nor should all of the blame be placed there. Introducing the interventions described above may have a tremendous beneficial effect in terms of reducing *Salmonella* on processed carcasses.

References

Barnhart, E. T., D. J. Caldwell, M. C. Crouch, J. A. Byrd, D. E. Corrier, and B. M. Hargis. 1999. Effect of lactose administration in drinking water prior to and during feed withdrawal on *Salmonella* recovery from broiler crops and ceca. Poult. Sci. 78:211–214.

Byrd, J. A., B. M. Hargis, D. J. Caldwell, R. H. Bailey, K. L. Herron, J. L. McReynolds, R. L. Brewer, R. C. Anderson, K. M. Bischoff, T. R. Callaway, and L. F. Kubena. 2001. Effect of lactic acid administration in the drinking water during preslaughter feed withdrawal on *Salmonella* and *Campylobacter* contamination of broilers. Poult. Sci. 80:278–283.

Cason, J. A., J. S. Bailey, and N. A. Cox. 1993. Location of *Salmonella typhimurium* during incubation and hatching of inoculated eggs. Poult. Sci. 72:2064–2068.

Cason, J. A., J. S. Bailey, and N. A. Cox. 1994. Transmission of *Salmonella typhimurium* during hatching of broiler chicks. Avian Dis. 38:583–588.

Cox, N. A., J. S. Bailey, J. M. Mauldin, and L. C. Blankenship. 1990. Research note: presence and impact of *Salmonella* contamination in commercial broiler hatcheries. Poult. Sci. 69:1606–1609.

Cox, N. A., J. S. Bailey, J. M. Mauldin, L. C. Blankenship, and J. L. Wilson. 1991. Research note: extent of salmonellae contamination in breeder hatcheries. Poult. Sci. 70:416–418.

Cox, N. A., M. E. Berrang, and J. A. Cason. 2000. *Salmonella* penetration of egg shells and proliferation in broiler hatching eggs: a review. Poult. Sci. 79:1571–1574.

Cox, N. A., C. L. Hofacre, R. J. Buhr, J. L. Wilson, J. S. Bailey, D. E. Cosby, M. T. Musgrove, K. L. Hiett, and S. M. Russell. 2003. Attempts to isolate naturally occurring *Campylobacter, Salmonella,* and *Clostridium perfringens* from the ductus deferens, testes and ceca of commercial broiler breeder roosters. Poultry Sci. (Suppl. 1) 82:26.

Goren, E., W. A. de Jong, P. Doornenbal, N. M. Bolder, R. W. Mulder, and A. Jansen. 1988. Reduction of *Salmonella* infection of broilers by spray application of intestinal microflora: a longitudinal study. Vet. Q. 10:249–255.

Hargis, B. M., D. J. Caldwell, R. L. Brewer, D. E. Corrier, and J. R. DeLoach. 1995. Evaluation of the chicken crop as a source of *Salmonella* contamination of broiler carcasses. Poult. Sci. 74:1548–1552.

Muira, S., G. Sato, and T. Miyamae. 1964. Occurrence and survival of *Salmonella* organisms in hatcher chick fluff in commercial hatcheries. Avian Dis. 8:546–554.

Northcutt, J. K., S. I. Savage, and L. R. Vest. 1997. Relationship between feed withdrawal and viscera condition of broilers. Poult. Sci. 76:410–414.

8

Validating HACCP for Small Plants

Michael Gregory and John Marcy

Introduction

Validation of critical limits or steps at a critical control point (CCP) in a Hazard Analysis Critical Control Point (HACCP) plan is the process of documenting that the limit is based on appropriate scientific principles and sound operating practices. Validation data demonstrate that the process examined produces safe food as expected. Applying the proper balance of science and common sense will result in a validation process and an HACCP monitoring plan that results in the production of foods that are safe by design. Because validation of critical limits is essential to the foundation of the HACCP plan, the process needs to be uncomplicated and easy to understand.

Validation: The FSIS Perspective

A discussion about validation must start with language from the Pathogen Reduction–Hazard Analysis Critical Control Point Final Rule (USDA FSIS 1996). The United States Department of Agriculture–Food Safety Inspection Service (FSIS) does not specify any method, means or tool to be used for validation. In fact, the language from the final rule is "FSIS is not prescribing that any particular validation method be used." While this lack of clear expectation may at times be problematic, it does, nonetheless, follow the very basic principle of the HACCP process that each plan is unique to that plant's production. Industry is expected to design, operate, and validate their plan. In order to implement an effective HACCP plan, the facility management must understand and be committed to the process as well as be involved at all levels and in every step of the plan's implementation.

"Well-Documented" Processes

A good place to start the validation process is to take advantage of what FSIS calls "well-documented" processes. These are techniques, process steps, and scientific principles or procedures that are sufficiently supported or accepted by longstanding use so that they need no further support when used. A classic example is the use of freezing to preserve meat and poultry products. There are regulatory standards in place that require certain time and temperature criteria be met for the products to be eligible for the mark of inspection (i.e., zero degrees within 72 hours of entering the freezer 9 CFR §381.66(f)(w)). In most cases, these well-documented processes can be used as "safe harbors" to demonstrate that the regulatory standard has been met.

FSIS Published Guidelines

FSIS publishes guidelines processors may adopt to achieve a regulatory performance standard and maintain the "safe harbor" approach. One of the first places to start research for these documents is found at the FSIS Web site (http://www.fsis.usda.gov/Regulations_&_Policies/Compliance_Guides_Index/index.asp) where these guidelines are published on line and kept up to date.

A few specific examples of documents routinely used by meat and poultry plant HACCP plan developers are listed below. There may be more information found in literature or on the FSIS Web site. All of these will provide the safe harbor and/or the "well-documented" support for the principle so long as the plant follows the process as outlined in the guidance documents. Again, deviation from any fundamental principle of these guidelines (i.e., changing the cooling time outlined) will invalidate the use of the guideline at least for that parameter and the plant must develop their own validation support data.

Examples of specific compliance guidelines available from the FSIS are listed below:

1. Compliance Guidelines To Control LISTERIA MONOCYTOGENES In Post-Lethality Exposed Ready-To-Eat Meat And Poultry Products. http://www.fsis.usda.gov/oppde/rdad/FRPubs/ 97–013F/LM_Rule_Compliance_Guidelines_May_2006.pdf.
2. Appendix A Compliance Guidelines For Meeting Lethality Performance Standards For Certain Meat And Poultry Products http://www.fsis.usda.gov/Frame/FrameRedirect.asp?main=http:// www.fsis.usda.gov/OPPDE/rdad/FRPubs/95–033F/95–033F_ Appendix_A.htm.

3. Appendix B Compliance Guidelines for Cooling Heat-Treated Meat and Poultry Products (Stabilization) http://www.fsis.usda.gov/OPPDE/rdad/FRPubs/95–033D_Appendix_B.htm.

4. Appendix A Guidance on Relative Humidity and Time/Temperature for Cooking/Heating and Applicability to Production of Other Ready-to-Eat Meat and Poultry Products http://www.fsis.usda.gov/OPPDE/rdad/FRPubs/95–033F/Appendix_A_guidance_95–033F.pdf.

5. Time-Temperature Tables For Cooking Ready-To-Eat Poultry Products http://www.fsis.usda.gov/OPPDE/rdad/FSISNotices/RTE_Poultry_Tables.pdf.

An example of another source of validation guidance and data is found at the University of Wisconsin (http://meathaccp.wisc.edu/ ;http://www.meatscience.org/meetings/WSC/2005/Presentations/wsc_2005_006_0000_Buege.pdf). This site provides excellent science for the production of jerky that will yield a safe food.

While HACCP plan managers may choose to pursue their own set of circumstances to achieve the regulatory standard, the small and very small plant operator may simply adopt these safe harbors as their critical limits. Validation of the process then is a simple matter of developing a minimum of 90 days of data from that process to demonstrate that it operates within those limits and as expected.

A novel use of a process or technique requires more validation and the operator of the establishment carries a larger burden of data development to support the new use of the technique. For example, chlorine as an antimicrobial rinse has long been accepted in meat and poultry plants. When this chemical was introduced as an on-line re-processing step it had to be validated for that use. Another example is steam, which has a long history as an antimicrobial agent. Yet when the beef industry chose to use it along with vacuum in a specific processing step, this process had to be validated. The key is whether the facility is using the process or technique in the classical sense or if there is a new approach being implemented in the process.

Validation Steps

The validation process is a matter of implementing the following steps:

1. Define the CCP limit or step to be validated.
2. Gather all scientific support that relates to the process.
3. Review the science and use only the information that applies.

4. Recap the science in some form of white paper or review sheet.
5. Seek assistance from experts (Extension, industry, FSIS, or trade associations).
6. Support the science with data from the process.

Defining the CCP Limit or Step to Be Validated

The CCP limit or step to be validated must be completely defined. Definitions should include the critical limits, monitoring procedures, monitoring locations, frequency of monitoring, corrective actions, and expected results. A flow diagram establishing the exact location of the step should also be included.

Gathering and Reviewing Supporting Scientific Literature, Data, and Resources

Facility managers typically use some form of published data that has direct application to their process as the foundation for validation procedures. The first question that generally arises is whether the processes described in the published data being cited are similar to the process in the plant. While this can be a somewhat subjective process, generally applicable published data can be found with applicable links to processes used today.

There are many sources of literature that can be researched by the small and very small facility managers. Many universities, for example, maintain a resource library of articles that may be used. HACCP Roundtable discussions such as those hosted by the University of Arkansas and Oklahoma State University (and many others) are ideal places to locate these libraries of articles. The Ohio State University has published a document that incorporates titles and links to many sources of literature that may be used as validation information.

Anyone with a minimum of Internet experience has probably already found that a simple Google search may return far more sources of information than can be evaluated in a short period of time. You can reduce the risk of unreliable sources by using Google Scholar to search for scholastic publications. The question then becomes, "How do I evaluate all of these articles?" A few simple guidelines are generally sufficient to narrow the list considerably.

The normal standard is to determine if the published information is from a source of scientific literature where there is a peer process employed before the articles are accepted and printed. The *Journal of Food Protection, Poultry Science,* the *Journal of Food Science,* and the *Journal of Dairy Science* are examples of excellent sources. Do not forget to look at articles that may be published by foreign scientists as well. Just be prepared to evaluate the validity of the source. An example is a scientific study that is performed by or for a government agency similar to USDA or FDA in a foreign country.

Once the source of the article or data is confirmed, look at the authors of the study or publication. Scientists who hold faculty appointments at a reputable college or university are generally accepted as experts in their field, particularly scientists who publish articles routinely. In addition, many of the studies are performed as part of grants from industry or government sources who want to know the scientific validity of a particular process or principle being considered in meat and poultry plants today. These studies then have direct links to the current thinking and application to the plants' needs. Experts in industry are also qualified in their particular field. Especially look for those industry experts who publish in peer-reviewed science journals, are sought as speakers or authors, or hold adjunct relationships to academic institutions. Industry experts may be uniquely qualified due to their close associations with the processes they speak or write about. Every plant should have identified at least one person qualified in toxicology who can assist in a crisis where a compound must be evaluated after it has found its way into the product. Chemists and microbiologists are also valuable experts in their field who may be contacted through trade associations or universities.

A source of validation data used and accepted in most HACCP plans are articles and publications from trade publications or the general press. These articles are usually not peer reviewed but may still be valid for the purpose. Use of these forms of information is generally limited in most HACCP validation files and is usually focused on one or a few aspects of the business. In addition, the articles' author's credentials can be traced either by attribution in the article or through industry or government sources to define the author as an expert in their field. Again, the small or very small plant owner need not be worried about finding this support. There are ample resources through trade associations, roundtables, and industry networking to assist them.

Meat and poultry companies have agreed and publicly announced through their trade associations that sharing of knowledge on food-safety principles and facts are not considered trade secrets. While a facility may have proprietary application of a much broader principle, most meat and poultry facility managers openly and regularly share how a principle or practice works, where information about the principle or practice may be published, and in certain limited situations may even allow competitors to observe the application in their plant. Sharing of information in this manner should be a common method of doing business for the small and very small plant operator.

The operator of a small and very small plant should already be taking advantage of any form of industry roundtable or trade association sponsored discussion group on HACCP. Several land grant universities host such discussions and are attended by industry, academic, and government officials. FSIS participates in these discussions to advise and provide guidance on current thinking. FSIS has increased support of roundtables and is hoping to create a database of questions asked at these meetings to be shared with those unable to attend these types of meetings.

There are several sources of predictive microbiology models that are instructive in evaluation of product parameters that affect pathogen growth or survival/kill. The best known is the Pathogen Modeling Program (PMP) from the Microbial Food Safety Research Unit (MFSRU) (http://ars.usda.gov/Main/site_main.htm?modecode=19-35-30-00) at the Agricultural Research Service Eastern Regional Research Center. However, the PMP does not contain all variables and the agreement with plant data is essential to begin to use one of these predictive models in validation of an HACCP plan. The MFSRU explains this precaution at http://www.ars.usda.gov/Services/docs.htm?docid=11919.

Development of a White Paper or Review Sheet

Scientific documentation, information from industry associations, plant data, predictive microbiological records, and expert advice must be assembled in a white paper or review sheet that briefly but clearly outlines the principles that apply to the validation process. The white paper should provide experts and FSIS official with adequate information to evaluate the merit of the proposed approach.

Validation: Expert Advice

FSIS is required to evaluate, and where appropriate, take regulatory control action when HACCP plans are not being properly implemented and documented. This includes the evaluation of material, data, and principles used to validate the plan. FSIS is also expected to provide guidance and advice when solicited by the operators of meat and poultry plants. The first source of information is the Technical Services Center located in Omaha, Nebraska. This resource is staffed by subject matter specialists of FSIS who are there to answer questions from any interested person, including academia, government, industry, and the public. Another source of information is the Enforcement, Investigations, and Analysis Officers (EIAOs), who perform Food Safety Assessments every day in meat and poultry plants. These FSIS employees have been through extensive training and have diverse experiences that make them a source of knowledge to be tapped. When you call for an EIAO officer's advice, you should be prepared for them to be guarded when giving advice. These officers will not tell the plant operator how to write the plan or how to use the validation material. They will, however, provide guidance and suggestions, based on their experience, about the validity or general applicability of the material to the plan. In most situations, the officer will provide expert advice and opinion about a question of design for an HACCP plan so long as the facility managers are not expecting FSIS to design their plan for them.

Finally, expert advice in any aspect of implementing the HACCP plan is available to the small and very small plant operator without breaking the bank. All land grant universities have extension specialists on staff to assist the public, including industry, as part of their charter. FSIS maintains a list of available experts from these schools on their Web site. Experts in HACCP implementation are also available for hire.

Validation: The Plant's Data

FSIS and HACCP principles require that the plant use data from their own process to validate their plan. Depending on the complexity of the process being validated and the use of the commonly accepted "well-documented" principles discussed earlier, the amount of data necessary will depend on the plan. If the plant operator is using a safe-harbor limit for their CCP (e.g., temperature guidelines published by FSIS) then, the data will generally be a simple study of the actual temperatures of meat or poultry exiting the cooking medium. If, however, the facility managers are using the current regulatory limit of 7 $\log_{10}$ for poultry or 6.5 $\log_{10}$ for red meat without accepting FSIS published guidelines for time and temperature, there will be considerably more validation data required to support their unique time and temperature combination. In fact, more detailed and scientifically complex studies in a laboratory may be needed to validate the principle.

Sources of plant data are obviously the data from implementing the HACCP plan. FSIS regulations expect that any new plan or a revised plan be validated within 90 days of implementation. If proper scientific documentation has been used to design the plan and the CCP, the validation file will then only have to be supported by this implementation data that demonstrates it is operating as expected. Typical statistical analyses are generally sufficient for this purpose. If the data reveals unanticipated deviations or that the process is not as stable as expected, the facility managers must modify their critical limit, change their process, or both to achieve the food-safety standard specified by the CCP. At no time should product be allowed to leave the plant operator's control whenever a CCP has been exceeded, even during this validation period, unless it has been evaluated by someone knowledgeable with the process and compliance with the plant's HACCP plan has been documented.

Small Business Regulatory Enforcement Fairness Act (SBREFA)

The SBREFA (http://www.sba.gov/advo/archive/sum_sbrefa.html) was passed in 1999 to give small businesses a greater voice in the development and enforcement of federal regulations. The Small Business Administration and the USDA-FSIS define a Very Small Plant as < 10 employees and annual sales < $2.5 million. A Small Plant is from 10 and to less than 500 employees. SBREFA is important to FSIS because of approximately 6,000 plants regulated by FSIS, all but 300 are small or very small.

There are two major parts of SBREFA: (1) Regulatory Development, and (2) the Office of the National Ombudsman (ONO), which is advised by regional Regulatory Fairness Boards. During regulatory development, SBREFA requires agencies to analyze and consider small business concerns when a proposed regulation will have a significant economic impact on a substantial number of small businesses. Small businesses are encouraged to participate in the rulemaking process.

The National Ombudsman and regional Fairness Boards provide another venue for small businesses to express their concerns by providing comments or contacting their state representatives. Through the provisions of SBREFA, the Small Business Administration (SBA) appoints a national ombudsman and creates 10 regional Fairness Boards, made up of small businesspersons. Small businesses may contact the national ombudsman or Fairness Boards about their complaints regarding agency regulatory compliance or enforcement decisions.

Regional Small Business Regulatory Fairness Boards were established in each of SBA's 10 regions to advise the Ombudsman on matters of concern to small business relating to the enforcement activities of agencies. The Regulatory Fairness Boards can (1) receive a copy of your appraisal form; (2) hold a follow-up meeting on your concern; (3) report on significant enforcement issues; and (4) reflect all concerns in their report to Congress. The boards cannot adjudicate your complaints directly or reverse agency decisions. Therefore, small businesses should continue exercising their rights and exhausting every option they believe is in their best interest.

Under the SBREFA, each agency must establish a policy to provide for the reduction, and under appropriate circumstances, for the waiver of civil penalties for violations of statutory or regulatory requirements by a small business. The language in this section was adopted from a statement and Executive memorandum issued by President Clinton in March 1995.

The SBREFA also expands the ability of small businesses in litigation with the government to recover attorney fees under the 1980 Equal Access to Justice Act. In administrative and judicial proceedings, if the government's demand is unreasonable when compared to the judgment or decision, then the small business is awarded attorney fees and other expenses related to defending against the action.

Conclusion

Validation of an HACCP plan, including the specific CCPs need not be complicated or a heavy burden for the small and very small meat or poultry plant operator. The process is a matter of following a few simple steps:

Define the CCP limit or step to be validated.
Gather all scientific support that relates to the process.
Review the science and use only the information that applies.
Recap the science in some form of white paper or review sheet.
Seek assistance from Extension, industry, FSIS, or trade associations.
Support the science with data from the process.
KNOW YOUR PLAN.

References

USDA-FSIS. 1996. Pathogen reduction—Hazard Analysis Critical Control Point Final Rule. At http://www.fsis.usda.gov/OPPDE/rdad/FRPubs/93–016F.pdf. Accessed February 1, 2008.

▌9 ▌

Campylobacter jejuni in Biofilms:
A Possible Mechanism of Survival Inside
and Outside the Chicken Host

Irene Hanning and Michael Slavik

Introduction

Campylobacter jejuni is a leading cause of foodborne bacterial diarrhea in the United States (CDC 2007). There are approximately 2.4 million cases of campylobacteriosis in the United States each year (Friedman 2000). *C. jejuni* is considered part of the normal flora within the gastrointestinal tract of chickens (Jeurissen et al. 1998) and can colonize the gut of the chicken, especially the ceca, at levels of 10^5 to 10^9 cfu g^{-1} (Berndtson et al. 1992). Raw poultry products are considered to be a major source of *C. jejuni* infections in humans. Release of the intestinal contents during processing can result in contamination of the raw product and poor handling or consumption of undercooked or raw products that are contaminated can result in infection (Rivoal et al. 1999).

Outside of the host, *C. jejuni* is extremely susceptible to multiple stresses and considered quite easy to kill. High and low salt conditions as well as drying are extremely biocidal and most strains perish quickly when exposed to a pH outside of 5.5 to 8.0 (Doyle et al. 1982; Reezal et al. 1998; Murphy et al. 2006). Host body temperatures of 37°C to 42°C are optimal for growth, outside this range, the rate of reduction in numbers is dependent on the temperature (Konkel et al. 1998; Chan et al. 2001). The anaerobic conditions of the gut are optimal because high levels of oxygen are toxic to *C. jejuni* (Murphy et al. 2006). Furthermore, the mixed microflora of the gut may provide amino acids and other metabolites for *C. jejuni,* which is unable to utilize 6 carbon sugars (Lee and Newell 2006). In response to stress outside the host, most pathogens engage in a stationary phase where growth and metabolism is halted and stress proteins are produced (Hengge-Aronis 1996). However, *C. jejuni* does not have the genetic ability to form a global stress response but instead increases surface carbohydrates and fatty acid membrane profiles to help protect the cell (Martinez-Rodriguez and Mackey 2005).

Given the highly susceptible nature, the strict growth requirements, and the lack of genetic survival mechanisms, the ability of *C. jejuni* to survive outside the host and become a leading cause of foodborne illness is perplexing. The possibility of *C. jejuni* in a biofilm form has been proposed as a mechanism of extended survival outside the host (Kalmokoff et al. 2006). Many bacteria have an ability to form a biofilm, an assemblage of bacteria encased in a sticky polymer. The assemblage could provide *C. jejuni* with protection from environmental insults including desiccation, sanitizers, and antibiotics (Gilbert et al. 1993; Chmielewski and Frank 2003). Furthermore, since oxygen diffusion is limited through the biofilm, an optimal atmosphere for microaerophilic *C. jejuni* could be created (Xavier and Foster 2007). *Campylobacter* species have been shown to grow and form biofilms in laboratory models (Somers et al. 1994; Buswell et al. 1998). Past studies also have shown this pathogen is capable of surviving in mixed biofilms under conditions that would not normally allow growth (Trachoo et al. 2002). Currently, laboratory models supply the only information available concerning *Campylobacter* and biofilms. The impact biofilms may have in the poultry industry is currently being investigated and may help answer questions concerning the survival of this fastidious pathogen.

Biofilms in General

The formation of a biofilm is a stepwise process shown in Figure 9.1 (Gilbert et al. 1993). The initial stage begins with a bacterium, the primary colonizer, attaching to a surface free of other organisms by means of chemotaxis or electrostatic interactions. The bacterium then actively secretes an exopolymer that tightly binds the cell to the surface. Secondary colonizers that are free-floating bacteria, as well as nutrients, are readily entrapped in the sticky exopolymer produced by the primary colonizer. The final stage is a mature biofilm made up of many cells that are reproducing, capturing nutrients, and entangling free-floating bacteria. The mature biofilm may also disperse portions of the community to colonize other surfaces.

Almost all biofilms are a mixture of species that can result in cooperation and provide advantages to the members (Branda et al. 2005). Biofilms can protect the members of the community from desiccation, antibiotics, phagocytosis, and sanitizers (Gilbert et al. 1993). The sticky exopolymer allows nutrients to be readily captured, and byproducts of nutrient degradation can be recycled by different

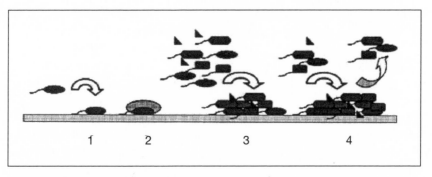

Figure 9.1. A schematic representation of biofilm formation. Primary colonizers, originally planktonic cells, are initially attracted and attach to a surface free of other bacteria (1). The secretion of exopolymers irreversibly adhere the cell to the surface (2). Secondary colonizers, formed by reproduction of the primary colonizer or captured from the environment, and nutrients ( ) become entrapped in the biofilm (3). In the mature biofilm (4), cells may be reproducing, recruiting planktonic cells, capturing nutrients and dispersing from the colony.

members in the community. The close proximity of cells provides opportunities for genetic exchange and the densely packed assemblage also facilitates sharing of secreted enzymes, iron-scavenging siderophores, and exopolymers (Xavier and Foster 2007).

It is now accepted that most bacterial cells found in natural settings are usually adhered to biotic and abiotic surfaces in the form of a biofilm, rather than as free-swimming cells (Branda et al. 2005). Biofilms can develop on almost any surface, from contact lenses, heart valves, and teeth to water pipes and food-processing surfaces (Chmielewski and Frank 2003; Fux et al. 2005). Biofilms are of major clinical importance as 60% of bacterial infections are considered to involve biofilm formation (Costerton et al. 1999). Biofilms can also be problematic in the food-processing environment. Biofilms have been reported to slow fluid-handling systems, hamper heat transfer, and even corrode metal surfaces (Chmielewski and Frank 2003). Growth of biofilms in the food-processing environment can lead to an increased risk for cross-contamination, which can reduce the shelf-life of products and increase the risk of foodborne illness. Since biofilms can be more resistant to sanitizers, it makes elimination from the processing plant a challenge.

Research indicates biofilms have the potential to impact all stages of the poultry process from the farm to the consumer. Biofilms on the farm may facilitate the spread of disease through flocks and in processing plants may act as sources of cross-contamination. Biofilms are difficult to remove and may not be eliminated using standard cleaning procedures (Chmielewski and Frank 2003). Therefore, identifying risk factors, environmental conditions, and surfaces that are ideal for biofilm formation are needed in order to control biofilms.

Biofilms on the Farm

It has been estimated that 90% of flocks in the United States are *Campylobacter* positive (Stern et al. 2001). Epidemiological studies have been undertaken to identify risk factors and determine sources of infection. Poor cleaning, biosecurity breaches, poor house maintenance, short empty periods, and the presence of other animals on the farm have all been suggested as risk factors (Hald et al. 1999). However, flock infection may not always be associated with environmental contamination (Payne et al. 1999). Detecting sources of *Campylobacter* infection to flocks may be difficult, and due to the susceptibilities and highly fastidious nature of the organism, carry-over between flocks is difficult to explain. Some studies suggest the risk factor for carry-over is relatively low and that a persistent reservoir, such as a biofilm, may be a more likely source (Shreeve 2002).

Desiccation is considered to be highly lethal to *C. jejuni*. Because biofilms protect bacteria from desiccation, it seems this strategy would be advantageous for *C. jejuni* when outside the host. But water is usually limited within the farm environment and some moisture must be present in order for biofilm formation to take place (Zottola and Sassahara 1994). The ability of *C. jejuni* to survive in water is well documented and *C. jejuni* has been shown to survive within an aquatic biofilm for up to 28 days (Buswell et al. 1998; Cools et al. 2003). Taken together, these studies suggest the possibility of a biofilm within a waterline could be a consistent source of *C. jejuni* to the flocks.

C. jejuni outbreaks associated with contaminated water have been reported. Furthermore, *C. jejuni* recovered from waterlines can be phenotypically and genotypically identical to strains recovered from birds (Pearson et al. 1996). Isolation of *Campylobacter* usually results from a break in chlorination or when nonchlorinated groundwater contaminates the system (Freidman et al. 2000). *C. jejuni* is very susceptible to chlorination and, therefore, the possibility of water contamination prior to stocking houses has been difficult to show (van der Giessen et al. 1998). Water contamination is actually thought to follow rather than precede flock colonization, which suggests the birds contaminate the waterlines (Lindblom et al. 1986; Engvall et al. 1986; Kazwala et al. 1990). Thus, a biofilm in the waterlines could spread *C. jejuni*, but probably would not be an initial source.

Nipple drinkers and watering troughs are other areas that have been shown to support the growth of a biofilm. Laboratory models using bacteria isolated from nipple drinkers show that *C. jejuni* can survive in biofilms for up to 7 days (Trachoo et al. 2002). Like waterlines, though, it has been suggested that nipple drinkers probably act as sources that enhance and promote infection rather than as an initial origin (Kazwala et al. 1990; Lindblom et al. 1986, Engvall et al. 1986). Initially, it is thought that the contamination of nipple drinkers occurs via fecal transfer from the beak to drinker. *C. jejuni* is most likely picked up on the beak when pecking at the litter. Nipple drinkers are somewhat cleaner than watering troughs as it has been shown that bacterial attachment to rubber is less than other surfaces and some types of rubber even inhibit attachment (Arnold 1998). Watering trough types of systems are less hygienic as they accumulate a heavy organic and bacterial load, which reduces the efficacy of the biocidal effect of chlorine (Poppe et al. 1991). This, in turn, may allow more opportunities for biofilm formation.

Coolers associated with airhandling units may be another likely area for biofilm formation. These cooling units have been shown to harbor *Pseudomonas aeruginosa,* an excellent biofilm former and model organism used in biofilm experiments (Walker et al. 2002; Toutain et al. 2004). *C. jejuni* could be entrapped and harbored within a preestablished biofilm and aerosolized. A dry aerosol would be considered lethal to *C. jejuni*, but when conditions are very humid, *C. jejuni* has been shown to travel considerable distances and retain viability (Newell et al. 2001). Research has also shown flocks closest to air-conditioning vents became positive for *C. jejuni* first (Refregier-Petton et al. 2001). But this conclusion would only explain infection during the summer when air conditioners are used.

Biofilms in Processing Plants

Unlike the farm environment, processing plants are very wet environments having areas that are ideal for biofilm formation (Lindsay et al. 1996; Gibson et al. 1999; Arnold and Silvers 2000). A survey of a poultry-processing plant revealed some types of rubber picker fingers, plastic curtains, and conveyor belts were heavily colonized by biofilms (Linsday et al. 1996). In the same study, stainless steel coupons applied within scald tanks and chutes also had extensive and confluent biofilms after 14 days. Laboratory models show biofilms can form on materials commonly used in poultry processing plants, including stainless steel, polyethylene, and conveyor belting (Arnold and Silvers 2000). These materials may support the biofilm formation of some poultry isolates, but of the materials found in a processing plant, *C. jejuni* has only been shown to attach to stainless steel and only under conditions needed for growth (Somers et al. 1994; Kalmokoff et al. 2006).

Because up to 100% of a flock may be *C. jejuni* positive and each bird within the flock may be carrying 10^{10} cfu of *C. jejuni* per gram of intestinal material (Lee and Newell 2006), it is easy to understand how equipment can be readily contaminated with *C. jejuni* if intestinal contents are released. The heavy use of water in processing plants is thought to contribute to the spread of *C. jejuni* through the plant (Rivoal et al. 1999). Multiple genotypic studies have shown that intestinal isolates are genotypically the same as those contaminating carcasses and that contamination of broilers during processing is principally due to rupture of the gastrointestinal tract (Rivoal 1999; Newell et al. 2001; Son et al. 2007). Newell et al. (2001) used *flaA* sequencing to show that strains of *C. jejuni* varied in the ability to survive the process, and only a few strains were able to survive the entire process, but none survived to contaminate birds processed on different days. These genotyping studies suggest that intestinal leakage is the cause of the majority of *C. jejuni* contamination within the processing plants and biofilms may only be a minor source of contamination.

Unlike most other bacteria in biofilms, *C. jejuni* within a biofilm has been shown to be more susceptible to sanitizers and environmental stress than free-swimming cells (Dykes et al. 2003). This would make survival of *C. jejuni* in biofilms most unlikely after daily cleaning and sanitization. However, this biofilm study was done using a monoculture, and most mature bacterial biofilms are composed of multiple species. It is possible that a small source of contamination could be from a mixed biofilm that supported *C. jejuni*. But considering that most of the contamination can be traced to the intestinal tract, it would be difficult to find and distinguish biofilm contamination from intestinal leakage. Because flocks are usually limited to colonization by three or fewer strains, sampling protocols may be limited to sampling five or fewer colonies per agar plate (Ayling et al. 1996; Newell et al. 2001). These sampling protocols would probably limit detection to the major sources of contamination. In order to pinpoint a biofilm source, a study would have to genotype a large number of isolates which to date has not been reported.

Biofilms on Raw Poultry Products

Early studies concluded that the location of bacteria on the chicken skin could protect the bacteria and allow them to survive through processing (Notermans 1975). These studies also suggested that the cells were able to survive various treatments because they were firmly attached. Subsequent research has found the bacterial attachment to raw products can be mediated by the production of viscous exopolymers; in other words, by forming a biofilm (Farber and Idziak 1982; Benedict et al. 1991). More recently, a study aimed at specifically identifying biofilms on raw chicken products found that the communities varied greatly in microbial species and that the community profile had a significant impact on shelf life (Boothe et al. 1999). It has been found that at the time of contamination, the physiological state and prior mode of growth, attached or planktonic cells, can impact subsequent survival of bacteria (Turnock et al. 2002). Bacteria that attach to raw poultry products early in the process and form biofilms may be protected from future cold storage stress.

Fastidous growth requirements do not allow *C. jejuni* to multiply under storage conditions of 4°C to 10°C (Blaser 1980). Although *Campylobacter* has been shown to survive better at 4°C than at room temperature, reductions in numbers still occur (Chan et al. 2001). The mechanisms of survival at cold temperatures are unclear, but since studies have found up to 98% of chicken carcasses obtained from retail are positive for *C. jejuni* contamination, survival during storage must occur (Jacobs-Reitsma 2002). Biofilms have been shown to provide *C. jejuni* with prolonged protection at 4°C (Buswell et al. 1998). Although research indicates *C. jejuni* could be surviving cold storage by using a biofilm strategy, it has not been specifically investigated.

Other evidence of biofilm formation on chicken carcasses has been shown in cell-signaling research. Cell density-dependent signaling systems are known as quorum sensing systems. Quorum sensing has been shown to play a role in both fresh meat spoilage and biofilm formation (Jay et al. 2003). Quorum sensing using autoinducer-2 (AI-2) is used by both Gram-negative and Gram-positive bacteria for interspecies communication (Schauder and Bassler 2001). The AI-2 language allows bacteria present in mixed populations, such as biofilms, to respond to changes in its own numbers and changes in numbers of other species. There is great interest in manipulating this cell signaling system as a method of preserving meat quality and disrupting biofilm formation (Smith et al. 2004). Cloak et al. (2002) were able to show AI-2 activity by *C. jejuni* inoculated in chicken broth. However, AI-2 signaling is also used to coordinate responses and react to stress in terms of growth, survival, and virulence (Sperandio et al. 1999). So it is unclear if there is a connection between AI-2 and biofilm formation by *C. jejuni*.

Biofilms in the Gut

In both chickens and humans, the epithelial cells of the gastrointestinal tract are covered by a mucus layer. The mucin forming this layer can be used by many intestinal bacteria as sources of carbon, nitrogen, and energy (Macfarlane et al. 2005). Biofilm formation can take place within the GI tract on food particles, on the mucosa layer and within the mucosa layer (Macfarlane and Macfarlane 2006). In fact, mucin has been shown to increase biofilm formation by *E. coli* (Bollinger et al. 2003). Biofilm formation by commensals in the gut is thought to be beneficial to both bacteria and host (Bollinger et al. 2006). Agglutination of the bacteria, promoted by secretory antibodies (sIgA), prevents bacteria from breaching the intestinal epithelial layer (Williams and Gibson 1972). Agglutination by sIgA may also promote the growth and retention of normal flora which could prevent colonization by pathogens (Bollinger et al. 2003). As for the bacteria, large and sticky aggregates of bacteria have a better chance of capturing food particles than free-swimming planktonic cells. Furthermore, because biofilms are typically composed of multiple species, the process of digesting can be a cooperative effort and, thus, the task accomplished more efficiently.

The mucus layer of the ceca and large intestine of the chicken provides an ideal habitat for *C. jejuni*. Near the surface of the epithelial cells, oxygen is present for host cell metabolism, but low enough for microaerophilic *C. jejuni* to grow. In the human gut, the mucus layer of the ileum and colon are typically colonized (Blaser 1995). In order to colonize a suitable area of any gastrointestinal tract, *C. jejuni* might be required to cross a biofilm of commensal bacteria and also penetrate the mucus layer. It was recently shown that *C. jejuni* upregulates motility in a monoculture biofilm (Kalmokoff et al. 2006). But, whether the upregulation in motility is specifically done to cross the biofilm and mucus is not known.

There is great interest in studying mucosal biofilms in the human digestive tract. Some studies indicate biofilm formation may play a significant role in the ability of bacteria to cause diseases such as inflammatory bowel disease (Swidsinski 2005). Other studies are interested in any protective effect biofilm formation of commensals may have against pathogen colonization (Bollinger et al. 2003). At this time, it is unclear what role a biofilm formation might play in *C. jejuni* colonization.

Conclusions and Future Endeavors

The ability of *C. jejuni* to form a biofilm in monoculture has been well documented. *C. jejuni* has been shown to form a biofilm on stainless steel, nitrocellulose fibers, polystyrene plates, glass, and polyvinyl chloride and in tissue culture flasks (Kalmokoff et al. 2006; Joshua et al. 2006; Reeser et al. 2007). However, these experiments show *C. jejuni* forms biofilms during growth conditions (Reeser et al. 2007). Collectively, these findings indicate a monoculture biofilm of *C. jejuni* outside the host would be nearly impossible because the conditions needed to initiate a biofilm would most likely not be found outside the host (Lee and Newell 2006). Therefore, with limited genetic survival abilities, it may be more advantageous for *C. jejuni* to remain motile and forgo biofilm formation in anticipation of finding a new host.

There is a possibility that *C. jejuni* incorporates into a preestablished biofilm as a secondary colonizer (Trachoo et al. 2001; Sanders et al. 2007). The concept of one species acting as a primary colonizer and promoting the biofilm formation of a secondary colonizer has been shown with *Pseudomonas fragi* promoting biofilm formation of the foodborne pathogen *Listera monocytogenes* (Sasahara and Zottola

1993). The idea also has been demonstrated with *Pseudomonas aeruginosa* promoting the biofilm formation of *E. coli* O157:H7 (Volden et al. 2007).

As a commensal in the chicken gut, *C. jejuni* is well adapted to life within a population of mixed bacterial species. In fact, utilizing amino acids and metabolites produced by other microflora is a necessity because *C. jejuni* lacks phosphofructokinase (Lee and Newell 2006). Within the biofilm, the densely packed cells share all secreted resources including the polymer (Xavier and Foster 2007). This sharing suggests that opportunities exist for "freeloaders" in a biofilm. *C. jejuni* is so highly adapted to life in the gut environment that outside the gut, it may have no choice but to exist as a "freeloader."

Research shows that biofilms have the potential to impact all stages of the poultry process from rearing on the farm, to processing in the plant, to surviving during storage and causing diseases. Most of the current research involving *C. jejuni* and biofilm formation is aimed at answering how this fastidious organism survives outside the host. But, as this biofilm research progresses, it is clear that biofilms are complex and may be used for more than just simply survival. As research continues involving *C. jejuni* and biofilms, it is possible that biofilm formation may play more roles in the life of *C. jejuni*.

Acknowledgments

This review was supported by the United States Department of Agriculture Food Safety Consortium.

References

Arnold, J. 1998. Development of bacterial biofilms during poultry processing. Poul. Av. Biol. Rev. 9:1–9.

Arnold, J., and S. Silvers. 2000. Comparison of poultry processing equipment surfaces for susceptibility to bacterial attachment and biofilm formation. Poult. Sci. 79:1215–1221.

Ayling, R., M. Woodward, S. Evans, and D. Newell. 1996. Restriction fragment length polymorphism of polymerase chain reaction products applied to the differentiation of poultry campylobacters for epidemiological investigations. Res. Vet. Sci. 60:168–172.

Bäckhed, F., R. Ley, J. Sonnenburg, D. Peterson, and J. Gordon. 2005. Host-bacterial mutualism in the human intestine. Science 307:1915–1920.

Benedict, R., F. Schultz, and S. Jones. 1991. Attachment and removal of *Salmonella* spp. on meat and poultry tissues. J. Food Safety 11:135–148.

Berndston, E., M. Tivemo, and A. Engvall. 1992. Distribution and numbers of *Campylobacter* in newly slaughtered broiler chickens and hens. Int. J. Food. Microbiol. 15:45–50.

Blaser, B. 1995. *Campylobacter* and related species. Pages 1948–1956 in Principles and practice of infectious disease. G. L. Mandell, J. E. Bennett, and R. Dolin, eds. Churchill Livingstone, Inc., New York, NY.

Blaser, M., H. Hardesty, B. Powers, and W. Wang. 1980. Survival of *Campylobacter fetus* subsp. *jejuni* in biological milieus. J. Clin. Microbiol. 11:309–313.

Bollinger, R., M. Everett, D. Palestrant, S. Love, S. Lin, and W. Parker. 2003. Human secretory immunoglobulin A may contribute to biofilm formation in the gut. Immunology 109:580–587.

Bollinger, R., M. Everett, S. Wahl, Y. Lee, P. Orndorff, and W. Parker. 2006. Secretory IgA and mucin-mediated biofilm formation by environmental strains of *Escherichia coli*: role of type 1 pili. Mol. Immunol. 43:378–387.

Boothe, D., J. Arnold, and V. Chew. 1999. Utilization of substrates by bacterial communities (biofilm) as they develop on stored chicken meat samples. Poult. Sci. 78:1801–1809.

Branda, S., S. Vik, L. Friedman, and R. Kolter. 2005. Biofilms: the matrix revisited. Trends Microbiol. 13:20–26.

Buswell, C., Y. Herlihy, C. Keevil, and S. Leach. 1998. The effect of aquatic biofilm structure on the survival of the enteric pathogen *Campylobacter jejuni* in water. Appl. Environ. Microbiol. 64:733–741.

CDC. 2007. Preliminary FoodNet data on the incidence of infection with pathogens transmitted commonly through food. MMWR 56:336–339.

Chan, K., H. Tran, R. Kanenaka, and S. Kathariou. 2001. Survival of clinical and poultry-derived isolates of *Campylobacter jejuni* at a low temperature. Appl. Environ. Microbiol. 67:4186–4191.

Chmielewski, R., and J. Frank. 2003. Biofilm formation and control in food processing facilities. Comp. Rev. Food Sci. Food Saf. 2:22–32.

Cloak, O., B. Solow, C. Briggs, C. Chen, and P. Fratamico. 2002. Quorum sensing and production of autoinducer-2 in *Campylobacter* spp., *Escherichia coli O157:H7,* and *Salmonella enterica* serovar Typhimurium in foods. Appl. Environ. Microbiol. 68:4666–4671.

Cools, I., M. Uyttendaele, C. Caro, E. D'Haese, H. Nelis, and J. Debevere. 2003. Survival of *Campylobacter jejuni* strains of different origin in drinking water. J. Appl. Microbiol. 94:886–892.

Costerton , J., P. Stewart, and E. Greenberg. 1999. Bacterial biofilms: a common cause of persistent infections. Science 284:1318–1322.

Doyle, M., and D. Roman. 1982. Sensitivity of *Campylobacter jejuni* to drying. J. Food Protect. 45:507–510.

Dykes, G., B. Sampathkumar, and D. Korber. 2003. Planktonic or biofilm growth affects survival, hydrophobicity and protein expression patterns of a pathogenic *Campylobacter jejuni* strain. Int. J. Food Microbiol. 89:1–10.

Engvall, A., A. Bergqvist, K. Sandstedt, and M. Danielsson-Tham. 1986. Colonization of broilers with *Campylobacter* in conventional broiler-chicken flocks. Acta.Vet. Scand. 27:540–547.

Farber, J., and E. Idziak. 1982. Detection of glucose oxidation products in chilled fresh beef undergoing spoilage. Appl. Environ. Microbiol. 44:521–524.

Friedman, C. R., J. Neimann, H. C. Wegener, and R. V. Tauxe. 2000. Epidemiology of *Campylobacter jejuni* infections in the United States and other industrialized nations. In *Campylobacter.* I. Nachamkin and M. Blaser, eds. ASM Press, Washington, DC.

Fux, C., J. Costerton, P. Stewart, and P. Stoodley. 2005. Survival strategies of infectious biofilms. Trends Microbiol. 13:34–40.

Gibson, H., J. Taylor, K. Hall, and J. Holah. 1999. Effectiveness of cleaning techniques used in the food industry in terms of the removal of bacterial biofilms. J. Appl. Microbiol. 87:41–48.

Gilbert, P., D. Evans, and M. Brown. 1993. Formation and dispersal of bacterial biofilms in vivo and in situ. J. Appl. Bacter. 74:67S–78S.

Hald, T., H. Wegener, and B. Jorgenson. Annual report of zoonoses in Denmark 1998. Ministry of Food, Agriculture, and Fisheries, Copenhagen, Denmark. 1999.

Hengge-Aronis, R. 1996. Regulation of gene expression during entry into stationary phase. Pages 1210–1233 in *Escherichia coli* and *Salmonella:* cellular and molecular biology. F. Neidhardt, R. Curtiss, J. Ingraham, E. Lin, K. Low, B. Magasanik, W. Reznikoff, and M. Riley, eds. ASM Press, Washington, DC.

Jacobs-Reitsma, W. 2002. *Campylobacter* in the food supply. Pages 467–481 in *Campylobacter.* I. Nachamkin and M. Blaser, eds. ASM Press, Washington, DC.

Jay, J., J. Vilai, and M. Hughes. 2003. Profile and activity of the bacterial biota of ground beef held from freshness to spoilage at 5–7 degrees C. Int. J. Food Microbiol. 81:105–111.

Jeurissen, S., E. Janse, N. van Rooijen, and E. Claassen. 1998. Inadequate anti-polysaccharide antibody responses in the chicken. Immunobiology 198:385–395.

Joshua, G., C. Guthrie-Irons, A. Karlyshev, and B. Wren. 2006. Biofilm formation in *Campylobacter jejuni.* Microbiology 152:387–396.

Kalmokoff, M., P. Lanthier, T. Tremblay, M. Foss, P. Lau, G. Sanders, J. Austin, J. Kelly, and C. Szymanski. 2006. Proteomic analysis of *Campylobacter jejuni* 11168 biofilms reveals a role for motility complex in biofilm formation. J. Bacter. 188:4312–4320.

Kazwala, R., J. Collins, J. Hannan, R. Crinion, and H. O'Mahony. 1990. Factors responsible for the introduction and spread of *Campylobacter jejuni* infection in commercial poultry production. Vet. Rec. 126:305–306.

Konkel, E., B. Kim, J. Klena, C. Young, and R. Ziprin. 1998. Characterisation of the thermal stress response of *Campylobacter jejuni.* Infect. Immun. 66:3666–3672.

Lee, M., and D. Newell. 2006. *Campylobacter* in poultry: filling an ecological niche. Avian Dis. 50:1–9.

Lindblom, G., E. Sjörgren, and B. Kaijser. 1986. Natural *Campylobacter* colonization in chickens raised under different environmental conditions. J. Hyg. Lond. 96:385–391.

Lindsay, D., I. Geornaras, and A. von Holy. 1996. Biofilms associated with poultry processing equipment. Microbios. 86:105–116.

Macfarlane, S., E. Woodmansey, and G. Macfarlane. 2005. Colonization of mucin by human intestinal bacteria and establishment of biofilm communities in a two-stage continuous culture system. Appl. Environ. Microbiol. 71:7483–7492.

Macfarlane, S., and G. Macfarlane. 2006. Composition and metabolic activities of bacterial biofilms colonizing food residues in the human gut. Appl. Environ. Microbiol. 72:6204–6211.

Martinez-Rodriguez, A., and B. Mackey. 2005. Physiological changes in *Campylobacter jejuni* on entry into stationary phase. Int. J. Food. Microbiol. 101:1–8.

Murphy, C., C. Carroll, and K. Jordan. 2006. Environmental survival mechanisms of the foodborne pathogen *Campylobacter jejuni.* J. Appl. Microbiol. 100:623–632.

Newell, D., J. Shreeve, M. Toszeghy, G. Domingue, S. Bull, T. Humphrey, and G. Mead. 2001. Changes in the carriage of *Campylobacter* strains by poultry carcasses during processing in abattoirs. Appl. Environ. Microbiol. 67:2636–2640.

Notermans, S., and E. Kampelmacher. 1975. Further studies on the attachment of bacteria to skin. Br. Poult. Sci. 16:487–496.

Payne, R., M. Lee, D. Dreesen, and H. M. Barnhart. 1999. Molecular epidemiology of *Campylobacter jejuni* in broiler flocks using random amplified polymorphic DNA-PCR and 23s rRNA-PCR and role of litter in its transmission. Appl. Environ. Microbiol. 65:260–263.

Pearson, A., M. Greenwood, R. Feltham, T. Healing, J. Donaldson, D. Jones, and R. Colwell. 1996. Microbial ecology of *Campylobacter jejuni* in a United Kingdom chicken supply chain: intermittent common source, vertical transmission, and amplification by flock propagation. Appl. Environ. Microbiol. 62:4614–4620.

Poppe, C., R. Irwin, S. Messier, G. Finley, and J. Oggel. 1991. The prevalence of *Salmonella enteritidis* and other *Salmonella* spp. among Canadian registered commercial chicken broiler flocks. Epidemiol. Infect. 107:201–211.

Reeser, R., R. Medler, S. Billington, B. Jost, and L. Joens. 2007. Characterization of *Campylobacter jejuni* biofilms under defined growth conditions. Appl. Environ. Microbiol. 73:1908–1913.

Reezal, A., B. McNeil, and J. Anderson. 1998. Effect of low-osmolarity nutrient media on growth and culturability of *Campylobacter* species. Appl. Environ. Microbiol. 64:4643–4649.

Refrégier-Petton, J., N. Rose, M. Denis, and G. Salvat. 2001. Risk factors for *Campylobacter* spp. contamination in French broiler-chicken flocks at the end of the rearing period. Prev. Vet. Med. 50:89–100.

Rivoal, K., M. Denis, G. Salvat, P. Colin, and G. Ermel. 1999. Molecular characterization of the diversity of *Campylobacter* spp. isolates collected from a poultry slaughterhouse: analysis of cross-contamination. Lett. Appl. Microbiol. 29:370–374.

Sanders, S., D. Boothe, J. Frank, and J. Arnold. 2007. Culture and Detection of *Campylobacter jejuni* within mixed microbial populations of biofilms on stainless steel. J. Food Prot. 70:1379–1385.

Sasahara, K., and E. Zottola. 1993. Biofilm formation by *Listeria monocytogenes* utilizes a primary colonizing microorganism in flowing systems. J. Food Prot. 56:1022–1028.

Schauder, S., and B. Bassler. 2001. The languages of bacteria. Genes Dev. 15:1468–1480.

Shreeve, J., M. Toszeghy, A. Ridley, and D. Newell. 2002. The carry-over of *Campylobacter* isolates between sequential poultry flocks. Avian Dis. 46(2):378–385.

Smith, J., P. Fratamico, and J. Novak. 2004. Quorum sensing: a primer for food microbiologists. J. Food Prot. 67:1053–1070.

Son, I., M. Englen, M. Berrang, P. Fedorka-Cray, and M. Harrison. 2007. Prevalence of *Arcobacter* and *Campylobacter* on broiler carcasses during processing. Int. J. Food Microbiol. 113:16–22.

Somers, E., J. Schoeni, and A. Wong. 1994. Effect of trisodium phosphate on biofilm and planktonic cells of *Campylobacter jejuni*, *Escherichia coli* O157:H7, *Listeria monocytogenes* and *Salmonella typhimurium*. Int. J. Food Microbiol. 122:269–276.

Sperandio, V., J. Mellies, W. Nguyen, S. Shin, and J. Kaper. 1999. Quorum sensing controls expression of the type III secretion gene transcription and protein secretion in enterohemorrhagic and enteropathogenic *Escherichia coli*. Proc. Natl. Acad. Sci. USA 96:15196–15201.

Stern, N., P. Fedorka-Cray, J. Bailey, N. Cox, S. Craven, K. Hiett, M. Musgrove, S. Ladely, D. Cosby, and G. Mead. 2001. Distribution of *Campylobacter* spp. in selected U.S. poultry production and processing operations. J. Food Prot. 64:1705–1710.

Swidsinski, A., J. Weber, V. Loening-Baucke, L. Hale, and H. Lochs H. 2005. Spatial organization and composition of the mucosal flora in patients with inflammatory bowel disease. J. Clin. Microbiol. 43:3380–3389.

Toutain, C., N. Caiazza, and G. O'Toole. 2004. Molecular basis of biofilm development by pseudomonads. Pages 43–63 in Microbial biofilms. M. Ghannoum and G. O'Toole, eds. ASM Press, Washington, DC.

Trachoo, N., J. Frank, and N. Stern. 2002. Survival of *Campylobacter jejuni* in biofilms isolated from chicken houses. J. Food Prot. 65:1110–1116.

Turnock, L., E. Somers, N. Faith, C. Czuprynski, and A. Wong. 2002. The effects of prior growth as a biofilm on the virulence of *Salmonella* Typhimurium for mice. Com. Immun. Microbiol. Infect. Dis. 25:43–48.

van de Giessen, A., J. Tilburg, W. Ritmeester, and J van der Plas. 1998. Reduction of campylobacter infections in broiler flocks by application of hygiene measures. Epidemiol. Infect. 121:57–66.

Volden, P., B. Klayman, and A. Camper. 2007. Page 67 in Characterization of *Escherichia coli* biofilm detachment in mixed species biofilms grown in capillary flow cells. ASM conferences on biofilms. Quebec City, Quebec, Canada..

Walker, S., J. Sander, J. Cline, and J. Helton. 2002. Characterization of Pseudomonas aeruginosa isolates associated with mortality in broiler chicks. Avian Dis. 46:1045–1050.

Williams, R., and R. Gibbons. 1972. Inhibition of bacterial adherence by secretory immunoglobulin A: a mechanism of antigen disposal. Science 177:697–699.

Xavier, J., and K. Foster. 2007. Cooperation and conflict in microbial biofilms. Proc. Natl. Acad. Sci. USA 104:876–881.

Xu, J., and J. Gordon. 2003. Inaugural article: Honor thy symbionts. Proc. Natl. Acad. Sci. USA 100:10452–1049.

Zottola, E., and K. Sasahara. 1994. Microbial biofilms in the food processing industry—should they be a concern? Int. J. Food Microbiol. 23:125–148.

∎ 10 ∎

Beef-Safety Research Conducted by the Food Safety Consortium

James L. Marsden

Introduction

When the Food Safety Consortium was founded in 1989, one of the identified areas of research was beef food safety. Research in beef safety was progressing at a slow pace. The national focus during that period was directed to the control of *Salmonella* in raw poultry products. One of the food-safety concerns relating to beef that was emerging during the late 1980s was an enteropathogenic strain of *E. coli*—*E. coli* O157:H7. The initial focus was on rapid detection of *E. coli* O157:H7, *Salmonella,* and other foodborne pathogens associated with meat and poultry products. Interventions to control microbiological contamination on beef carcasses were under development, but rarely used in commercial beef-processing plants. Much of the research on beef food-safety interventions involved organic acid sprays to reduce microbiological contamination during the slaughter process.

An event occurred in December 1992 and January 1993 that greatly impacted beef-safety research and establish the importance of the Food Safety Consortium. An outbreak of *E. coli* O157:H7 in the Pacific Northwest sickened hundreds and resulted in the deaths of four children. The cause of the outbreak was undercooked ground beef served in Jack-in-the-Box restaurants. The outbreak resulted in a public outcry for more stringent meat inspection and the government responded. The USDA expedited the mandatory implementation of the Hazard Analysis and Critical Control Point System and enacted microbiological testing requirements. In 1994, the Food Safety Inspection Service stated that raw ground beef products contaminated with *E. coli* O157:H7 are adulterated within the meaning of the Federal Meat Inspection Act (FMIA), unless the ground beef was further processed. In 1999, that definition was expanded to include nonintact products that had either been mechanically tenderized through blade or needle tenderization or enhanced product that had been injected with a brine solution. Subsequently, the FSIS issued a directive in 2002 that required a reassessment of HACCP plans in facilities that produce nonintact beef products. The USDA's policies and the need to address the overall problem of *E. coli* O157:H7 forced the beef-safety researchers to unite and work together with government public health agencies, consumer groups and technology providers to develop and put into practice systems to eliminate this pathogen in raw beef products.

During the mid-1990s, the USDA held several public meetings on *E. coli* O157:H7 that were attended by industry and consumer groups, including parents who had lost children to the disease. Representatives from the FSC participated fully in the public process. Initially, there was a great deal of mistrust and animosity. To their credit, industry representatives, researchers, and consumers learned to work together to agree on and advance solutions. This unprecedented alliance allowed the USDA to establish consensus and act quickly to modify policies and regulations regarding food-safety technologies.

A Blue Ribbon Task Force consisting of experts from industry and academia evaluated the vectors of *E. coli* O157:H7 contamination in beef products and determined that the slaughter process was the most likely point of contamination and also the point in the process where interventions could be most effectively applied. Beef carcasses and muscle tissue are initially sterile. They become contaminated from the hide of the animal, from cross-contamination during the evisceration process and from aerosolized environmental contamination. Low levels of contamination will increase if the carcass is not rapidly chilled. Carcass contamination is transferred into meat products, including cuts and trimmings destined for ground beef, during the fabrication of the carcass.

The Task Force report established a blueprint for future research. Areas of research focus that were identified included preharvest food safety, beef slaughter interventions, processing interventions, pasteurization technologies, including irradiation, improved sampling and testing methodologies, and consumer and food-service cooking and handling procedures.

The Food Safety Consortium had already established a solid beef research program at Kansas State University, which allowed for an immediate response to the *E. coli* O157:H7 crisis. The FSC's research efforts were directed to address each of the critical needs.

Carcass Decontamination Strategies

In 1993, the only approved beef carcass interventions were knife trimming and treatment with organic acids. FSC researchers evaluated the effectiveness of these interventions and sought to identify and validate additional strategies.

In order to prevent carcass contamination and eliminate it if it does occur, FSC researchers initiated two visionary projects in 1994. The first was an evaluation of a system for removing physical contaminants, including fecal material that may contain *E. coli* O157:H7 from the carcass surface. This technology—steam vacuuming—was intended to act as an adjunct to knife trimming and a means of achieving the USDA's zero tolerance for physical defects on beef carcasses. Phebus et al. (1997) reported knife trimming alone and in combination with other treatments reduced inoculated pathogen populations. FSC researchers determined that by combining sanitary knife trimming with the use of steam vacuuming, physical defects can be completely removed from beef carcasses and that a corresponding decrease in microbiological populations is achieved.

The second project involved a steam pasteurization treatment for the entire carcass. This project was carried out in conjunction with the Cargill Meat Group and Frigoscandia, a major food equipment manufacturer.

The results of these research projects were published (Nutsch et al. 1998; Phebus et al. 1997). Both projects were successful in providing the beef industry with important intervention technologies at a time when alternatives to prevent or control carcass contamination did not exist. Both steam vacuuming and steam pasteurization were widely adopted in the beef industry as a result of the Food Safety Consortium research projects. The thermal pasteurization of carcasses provided the beef industry with a valuable tool for controlling *E. coli* O157:H7 at a critical point in time. The combination of sanitary slaughter practices and thermal pasteurization greatly reduces the risk of pathogen contamination on beef carcasses.

Other beef carcass interventions evaluated by FSC researchers included knife trimming, organic acid treatments, chlorine compounds, ozone, tri-sodium phosphate, lactoferrin, peracectic acid, UV light, and bacteriocins.

Many beef slaughter plants employ a combination of thermal pasteurization, steam vacuuming, and one or more chemical treatments. The combined results of these treatments are carcasses that are both visually and microbiologically clean.

Irradiation Pasteurization

FSC researchers conducted extensive studies on pulse electron beam irradiation. This work was carried out in conjunction with Sandia National Laboratories and resulted in several publications. The research documented the pasteurization effect of irradiation on pathogens, including *E. coli* O157:H7 and *Salmonella* when the proper dose of electron beam energy is applied to raw ground beef patties. In addition, the research identified optimum packaging conditions to allow for irradiation pasteurization without compromising product quality.

The studies on pasteurization using pulse electron beam irradiation included ground beef, whole muscle beef products, and boneless pork chops. Effects of irradiation on color, oxidative rancidity, flavor, texture, and aroma were determined in five separate studies. Some of the conclusions of these comprehensive studies were that irradiation had minimal effects on flavor, texture, aroma, and color of beef or pork products. In irradiated pork, the muscle color was enhanced and was considered more "intense."

Additional studies evaluated the effect of packaging on the acceptability of irradiated beef and pork products. When the process was conducted in an aerobic package, there was a detrimental effect on oxidation and color. When the irradiation process was conducted in an anaerobic environment under controlled conditions, consumer panelists found no difference between irradiated and control pork chops.

The FSC research on irradiation pasteurization was published in a series of journal articles. This work helped provide a basis for optimizing irradiation process with respect to product and packaging requirements.

Research on Microbiological Sampling and Testing

Another breakthrough that occurred during the period was the use of microbiological testing to validate food-safety technologies and verify their effectiveness in practice. In 1994, the USDA enacted the Pathogen Reduction and HACCP regulation. One of the provisions of the regulation established a requirement for microbiological testing of carcasses to verify process control. This mandatory program involves testing for generic *E. coli*. The ongoing FSC projects involving carcass sampling procedures and the rapid detection of microbiological pathogens provided valuable input into microbiological testing strategies. In addition to mandatory carcass testing requirements, the beef industry has voluntarily enacted widespread testing for the pathogen—*E. coli* O157:H7. Today, pathogen testing is widely used in the beef industry to verify the safety of raw materials for ground beef and further processed products.

FSC researchers were involved in the training of industry and government personnel in carcass sampling and microbiological testing procedures related to the implementation of the HACCP system. Additional work has been conducted to validate rapid microbiological testing procedures for *E. coli* O157:H7, *Sal-monella,* and *Listeria monocytogenes.*

In addition, FSC researchers also assisted in the development and validation of PCR-based testing systems for the rapid detection of microbiological pathogens in food products.

The Safety of Nonintact Beef Products

One beef-safety issue relating to *E. coli* O157:H7 that surfaced in the 1990s was the risk associated with nonintact beef products. The safety of intact steaks and roasts is assured because microbiological contaminants are limited to the exterior surface of these products and the cooking process readily inactivates bacteria on the surface of the muscle. The interior muscles are free of microbiological contamination. Therefore, intact steaks and roasts may be cooked to a rare degree of doneness without the risk that *E. coli* O157:H7, *Salmonella,* or other pathogens will survive the cooking process and pose a risk to consumers. However, industry practices change over the years, and by the mid-1990s, most steaks and roasts were processed in a manner that allowed for the possible translocation of bacteria from the surface of the cut into the interior muscle. The technologies that resulted in "nonintact" steaks and roasts included blade tenderization, needle injection, marination, and section and forming.

A survey conducted by the North American Meat Processors Association determined that a very low percentage of steaks destined for food-service applications were blade tenderized and therefore would be considered "nonintact." The Food Safety Inspection Service of the U.S. Department of Agriculture defines nonintact beef products as ground beef; beef injected with solution; beef that has been mechanically tenderized by needling, cubing, or pounding devices, and beef that has been reconstructed into formed entrees (FSIS-USDA 1999). Whole muscle cuts (e.g., chucks, ribs, tenderloins, strip loins, top sirloin butts, rounds) may be treated with these technologies to increase tenderness or to add ingredients for quality purposes. The treatments often occur before fabrication into steaks and may include solid-needle or hollow-needle tenderizing, in which a solution is pumped into the whole muscle. In the latter case, the solution typically is recirculated, refrigerated, and treated to ensure quality.

In 2001, the National Advisory Committee on Microbiological Criteria for Foods (NACMCF) was asked by the FSIS to address the potential public health risk of *E. coli* O157:H7 in tenderized, nonintact beef steaks and roasts. In its final report, the committee indicated that based on the available data, nonintact, blade-tenderized beef steaks did not present a greater risk to consumers if the meat is oven broiled to an internal temperature of 60°C. The committee could not make a similar conclusion for nonintact beef roasts, based on an absence of available data.

The Food Safety Consortium research group at Kansas State University addressed the problem of nonintact beef products through a series of projects designed to determine the risk and to quantify the translocation of microorganisms from the outside surface of beef subprimals into the interior of the muscle. Additional research was conducted to evaluate various cooking methods with respect to the inactivation of microbiological contamination both on the surface and inside various beef cuts.

In a study examining translocation, Food Safety Consortium researchers at Kansas State University found that after a single pass through a blade tenderizer, 3–4% of *E. coli* O157:H7 on the surface of inoculated subprimals was transferred to the interior muscle tissue. As part of the same study, researchers examined the effectiveness of cooking procedures on the destruction of *E. coli* O157:H7 in mechanically tenderized beef steaks. Nonintact subprimals inoculated with 10^7 CFU/cm^2 *E. coli* O157:H7 were broiled to endpoint temperatures ranging from 49°C to 77°C. Based on this project, it was concluded that a target temperature of at least 60°C provides the necessary thermal destruction required to eliminate the public health risk from *E. coli* O157:H7. In this study, even though the steaks were rapidly chilled in ice water after broiling to a specific endpoint temperature, the internal temperature continued to rise by approximately −11°C. This increase in temperature would serve as an additional margin of safety, especially in a foodservice setting.

As a result of extensive research on the nonintact beef products, FSC researchers determined that the risks associated with blade tenderization, needle injection, and other technologies are dependent on the following factors:

1. The effectiveness of carcass intervention strategies applied during the slaughter process. According to a report presented at the Non-Intact Products Processing Workshop, "Validated technologies, including steam and hot water pasteurization and USDA's zero tolerance policy for fecal contamination on carcasses, greatly reduce the likelihood that high levels of enteric pathogens will be present on the surface of subprimal meat cuts prior to needling or mechanical tenderization."
2. Possible application of an approved antimicrobial to subprimals prior to tenderization.
3. A critical control point addressing ambient and/or product temperature.
4. Thermal process that results in an internal temperature of 60°C.

The results of the Food Safety Consortium research on the safety of nonintact beef products is consistent with the National Advisory Committee on Microbiological Criteria for Foods report that indicated that based on the available data, nonintact, blade-tenderized beef steaks did not present a greater risk to consumers if the meat is oven broiled to an internal temperature of 60°C.

Processing Interventions for Raw Beef Products

A number of processing interventions have been evaluated by FSC researchers. These include acidified sodium chlorite, ozone, UV, advanced oxidation technologies, pressurized CO_2, and chlorine dioxide. Research focus was placed on technologies that can be applied to beef trimmings prior to grinding and to beef subprimals prior to blade tenderizations or needle injection. Two of the technologies that were validated by FSC researchers were the application of acidified sodium chlorite to raw beef trimmings and the application of a UV-based advanced oxidation technology to beef subprimals. Both technologies have been adopted by the industry and are in commercial use.

Multiple Hurdles

FSC research has shown that the effectiveness of interventions is enhanced by staging "multiples hurdles" in the farm to table process. Although no single intervention completely eliminates the risk of E. coli O157:H7 and other pathogens in raw ground beef and processed raw beef products, the combination of multiple interventions can greatly decrease the overall risk. For example, a preharvest food-safety program combined with the application of steam vacuuming and steam pasteurization in a sanitary slaughter process reduces the risk of microbiological contamination on the finished carcass. The risk is further reduced through the application of controlled carcass chilling and a chemical intervention applied postchill. An additional intervention may be applied to trimmings prior to grinding or to subprimals prior to blade tenderization to provide further reduction in risk. In theory, these steps in combination with proper handing and cooking by consumers of restaurants should reduce the risk of pathogen contamination in the cooked, finished product to very low levels. This approach is termed "Multiple Hurdles" and is used widely throughout the beef industry.

Summary

Although major improvements have been achieved in addressing the problem of E. coli O157:H7 in raw ground beef and processed beef products, fifteen years after the Jack-in-the-Box outbreak, public health and regulatory problems continue to occur. The reason is in the extreme difficulty in eliminating a naturally occurring pathogen from a raw product. Prior to the 1993 outbreak, the presence of pathogens in raw meat products was considered a natural occurrence. An example is the expected presence of *Salmonella* and *Campylobacter* in raw poultry. Pathogens are destroyed during the cooking process. Without lethality treatment such as cooking or irradiation, eliminating pathogens from raw meat products poses a monumental challenge.

In retrospect, the researchers and the beef industry on the whole acted in an organized and effective manner to reduce the risk of *E. coli* O157:H7. Research efforts were initially directed to the most critical point in the farm-to-table continuum —the slaughter process. Within a year after the 1993 outbreak, a thermal pasteurization system for beef carcasses had been developed and was operating in several beef slaughter plants. Improved technologies for removing physical defects from beef carcasses and chemical antimicrobial treatments were developed, approved, and put into practice. Today, virtually every beef slaughter plant employs one or more of these technologies. After addressing slaughter-related technology needs, the industry turned its attention to the development of food-safety technologies for further processed products and to preharvest research. Progress has been made in these areas as well.

Statistics from the USDA and the U.S. Centers for Disease Control and Prevention show a major improvement in the reduction of incidence of *E. coli* O157:H7 in raw ground beef between the years 2001 and 2004. This reduction in incidence was manifested in fewer cases and outbreaks of the disease associated with beef products. Unfortunately, that positive trend is starting to reverse. The USDA is finding that an increasing number of *E. coli* O157:H7 positives, and cases and outbreaks are on the rise.

Clearly, the Food Safety Consortium and the beef industry must continue their efforts and take solutions to the problem of *E. coli* O157:H7 to a new level. Innovative technologies are under development to prevent contamination during the slaughter process. The problem of contamination in nonintact beef products is being addressed through research and technology development. Research conducted by the Food Safety Consortium will continue to lead the way to safer beef products and ultimately to better solutions to the problems associated with microbiological contamination in foods.

References

Luchsinger, S. E., D. H. Kropf, E. Chambers IV, C. M. Garcia Zepeda, M. C. Hunt, S. L. Stroda, M. E. Hollingsworth, J. L. Marsden, and C. L. Kastner. Patties and whole muscle beef. Journal of Sensory Studies 12:105–126.

Luchsinger, S. E., D. H. Kropf, C. M. Garcia Zepeda, E. Chambers IV, M. E. Hollingsworth, M. C. Hunt, J. L. Marsden, C. L. Kastner, and W. G. Kuecker. 1996. Sensory analysis and consumer acceptance of irradiated boneless pork chops. Journal of Food Science 61(6):1261–1266.

Luchsinger, S. E., D. H. Kropf, C. M. Garcia Zepeda, M. C. Hunt, J. L. Marsden, E. J. Rubio Cañas, C. L. Kastner, W. G. Kuecker, and T. Mata. 1996. Color and oxidative rancidity of gamma and electron beam-irradiated boneless pork chops. Journal of Food Science 61(5):1000–1005, 1093.

Luchsinger, S. E., D. H. Kropf, C. M. Garcia Zepeda, M. C. Hunt, S. L. Stroda, J. L. Marsden, and C. L. Kastner. 1997. Color and oxidative properties of irradiated ground beef patties. Journal of Muscle Foods 8:445–464.

Nutsch, A. L., R. K. Phebus, M. J. Riemann, J. S. Kotrola, R. C. Wilson, J. E. Boyer Jr., and T. L. Brown. 1998. Steam pasteurization of commercially slaughtered beef carcasses: evaluation of bacterial populations at five anatomical locations. J. Food Prot. 61:571–577.

Nutsch, A. L., R. K. Phebus, M. J. Riemann, D. E. Schafer, J. E. Boyer Jr., R. C. Wilson, J. D. Leising, and C. L. Kastner. 1997. Evaluation of a steam pasteurization process in a commercial beef processing facility. J. Food Prot. 60:485–492.

Phebus, R. K., A. L. Nutsch, D. E. Schafer, C. R. Wilson, M. J. Riemann, J. D. Leising, C. L. Kastner, J. R. Wolf, and R. K. Prasai. 1997. Comparison of steam pasteurization and other methods for reduction of pathogens on surfaces of freshly slaughtered beef carcasses. J. Food Prot. 60:476–484.

USDA-FSIS. 1999. Beef products contaminated with Escherichia coli O157:H7. Fed. Regist. 64(11):2803–2805.

▌11▐

Occurrence of *Listeria monocytogenes* in Raw and Ready-to-Eat Foods and Food-Processing Environments and Intervention Strategies for Control

Bwalya Lungu, Steven C. Ricke, and Michael G. Johnson

Introduction

Listeria monocytogenes is an intracellular pathogen that causes listeriosis. *L. monocytogenes* along with *Salmonella* and Toxoplasma is responsible for more than 75% of foodborne deaths caused by known pathogens. Specifically, the Centers for Disease Control and Prevention reports that *L. monocytogenes* accounts for an estimated 2,500 illnesses and 500 deaths per year (CDC 2005). Unlike infection with other common foodborne pathogens that rarely result in deaths, human listeriosis is associated with a mortality rate of approximately 20–30%. The fatality rate in untreated patients or those treated late can rise to 70%. *L. monocytogenes* mainly affects the elderly, neonates, pregnant women, immune compromised individuals, and people on immune suppressant medication such as organ transplant and cancer patients (CDC 2005; Mead et al. 1999; Altekruse et al. 1997). The emergence of listeriosis could be the result of changes in social and economic patterns. During the past 50 years improvements in medicine, public health, sanitation, and nutrition have resulted in increased life expectancy, particularly in developed countries. The U.S. population is aging as its baby boomer population reaches a mature stage. By 2010, it is predicted that more than 26% of the U.S. population will be aged between 45 and 64. Cancer is also one of the leading causes of death in the United States (ACCT 2001; Anonymous 2006a; 2006b). In addition, new HIV/AIDS cases are reported each year. *L. monocytogenes* is an opportunistic human pathogen of high public concern that causes listeriosis, a disease that mainly affects the immunocompromised, the elderly, infants, and pregnant women. Due to a rise in the number of cases of immunocompromised individuals as a result of the emergence of diseases like AIDS, intensive cancer therapies, immunosuppressive drug therapies, organ transplants, and the rise in number of elderly people from the aging baby boomer population, *L. monocytogenes* has become a pathogen of serious concern.

Symptoms of listeriosis vary depending on the individual but may include fever, diarrhea, septicemia, and meningitis, which develops predominantly in newborns and the elderly. Other symptoms include meningioencephalitis, endocarditis, endophthalmitis, osteomylitis, brain abscesses, and peritonitis (Mead et al. 1999; Painter and Slutsker 2007). In pregnant women, *L. monocytogenes* causes mild flu-like symptoms but may also cause more severe complications including preterm labor, chorioamnionitis, spontaneous abortion, and stillbirth. *L. monocytogenes* crosses the placenta, resulting in abortion or other serious problems such as early and late onset disease in neonates with the most severe manifestation being granulomatosis infantiseptica (reviewed in DiMaio 2000). However, the mechanisms by which *L. monocytogenes* targets and crosses the maternofetal barrier are largely unknown.

The most common route of transmission to humans is by consumption of food contaminated by *L. monocytogenes,* which is supported by well-documented foodborne outbreaks (Aureli et al. 2000; Graves et al. 2005; Jacquet et al. 1995; Dalton et al. 1997; Miettinen et al. 1999; Schlech et al. 1993; Gilbert et al. 1993; Mead 1999). In 1999 and 2002 there were two massive recalls of 26 to 27 million pounds each of RTE products due to contamination by *L. monocytogenes.* Products from those two major recalls were reported by the CDC to be associated with 73 and 53 illnesses and 16 and 8 deaths, respectively (Anonymous 1999, 2002; Mead 1999; Graves et al. 2005). Approximately 2 to 6% of investigated healthy people are asymptomatic carriers of *L. monocytogenes* (Jensen 1993; Schuchat et al. 1993) and if they work in food-service establishments or in hospitals, they could transmit the bacteria to other susceptible individuals. Results of an investigation of a California outbreak in 1985 revealed that community-acquired outbreaks could be amplified through secondary transmission by fecal carriers (Mascola et al. 1992). In addition to the high death rate among the susceptible population, the minimum infective dose of listeriosis is not known and the severity of the disease is dependent on the immune competence of the individual; therefore, the USDA-FSIS has established a zero tolerance level for this organism in RTE foods. The severity and high fatality rates of the disease make it imperative that this human pathogen be controlled.

L. monocytogenes, a Gram-positive non-spore-forming bacterium, is widely distributed in nature, can survive a wide variety of environments, is microaerobic and psychrotrophic, and can grow in human phagocytes. *L. monocytogenes* is resistant to adverse environmental conditions including low temperature, irradiation, nutrient deprivation, oxidative stress, osmotic stress, and acid stress (Ferreira et al. 2001, 2003; Lou and Yousef 1996, 1997; Herbert and Foster 2001; Mendonca et al. 2004). In addition the organism can survive in or on foods for very long periods of time (reviewed in Farber and Peterkin 1991). All of these traits make *L. monocytogenes* a major concern for both public health and the food-processing industry. Therefore, our objectives were to (1) review the prevalence of *L. monocytogenes* in pre- and postharvest food-production systems as well as in the food-processing environment, and (2) summarize the progress that has been made to control this pathogen.

Risk Impact of *L. monocytogenes*

L. monocytogenes cannot be completely eliminated from the processing environment of RTE food products and a survey conducted by the USDA-FSIS showed that 1 to 10% of retail RTE products were contaminated with *L. monocytogenes* (Levine et al. 2001). At present there is a zero tolerance policy for *L. monocytogenes* in cooked RTE foods requiring less than one viable bacterial cell per 25 g sample. An acceptable non-zero level in CFU/g of *L. monocytogenes* contamination in cooked/RTE products cannot be set by regulatory agencies because the infectious dose varies greatly depending on the immune status of the individual ingesting the *Listeria* contaminated food. The USDA *L. monocytogenes* risk-assessment report states that certain foods such as frankfurters may pose a greater public health risk and this finding is significant because 20 billion frankfurters are consumed annually in the United States (Anonymous 2001). Chen et al. (2003) recently suggested that low levels of *L. monocytogenes* might be a low risk to immunocompetent people. They argue the zero tolerance policy does not distinguish between foods contaminated at high and low levels. They also suggest that a management strategy focusing on the concentration of *L. monocytogenes* rather than only its presence may have a greater impact on the improvement of public health by facilitating the development of control measures to limit the levels of *L. monocytogenes* in foods.

Interestingly, *Listeria* infections are mainly reported in industrial countries, probably due to the food-processing, distribution, and detection methods used in those countries. It is sometimes difficult to determine the specific source and route of *L. monocytogenes* foodborne outbreaks because this organism can have an incubation period ranging from two days to five weeks. This wide range in incubation periods presents a serious challenge in accurately determining which foods are potentially more contaminated with this pathogen. For example in 1985, in southern California at least 142 cases of listeriosis resulting in 48 deaths (20 fetuses, 10 neonates, and 18 nonpregnant adults) were reported (Linnan et al. 1988). In this outbreak the source was traced to Mexican-style soft cheese that was contaminated due to failure or misuse of the milk pasteurization equipment. An outbreak of *L. monocytogenes* in the New England states was linked to pasteurized milk (Fleming et al. 1985), suggesting it was not destroyed by normal milk pasteurization mandated by the FDA milk pasteurization ordinance.

Environmental Ecology of *L. monocytogenes*

L. monocytogenes is ubiquitous in the environment and can be found on decaying vegetation, in soil, human and animal feces, sewage, silage, food-processing environments, milk of normal and mastitic cows, and water (Weis and Seeliger 1975; Welshimer and Donker-Voet 1971; Botzler et al. 1974; Yoshida et al. 2000; Thimothe et al. 2002; Lawrence and Gilmour 1994; Fenlon et al. 1996; Garrec et al. 2003; Sanaa et al. 1993). *L. monocytogenes* has been isolated from cattle, sheep, goats, poultry, and fish, but infrequently from wild animals (Nightingale et al. 2004; Hofer 1983; Løken et al. 1982; Grønstøl 1979; Thimothe et al. 2002; Weis and Seeliger 1975; Fenlon 1985; Yoshida et al. 2000). Asymptomatic animals can carry the pathogen and contaminate foods of animal origin such as poultry, red meats, and dairy products. Vegetables can become contaminated from the soil or manure used as fertilizer. The use of animal manures as fertilizers in vegetable and other agricultural fields presents a potential dissemination route for *L. monocytogenes* in the environment. Fecal material is considered to be a potential source of contamination of animal carcasses and raw milk by *L. monocytogenes*. Skovgaard and Morgen (1988) found that 62% of feed samples and 51% of fecal samples from dairy farms harbored *L. monocytogenes*, while 33% of poultry fecal samples and 47% of poultry neck-skin samples were contaminated with *L. monocytogenes*. An epidemiological study conducted by Husu (1990) observed that more cows excreted *L. monocytogenes* in the winter with 16.1% of cows testing positive in December, with a strong positive correlation between the presence of *L. monocytogenes* and the feeding of silage. An investigation of fecal samples from chickens, sheep and cattle from abattoirs in Turkey revealed a prevalence of 4.36, 0.58 and 1.53%, respectively (Kalender 2003). Jiang et al. (2004) found that *L. monocytogenes* persisted in unautoclaved soil containing bovine manure for up to three weeks at 21°C. However, longer survival times were observed in soils kept at temperatures as low as 5°C. Persistent *L. monocytogenes* cells could subsequently be transmitted to fresh produce or to fieldworkers, especially during the cold months of the year.

It has been hypothesized that the main source of *L. monocytogenes* contamination of food products is live animals (Skovgaard and Norrung 1989; Autio et al. 2000). In particular, swine farms have been considered to be a primary source of *Listeria* species detected in slaughterhouses (Skovgaard and Norrung 1989; Van Renterghem et al. 1991; Salvat et al. 1995; Nesbakken et al. 1994). Several studies have shown that *L. monocytogenes* is present in swine operation environments and is asymptomatically carried by pigs (Skovgaard and Norrung 1989; Adesiyun and Krishnan, 1995; Fenlon et al. 1996). Beloeil et al. (2003) recovered *L. monocytogenes* from piggery pens as well as dry and wet feed samples. *L. monocytogenes* was isolated from more wet feed samples than dry feeds, suggesting that wet feeds likely increase pig infection. In addition, pigs fed contaminated feed continued to shed *L. monocytogenes* in their fecal matter and kept swine facilities continually contaminated with *L. monocytogenes*. In another study conducted in Yugoslavia, 45% percent of the pigs examined harbored *L. monocytogenes* in the tonsils and 3% of those were fecal excretors (Buncic 1991). In a study conducted by Skovgaard and Norrung (1989), *L. monocytogenes* was found in both pig feces and minced pork produced from those pigs suggesting that fecal contamination may also play a role in the dissemination and persistence of *L. monocytogenes* to raw meats.

However, pigs are not the only animals that have been found to harbor *L. monocytogenes*. Buncic (1991) reported that 29% of cattle retropharyngeal swabs and 19% of fecal matter samples harbored *L. monocytogenes*. Nightingale et al. (2004) conducted a case control study to determine the transmission and ecology of *L. monocytogenes* on farms. The farms involved in the study included 24 case farms with at least one recent case of listeriosis and 28 matched control farms with no listeriosis cases. Their study reported that the overall prevalence of *L. monocytogenes* in cattle farms (24.4%) was similar to that observed in control farms (20.2%). However, they observed that the prevalence of *L. monocytogenes* in small-ruminant (goat and sheep) farms (32.9%) was not only higher than in cattle farms but was also significantly higher than those observed in control farms (5.9%). Therefore, it is likely that these small-ruminant farm animals play a significant role in transmission and dispersal of *L. monocytogenes* within their farm environments. They observed that *L. monocytogenes* isolated from clinical cases and fecal samples were more frequent in the environmental samples than the feed samples, suggesting that infected animals may contribute to *L. monocytogenes* dispersal into the farm environment. They concluded that the epidemiology and transmission of *L. monocytogenes* differs between small-ruminant and cattle farms. In addition, the bovine farm ecosystem maintains a high prevalence of *L. monocytogenes*, including subtypes linked to human listeriosis cases and outbreaks. They also concluded that cattle more than likely contribute to amplification and dispersal of *L. monocytogenes* into the farm environment (Nightingale et al. 2004). Silage, sewage sludges, and wastewater are also potential reservoirs of *L. monocytogenes*. The prevalence of *L. monocytogenes* in these environments demonstrates the ability of this pathogen to survive in low oxygen environments (De Luca et al. 1998). Garrec et al. (2003) found *L. monocytogenes* in dewatered sludges and tank-stored sludges at 73 and 80%, respectively. Consequently, the prevalence of *L. monocytogenes* in bovine and other farm ecosystems presents a challenge to the food industry where the USDA-FSIS has imposed a zero tolerance of *L. monocytogenes* on RTE foods. Meat-processing plants could be contaminated with *L. monocytogenes* from raw meat products, and some *L. monocytogenes* cells may persist within the plant environment and thus recontaminate processed RTE meat products.

In meat-processing plants, the contamination of RTE meat products usually occurs when the food product comes in contact with a contaminated surface between the cooking and packaging steps, usually during product slicing. The surfaces of equipment used for food handling or processing are potential sources of contamination especially when improperly cleaned or sanitized (Dunsmore 1981). Therefore, the focus of food-processing plants needs to be prevention of recontamination of RTE food products. *L. monocytogenes* is often found around wet areas and cleaning aids such as floors, drains, wash areas, ceiling condensate, mops and sponges, brine chillers, and peeler stations (Marriott 1999). Lawrence and Gilmour (1994) detected *L. monocytogenes* in various sites within the poultry-processing environment including stainless steel surfaces, conveyor belts, door handles, floors, squeegees, worker gloves, and drains. A high incidence of *L. mono-cytogenes* was observed for raw product. Although no *L. monocytogenes* were isolated from cooked products during their study, the presence of *L. monocytogenes* in the plant environment increases the risk for post-process recontamination. Somers and Wong (2004) found that the cleaning efficacy of two detergent and sanitizer combinations on biofilms formed on a variety of materials in the presence of RTE meat residue was surface dependent and decreased with residue-soiled surfaces. They also observed that biofilms that developed on brick or conveyor material were more resistant to the cleaning action of the detergents and sanitizers. *L. monocytogenes* can grow or survive in planktonic form or as a biofilm (Lunden et al. 2003; Holah et al. 2002). These two phenotypes may behave differently and may consequently impact food industry control measures differently. In the natural or food-processing environment, ideal conditions necessary for growth and survival of bacterial cells are not always present. Therefore, the ability of *L. monocytogenes* to adapt to varying external stresses is important for its survival or growth in the environment as well as in contaminated foodstuffs and host organisms.

Prevalence of *L. monocytogenes* in Poultry Products and Processing Facilities

Historically, broiler chickens have represented the most common avian host for *L. monocytogenes*. Since there are many similiarities between processing procedures for chickens and turkeys, products from chickens and turkeys exhibit a similar risk for *L. monocytogenes* contamination. Listeric meningitis caused by *L. monocytogenes* has been linked with the consumption of turkey products and frankfurters (Hatkin and Philips 1986; Barnes et al. 1989). Slaughterhouse surveys of poultry samples held at 4°C revealed *L. monocytogenes* contamination on 45% of fresh turkey wings, 28.3% fresh legs, and 23.3% of fresh tails. These surveys also showed that turkey parts became increasingly more contaminated (2 to 2.5 fold) as stages of processing increased (Genigeorgis et al. 1990). In a separate survey, *Listeria* counts as high as 10^4 CFU/g of meat were found in two of the 509 raw turkey products sampled by Wilson (1995). In a study conducted by Wallace et al. (2003), the rates of recovery of *L. monocytogenes* from three of the 12 plants tested was significant, yielding recovery rates of 1.5% (plant 367), 2.2% (plant 439), and 16% (plant 133) over a two-year period. From *Listeria* positive plants, using extended refrigerated storage at 4 and 10°C, *L. monocytogenes* was recovered from frankfurters at rates of 6.54 and 6.81%, respectively, for all types of frankfurters. Conversely, when both the positive and negative plants tested were taken into consideration, the rates of *L. monocytogenes* recovery decreased to 1.64 and 1.59%, respectively, for the 4 and 10°C stored frankfurters. Overall, the recovery rates did not change appreciably over time and there was no significant difference in *L. monocytogenes* recovery rates with respect to frankfurter storage temperature (4 or 10°C). Molecular subtyping of multiple *L. monocytogenes*–positive isolates from the plants revealed that 90% of the 1,105 isolates tested were serotype 1/2a. However, in some cases recovery of more than one serotype from a given plant also occurred.

Rorvik et al. (2003) investigated the prevalence of *L. monocytogenes* in poultry products in Norway. They reported a high prevalence of *L. monocytogenes* in the test products with specific strains persisting in the processing environment; however, no listeriosis cases were linked to any of the *L. monocytogenes* strains isolated. It appeared that broilers were contaminated during the slaughter process. Their study suggested that plant practices need to include preventive measures to avoid contamination as well as post-process recontamination of poultry products. In a study conducted by Huff et al. (2005) results showed that when turkey poults were air-sac inoculated with the human pathogenic strain Scott A of *L. monocytogenes* at high-challenge doses, workers could later isolate this pathogen in 48 to 59% of the knee joint fluids of such birds. Low-challenge doses resulted in the isolation of *L. monocytogenes* from 11% of the knee joint fluids. Their results suggested that this pathogen once it gains entry to this animal can colonize the turkey knee joint and represent a previously unrecognized and potentially serious source of pathogen cross contamination in the processing plant.

Prevalence of *L. monocytogenes* in Red Meat Products and Processing Facilities

Sources of in-plant carcass microbial contamination during slaughter include those associated with processing practices, slaughter plant facilities, and plant personnel (Lawrence and Gilmour 1994; Rorvik et al. 2003). Rivera-Betancourt et al. (2004) reported *L. monocytogenes* to be prevalent on the hides of cattle as well as in processing plant facilities. Samadpour et al. (2006) isolated *L. monocytogenes* from 3.5% of ground beef samples. Skovgaard and Morgen (1988) found that 28% of minced beef samples from retail sources harbored *L. monocytogenes*. Buncic (1991) did not isolate *L. monocytogenes* from deeper muscle tissue of beef carcasses but noted that 69% of minced meat (pork and beef mixture), 19% of raw dry sausages, and 21% of vacuum packaged smoked sausages were contaminated with *L. monocytogenes*. However, smoked sausages heated to an internal temperature of 70 to 75°C after the smoking process were not contaminated. Coillie et al. (2004) found a high prevalence of *L. monocytogenes* in minced meat.

Bison or buffalo meat is an emerging meat preference for consumers in U.S. and European markets (Li et al. 2004). A study conducted by Li et al. (2004) found that the overall prevalence of *Listeria* species in bison was 18.3%. The presence of *Listeria* spp. at pre-dehiding, postevisceration, post-USDA inspection, or postwashing was 42.24, 18.1, 6.03, or 1.72%, respectively. Significant reductions in the proportion of carcass samples positive for *Listeria* spp. were observed between predehiding and postevisceration as well as between post evisceration and post-USDA inspection points. Of the 355 carcasses tested, only 1.13% tested positive for *L. monocytogenes*.

Prevalence of *L. monocytogenes* in Milk-Processing Facilities

Raw milk, unpasteurized milk, or improperly pasteurized milk and milk products present some of the most common causes of sporadic illnesses by *L. monocytogenes*. However, datum regarding the national prevalence of bacterial pathogens including *L. monocytogenes* in these food products is not available. Milk contamination can occur via a number of ways including bacterial shedding from the udder. A number of studies have reported recovery of *L. monocytogenes* from milk or milk products. Hassan et al. (2000) demonstrated that *L. monocytogenes* was prevalent in milk filters in New York dairy herds. Jayarao and Henning (2001) found that 4.6% of bulk milk tanks from South Dakota and Minnesota were contaminated with *L. monocytogenes*. In another study, Van Kessel et al. (2004) collected milk samples from farms across 21 states and recovered *L. monocytogenes* from 6.5% of bulk milk tanks tested.

Growth and Survival of *Listeria monocytogenes* in Anaerobic or Modified Atmospheres

Food-packaging techniques as well as experimental conditions that have reduced oxygen as part of the control or inhibition of this pathogen have been used as commercial preservation systems for meat and vegetable products. Vacuum packaged or modified atmosphere packaging has been used for shelf-life extension of food products, including vegetables and meats. Although this packaging excludes the growth of aerobic bacteria, anaerobic and facultative anaerobes are not suppressed. Any prior treatments applied to these foods need to be sufficient to eliminate foodborne pathogenic bacteria such as *L. monocytogenes* that are capable of proliferating under low oxygen conditions.

Knabel et al. (1990) reported recovering severely heat injured cells in tryptic soy broth with yeast extract (TSBYE) within sealed tubes for thermal death time as a result of the formation of reduced (low oxygen) conditions in the depths of the TSBYE. They also reported a significantly increased recovery of *L. monocytogenes* cells under strict anaerobic versus aerobic conditions. They noted that cells grown at 43°C and recovered under strict anaerobic conditions exhibited an increased heat resistance at 62.8°C compared to those that were heat shocked at the same temperature or grown at lower temperatures. In fact, cells grown at 43°C and recovered anaerobically yielded a six-fold increase in heat resistance at 62.8°C compared to aerobically grown cells held at 37°C. Based on their results, it appears that high levels of *L. monocytogenes* would survive the minimum low temperature, long-term pasteurization treatments recommended by the FDA for milk pasteurization. With the aid of a novel strictly anaerobic recovery and enrichment system that incorporated lithium for the detection of heat-injured *L. monocytogenes* in pasteurized milk, Mendonca and Knabel (1994) recovered increased levels of severely heat-injured cells. Consequently, it appears that anaerobiosis may help resuscitate heat-injured *L. monocytogenes*. In a study conducted by George et al. (1998), the heat resistance of *L. monocytogenes* and other Gram-negative foodborne pathogens increased eight-fold under anaerobic conditions. Heat resistance was highest when anaerobic mixtures were used. Moderate heat resistance was observed when low oxygen concentrations (0.5 to 1%) were tested. Their data suggest that the use of anaerobic mixtures or vacuum packaging could potentially increase the risk of pathogenic bacteria surviving the heat treatment applied to minimally processed foods.

Barakat and Harris (1999) found that cooked poultry products inoculated with *L. monocytogenes*, packaged in 44:56 CO_2-N_2 and incubated at 3.5, 6.5, or 10°C for up to five weeks allowed growth of *L. monocytogenes*. Oxygen concentrations when detected were less than 1% and the pH of the chicken product was 6.3. The addition of lactate prolonged the lag phase but did not prevent growth of *L. monocytogenes* under their test conditions. In addition the presence of naturally occurring microbiota did not prevent the growth of *L. monocytogenes*. It appears that the use of modified atmospheres may increase the risk of survival of *L. monocytogenes* in poultry products at low temperatures. Ingham et al. (2006) observed that beef jerky dried to a_w values that were less than 0.87 with vacuum packaging did not support the growth of bacterial pathogens including *L. monocytogenes* over a four-week period with an incubation temperature of 21°C. However these conditions still permitted the survival of high numbers of viable *L. monocytogenes* cells during the duration of their study. Wimpfheimer et al. (1990) observed that *L. monocytogenes* Scott A failed to grow on minced raw chicken in a modified atmosphere (75% CO_2 : 25% N_2 lacking oxygen) at 4, 10, and 27°C. Conversely, *L. monocytogenes* Scott A increased by approximately 6 log CFU/g at 4°C in 21 days in amended modified atmospheres (72.5% CO_2, 22.5% N_2, 5% O_2) containing oxygen which more closely resembled commercial packaging atmospheres used. However, this atmosphere inhibited the growth of other competitive spoilage organisms by more than 4 log CFU/g at 4°C compared with those that were incubated in air. They also concluded that growth of *L. monocytogenes* in aerobic modified atmospheres was not affected by initial inoculum or by the initial level of aerobic competitive spoilage organisms. While modified atmosphere packaging substantially inhibited aerobic spoilage bacteria, it did not inhibit *L. monocytogenes* growth.

In a study conducted over a two-year period in Ireland, Francis and O'Beirne (2006) reported that fresh-cut vegetables packaged under modified atmosphere can support the growth of numerous *Listeria* species including *L. monocytogenes*. They noted a contamination rate of 9.58% with *Listeria* species in modified-atmosphere-packed fresh-cut fruits and vegetables. Overall they observed a 2.9% recovery of *L. monocytogenes* from the food samples tested and that dry coleslaw mix appeared to have the highest incidence (20%) of *L. monocytogenes*. They concluded that contamination of products was more frequent during the summer and autumn months. They successfully identified four pulsed field gel electrophoresis (PFGE) profiles and found that half of the isolates belonged to one molecular profile. Chen and Hotchkiss (1993) documented the growth of *L. monocytogenes* in cottage cheese under modified atmosphere packaging and concluded that *L. monocytogenes* was unable to grow in cottage cheese packaged with carbon dioxide and stored at 4°C but grew by one log CFU/g at 7°C. However, when cottage cheese was held in conventional packaging materials, *L. monocytogenes* grew by 3 log CFU/g. Low refrigeration temperatures appear to be essential to inhibit the growth of *L. monocytogenes* in cheese under modified-atmosphere conditions. However, the effectiveness of modified atmospheres may depend not only on low temperature but also on low pH.

Current Strategies for Limiting Postharvest Contamination of *L. monocytogenes*

Since *L. monocytogenes* has the ability to survive different physicochemical conditions such as refrigeration temperatures, low pH, high salt concentrations, and high temperatures, it appears that combinations of lower temperatures and lower pH best inhibit growth of this pathogen under low oxygen or modified-atmosphere conditions. The food type and cell growth phase may also play significant roles in enhancing the effectiveness of any intervention strategy to control *L. monocytogenes* using low oxygen or modified atmospheres. Therefore, considerable research has been conducted over the past few years to further develop and optimize intervention technologies to limit post-process recontamination of food products by *L. monocytogenes*. Research on potential methods includes the development and application of chemical, biological, and physical agents and processes. The food industry currently utilizes steam, hot water, radiant heat, and high-pressure processing as pre- and postpackaging lethal treatments against microbial pathogens. Murphy et al. (2006) tested the efficacy of a combined treatment of steam surface pasteurization with an organic acid solution (2% acetic acid, 1% lactic acid, 0.1% propionic acid, and 0.1% benzoic acid) against *L. monocytogenes* on fully cooked frankfurters during vacuum packaging. They showed a 3-log reduction of *L. monocytogenes* when frankfurters were treated with steam for 1.5 seconds. A combination of organic acid treatment with steam pasteurization further inhibited the growth of surviving *L. monocytogenes* cells for 19 or 14 weeks when frankfurters were stored at 4 or 7°C, respectively. Hayman et al. (2004) showed that high-pressure processing at 600 MPa, 20°C for 180 seconds can extend the shelf life of RTE meats and reduce *L. monocytogenes* numbers by more than 4-log CFU/g. Jofre et al. (2007) showed high-pressure processing of actively packaged ham slices reduced *L. monocytogenes* populations by about 4-log CFU/g in the presence of bacteriocins. In contrast, *L. monocytogenes* populations increased to about 6.5-log CFU/g following treatment of cooked ham with high-pressure processing in the presence or absence of lactate. Montero et al. (2007) showed that the combined treatment of cold-smoked dolphinfish with 2.93% salt, 82 ppm phenol, and high-pressure treatment of 300 MPa at 20°C for 15 minutes kept *L. monocytogenes* counts under the detection limit. Grant and Patterson (1995) observed D-values of 90.0–97.5 s, 34.0–53.0 s, or 22.4–28.0 s at 60, 65, or 70°C, respectively, for *L. monocytogenes* in cook-chill roast beef and gravy stored for 14 days at 2–3°C. However, after irradiating the food samples at 0.8 kGy, D-values were reduced to 44.0–46.4 s, 15.3–16.8 s, or 5.5–7.8 s, respectively suggesting the occurrence of radiation-induced heat sensitization in *L. monocytogenes*. Sommers and Fan (2003) reported that gamma irradiation at a dose of 1.5 kGy in combination with a 0.125% sodium diacetate solution reduced *L. monocytogenes* by more than 9 log CFU/g. They also found that post-irradiation growth of *L. monocytogenes* in beef bologna emulsion at 9°C was dependent on both sodium diacetate concentration and ionizing radiation dose.

O'Bryan et al. (2006) concluded from the published refereed research up to early 2006 on thermal inactivation of *Listeria* and other pathogens in poultry and meat products that heat resistance in pathogenic bacteria including *L. monocytogenes* is affected by several factors such as age of culture, growth conditions, and pH. In addition different strains of the same organism have different responses to heat. Murphy et al. (2001) examined the effects of an air temperature of 149°C, air velocity of 7.1 to 12.7 m³/min, and wet bulb temperature of 39 to 98°C on thermal inactivation of *Salmonella* spp. and *L. innocua*, strain M1, a surrogate strain of *L. monocytogenes* in chicken breast patties. The chicken patties were processed to an internal temperature of 55 to 80°C. They concluded that the thermal lethality of *Listeria* and *Salmonella* increased with increasing product temperature and wet bulb temperature. Murphy et al. (2003) used thermal inactivation (in-package pasteurization with hot water for 50 minutes) to reduce *L. monocytogenes* on ready to eat turkey breast meat products by 7 log CFU/g. Murphy et al. (2004) observed that the addition of sodium lactate to ground chicken thigh and leg meat increased the D-values of *L. monocytogenes* by as much as 75%. However, McMahon et al. (1999) reported that adding sodium lactate to ground beef increased the sensitivity of *L. monocytogenes* to heat. In order to reduce or eliminate pathogen contamination of cooked RTE meat and poultry products, more emphasis is now being placed on thorough cooking of these food products. Pradhan et al. (2007) developed a computer model to predict heat and mass transfer and thermal inactivation kinetics for *Listeria* inoculated into irregular shaped chicken breast products and heated by convection in a pilot plant scale air-steam impingement oven.

The ubiquitous nature of *L. monocytogenes* and its ability to survive adverse conditions make it extremely difficult to control and eradicate in the food-processing environment (Samelis and Metaxopoulos 1999; Ferreira et al. 2001; Lou and Yousef

1996, 1997; Herbert and Foster 2001; Mendonca et al. 2004). The current sanitation strategies and hygienic practices applied in the food-processing plants are often insufficient to prevent post-process recontamination of RTE food products. Therefore, post packaging hurdle technologies are needed to inhibit *L. monocytogenes* during storage. The incorporation of generally recognized as safe (GRAS) chemicals including organic acids and their salts as safety barriers in processed RTE meat and poultry products is heavily utilized in the food industry. Several studies have shown the effectiveness of organic acids and their salts against *L. monocytogenes*. However, the effect of these chemicals likely depend on various factors including chemical concentration, food type, the presence of other inhibitory compounds, and the growth phase state of the organism. Lungu and Johnson (2005a, 2005b) tested the efficacy of 6% sodium lactate, 6% sodium diacetate, or 6% potassium sorbate alone or in combination with nisin and/or zein coatings against *L. monocytogenes* on the surface of full fat turkey frankfurters. Their studies showed that sodium diacetate when used alone or in combination with nisin or zein coatings was very effective against *L. monocytogenes* while sodium lactate and potassium sorbate were both ineffective. Lu et al. (2005) observed that the maximum population of *L. monocytogenes* was decreased and generation time and lag phase were increased following surface treatments of frankfurters with 6% sodium diacetate, 6% sodium lacatate-sodium diacetate-potassium benzoate, 3% sodium diacetate-potassium benzoate, or 6% sodium diacetate-potassium benzoate solutions at 1.1°C. However, *L. monocytogenes* survived at higher refrigeration temperatures even in the presence of organic acids. Samelis et al. (2001) observed no significant increases in the growth of *L. monocytogenes* on sliced pork bologna stored at 4°C in vacuum packages when the products were treated with 2.5 or 5% acetic acid, 5% sodium diacetate, or 5% potassium benzoate for 120 days. However, treatment with 5% potassium sorbate or 5% lactic acid allowed growth of *L. monocytogenes* after 50 or 90 days, respectively. Sodium lactate and sodium acetate permitted growth of *L. monocytogenes* at earlier days of storage. The results obtained by Oh and Marshall (1995) demonstrated that *L. monocytogenes* that was attached to stainless steel exhibited increased resistance to acetic acid and monolaurin and increased resistance was observed with increase in culture age.

Future Prospects for Limiting *L. monocytogenes* in Postharvest Meat Production

As more research is conducted it has become clearer that *L. monocytogenes* has the capability to adjust to environmental conditions normally considered adverse. For example, *L. monocytogenes* has the potential to adapt to a low pH using a mechanism termed the acid tolerance response (ATR). The ATR is an attempted response by a bacterial cell to sustain its survival or growth in a low pH environment. O'Driscoll et al. (1996) showed that the ATR of *L. monocytogenes* was induced following exposure to mild acid (pH 5.5) and this further protected the pathogen from severe acid stress (pH 3.5). However, the susceptibility to severe acid stress was also found to be growth-phase dependent with stationary-phase cells showing a natural resistance to a pH challenge of 3.5. Exponential-phase cells required adaptation at pH 5.5 prior to survival at a more severe acid stress (pH 3.5). Adaptation to acid was also shown to require protein synthesis (O'Driscoll et al. 1996, 1997). Bonnet et al. (2006) noted that ATR-induced *L. monocytogenes* cells exhibited a decreased proton motive and an increased resistance to nisin. Podolak et al. (1996) found that 1.5% fumaric acid (pH 2.52) was more effective in reducing *L. monocytogenes* on the surface of beef cubes than 1% acetic acid (pH 2.93) or 1% lactic acid (pH 2.77). Fumaric acid reduced the CFU/cm^2 of *L. monocytogenes* by 0.80 while lactic acid and acetic acid only reduced the count by 0.53 and 0.40 CFU/cm^2, respectively. Furthermore, *L. monocytogenes* is capable of surviving and growing in liquid medium with up to 16% NaCl for 33 days at refrigerator temperatures (Hudson 1992). Garner et al. (2006) showed that exposure of *L. monocytogenes* to salt and organic acids increased the ability of this pathogen to invade Caco-2 cells but decreased its ability to survive gastric stress. They suggested that virulence associated characteristics that determine the infectious dose of *L. monocytogenes* are likely affected by food-specific properties such as pH or the presence of salt or organic acids.

Exposure of microorganisms to sublethal levels of a first or second preservative stress such as acids in the sequence of a food-processing scheme may allow *L. monocytogenes* to survive later preservative steps that would normally inhibit or kill the pathogen. Lou and Yousef (1997) studied the effects of "stress hardening" in *L. monocytogenes* with sublethal/lethal levels, and they revealed that incubations of *L. monocytogenes* cells in mild conditions of acid pH, ethanol, H_2O_2, or NaCl significantly increased their resistance to normally lethal conditions when using the same stress. Since organic acids are currently relied upon heavily in the food industry to inhibit microbial spoilage, the potential for development of "stress hardening" may be significant given the potential for cross protection by stress resistance to more than a single intervention. Therefore, multiple hurdle technologies have been examined for potential implementation (Ricke et al. 2005). Typically, more than one preservation method is used to inhibit growth and survival of pathogenic and spoilage organisms. Hurdle technology utilizes multiple preservative barriers in a sequential manner by combining different preservation methods to inhibit microbial growth.

The use of bacteriocins has become attractive to the food industry because the industry is facing both increasing consumer demand for natural, non-toxic products and increasing concern about foodborne diseases. Nisin is an antibacterial biopeptide produced by *Lactococcus lactis* that has hydrophobic properties and exhibits a broad spectrum of inhibitory activity against the vegetative cells and spores of Gram-positive bacteria. Several studies have shown that the presence of bacteriocins is an effective approach to reducing contamination and growth of *L. monocytogenes* on meats and poultry (Ming et al. 1997; Lungu and Johnson 2005a, 2005b; Janes et al. 2002; Sivarooban et al. 2007). Nisin is added to a variety of foods throughout the world and has received GRAS status in the United States (CFR Title 21 (184.1538)). It has been used with certain cheese products, shelf-life extension of milk and inactivation of thermophilic bacteria in canned foods (Stevens et al. 1991; Montville et al. 1992; Liu and Hansen 1990). Nisin has been shown to inhibit Gram-positive organisms (Lungu and Johnson 2005a), and when combined with chelating agents it can also inhibit Gram-negative bacteria such as *Salmonella* (Natrajan and Sheldon 2000). Generally nisin is more effective at lower temperatures under acidic conditions against low spore loads (Montville et al. 1992; Rogers and Montville 1994). Spores are less sensitive to nisin than are vegetative cells. Although bacteriocins have applications in many food systems, foods should not be preserved by bacteriocins alone but rather bacteriocins should be used in combination as part of a system with multiple hurdles such as low temperature and pH.

Over the years concerns over environmental pollution shifted the research focus to the production of biodegradable films as a weapon in the fight against microbial contamination and growth in food systems (Ko et al. 2001). Biodegradable food-grade films have been synthesized from proteins and examples of these include casein, collagen, corn zein, gelatin, soy-protein, and wheat gluten (Padgett et al. 1998). Edible films can be used as barriers against oxygen penetration, prevent moisture loss in foods, as carriers for antioxidants and antimicrobial agents, to protect against surface discolorations and retain quality during storage (Howard and Dewi 1995, 1996; Zhuang et al. 1996). Several researchers have shown the effectiveness of edible films impregnated with bacteriocins and other chemical antimicrobials against microbial growth in poultry and meats (Carlin et al. 2001; Janes et al. 2002; Cagri et al. 2002; Lungu and Johnson 2005a, 2005b; Sivarooban et al. 2007). Sivarooban et al. (2007) noted that the combination of nisin and grape seed extract inhibited *L. monocytogenes* in tryptic soy broth or turkey frankfurters to undetectable levels after 15 hours or 21 days, respectively. Janes et al. (2002) showed that zein films containing nisin or 1% calcium propionate prevented the growth of *L. monocytogenes* during aerobic incubation on ready-to-eat chicken at 4°C but not at 8°C. However, nisin alone or a combination of nisin and calcium propionate was ineffective against *L. monocytogenes* at both temperatures.

Conclusions

Despite the advances that have been made to control *L. monocytogenes*, this pathogen continues to persist in the food-processing environment where it subsequently recontaminates RTE processed food products. Therefore, *L. monocytogenes* remains a problem for both public health and the food industry. It is imperative to achieve a better understanding of the long-term survival of *L. monocytogenes* in foods, and on the food-processing surfaces, as well as understand the practical processing conditions that may prevent survival and persistence of this pathogen. The ubiquitous nature of *L. monocytogenes* in the environment makes it especially difficult to control in both raw and RTE food products. *L. monocytogenes* contamination of processed RTE food products such as milk, milk products, and processed meat or poultry products usually occurs by cross-contamination of the finished product within the food-processing environment.

The extensive research that has been conducted to devise novel physical, chemical, and biological control methods has been done under aerobic conditions, and in most cases robust cells grown and kept in complex media conditions have been used (reviewed in Lungu et al. 2009). However, *L. monocytogenes* is a facultative pathogen that commonly encounters anaerobic conditions as well as extreme environmental stresses in the form of nutrient-deprivation, chemical stress, low pH, extreme temperatures, and osmotic stress. Therefore when devising any new intervention strategy, these factors need to be taken into account for any inhibitory or control method to be successful.

L. monocytogenes readily adapts to different environments; therefore, the combination of physical, chemical, and biological methods in a multiple hurdle system to inhibit *L. monocytogenes* growth in foods as well as in food-processing facilities needs more attention. Synergistic action within the multiple hurdles may diminish the chances of *L. monocytogenes* being able to respond effectively to multiple control methods in one system, and may also reduce the chances for the bacteria to develop resistance to any one method especially with regard to the use of natural or chemical antimicrobials. More work characterizing the genetic and cellular mechanisms by which *L. monocytogenes* acquires resistance to currently used control methods is still needed. Understanding the cellular and molecular mechanisms that govern growth, survival, and persistence of *L. monocytogenes* may provide the knowledge that will enable scientists and the industry to come up with better control strategies against this pathogen.

Acknowledgments

This work was supported in part by a special grant from the USDA-CSREES for the Food Safety Consortium. The authors wish to thank Dr. Arun Muthaiyan and Julia Sonka, Center for Food Safety, University of Arkansas, for help in preparing this review.

References

ACCT (Arkansas Cancer Control Taskforce). 2001. The Arkansas cancer plan 2001–2005: a framework for action. At www.healthyarkansas.com/disease/cancerplan.pdf. Accessed August 2007.

Adesiyun, A. A., and C. Krishnan. 1995. Occurrence of *Yersinia enterocolitica* O:3, *Listeria monocytogenes* O:4 and thermophilic *Campylobacter* spp in slaughter pigs and carcasses in Trinidad. Food Microbiol. 12:99–107.

Altekruse, S. F., M. L. Cohen, and D. L. Swerdlow. 1997. Emerging foodborne diseases. Emerg. Infect Dis. At www.cdc.gov/ncidod/eid/vol3no3/cohen.htm. Accessed December 2006.

Anonymous. 1999. FSIS. At http://www.fsis.usda.gov/OA/speeches/1999/cw_lm.htm. Accessed August 2007.

Anonymous. 2001. Draft assessment of the relative risk to public health from foodborne *Listeria monocytogenes* among selected categories of ready-to-eat foods. At http://www.foodsafety.gov/~dms/lmrisk.html. Accessed August 2007.

Anonymous. 2002. Update: Listeriosis outbreak investigation. At http://www.cdc.gov/od/oc/media/pressrel/r021121.htm. Accessed August 2007.

Anonymous. 2006a. Changing demographics and their implications for Louisiana's tourism industry. At http://www.latour.lsu.edu/pdfs/demographics.pdf. Accessed August 2007.

Anonymous. 2006b. Cancer in Arkansas. At www.arkansascancercoalition.org/cancer_ar.asp. Accessed August 2007.

Aureli, P., G. C. Fiorucci, D. Caroli, G. Marchiaro, O. Novara, L. Leone, and S. Salmaso. 2000. An outbreak of febrile gastroenteritis associated with corn contaminated by *Listeria monocytogenes*. N. Engl. J. Med. 342:1236–1241.

Autio, T., T. Sateri, M. Fredriksson-Ahomaa, M. Rahkio, J. Lunden, and H. Korkeala. 2000. *Listeria monocytogenes* contamination pattern in pig slaughterhouses. J. Food Prot. 23:1438–1442.

Barakat, R. K., and L. J. Harris. 1999. Growth of *Listeria monocytogenes* and *Yersinia enterocolitica* on cooked modified-atmosphere-packaged poultry in the presence and absence of a naturally occurring microbiota. Appl. Environ. Microbiol. 65:342–345.

Barnes, R. P., P. Archer, J. Strack, and G. R. Istre. 1989. Listeriosis associated with the consumption of turkey franks. Morbid. Mortal. Weekly Rep. 38:267–268.

Beloeil, P. A., P. Fravalo, C. Chauvin, C. Fablet, G. Salvat, and F. Madec. 2003. *Listeria* spp. contamination in piggeries: comparison of three sites of environmental swabbing for detection and risk factor hypothesis. J. Vet. Med. 50:155–160.

Bonnet, M., M. M. Rafi, M. L. Chikindas, and T. J. Montville. 2006. Bioenergetic mechanism for nisin resistance, induced by the acid tolerance response of *Listeria monocytogenes*. Appl. Environ. Microbiol. 72:2556–2563.

Botzler, R. G., A. B. Cowan, and T. F. Wetzler. 1974. Survival of *Listeria monocytogenes* in soil and water. J. Wildl. Dis. 10:204–212.

Buncic, S. 1991. The incidence of *Listeria monocytogenes* in slaughtered animals, in meat, and in meat products in Yugoslavia. Int. J. Food Microbiol. 12:173–180.

Cagri, A., Z. Ustunol, and E. T. Ryser. 2002. Inhibition of three pathogens on bologna and summer sausage using antimicrobial edible films. J. Food Sci. 67:2317–2324.

Carlin, F., N. Gontard, M. Reich, and C. Nguyen-The. 2001. Utilization of zein coating and sorbic acid to reduce *Listeria monocytogenes* growth on cooked sweet corn. J. Food Sci. 66:1385–1389.

CDC. 2005. Listeriosis. http://www.cdc.gov/ncidod/dbmd/diseaseinfo/listeriosis_g.htm. Accessed August 2007.

Chen, J. H., and J. H. Hotchkiss. 1993. Growth of *Listeria monocytogenes* and *Clostridium sporogenes* in cottage cheese in modified atmosphere packaging. J. Dairy Sci. 76:972–977.

Chen, J. H., W. H. Ross, V. N. Scott, and D. E. Gombas. 2003. *Listeria monocytogenes:* low levels equals low risk. J. Food Prot. 66:570–577.

Code of Federal Regulations. 2001. CFR Title 21 (184.1538), Office of Federal Register, U.S. Government Printing Office, Washington, DC.

Coillie, E. V., H. Werbrouck, M. Heyndrickx, L. Herman, and N. Rijpens. 2004. Prevalence and typing of *Listeria monocytogenes* in ready-to-eat food products on the Belgian market. J. Food Prot. 67:2480–2487.

Dalton, C. B., C. C. Austin, J. Sobel, P. S. Hayes, W. F. Bibb, L. M. Graves, B. Swaminathan, M. E. Proctor, and P. M. Griffin. 1997. An outbreak of gastroenteritis and fever due to *Listeria monocytogenes* in milk. N. Engl. J. Med. 336:100–105.

De Luca, D., F. Zanetti, P. Fateh-Moghadm, and S. Stampi. 1998. Occurrrence of *Listeria monocytogenes* in sewage sludge. Zentralbl Hyg. Umweltmed. 201:269–277.

DiMaio, H. 2000. *Listeria* infection in women—Listeriosis and pregnancy: food for thought. Primary Care Update for OB/GYNS. 7:40–45.

Dunsmore, D. G. 1981. Bacteriological control of food equipment surfaces by cleaning systems. I. Detergent effects. J. Food Prot. 44:15–20.

Farber, J. M., and P. I. Peterkin. 1991. *Listeria monocytogenes*, a food-borne pathogen. Microbiol. Rev. 55:476–511.

Fenlon, D. R. 1985. Wild birds and silage as reservoirs of *Listeria* in the agricultural environment. J. Appl. Bacteriol. 59:537–543.

Fenlon, D. R., J. Wilson, and W. Donachie. 1996. The incidence and level of *Listeria monocytogenes* contamination of food sources at primary production and initial processing. J. Appl. Bacteriol. 81:641–650.

Ferreira, A., C. P. O'Byrne, and K. J. Boor. 2001. Role of σ^B in heat, ethanol, acid and oxidative stress resistance and during carbon starvation in *Listeria monocytogenes*. Appl. Environ. Microbiol. 67:4454–4457.

Ferreira, A., D. Sue, C. P. O'Byrne, and K. J. Boor. 2003. Role of *Listeria monocytogenes* σ^B in survival of lethal acidic conditions and in the acquired acid tolerance response. Appl. Environ. Microbiol. 69:2692–2698.

Fleming, D. W., S. L. Cochi, K. L. MacDonald, J. Brondum, P. S. Hayes, B. D. Plikaytis, M. B. Holmes, A. Audurier, C. V. Broome, and A. L. Reingold. 1985. Pasteurized milk as a vehicle of infection in an outbreak of listeriosis. N. Engl. J. Med. 312:404–407.

Francis, G. A., and D. O'Beirne. 2006. Isolation and pulsed-field gel electrophoresis typing of *Listeria monocytogenes* from modified atmosphere packaged fresh-cut vegetables collected in Ireland. J. Food Prot. 69:2524–2528.

Garner, M. R., K. E. James, M. C. Callahan, M. Wiedmann, and K. J. Boor. 2006. Exposure to salt and organic acids increases the ability of *Listeria monocytogenes* to invade Caco-2 cells but decreases its ability to survive gastric stress. Appl. Environ. Microbiol. 72:5384–5395.

Garrec, N., F. Picard-Bonnaud, and A. M. Pourcher. 2003. Occurrence of *Listeria* spp. and *Listeria monocytogenes* in sewage sludge used for land application: effect of dewatering, liming and storage in tank on survival of *Listeria* species. FEMS Immunol. Med. Microbiol. 35:275–283.

Genigeorgis, C. A., P. Danca, and D. Dutulescu. 1990. Prevalence of *Listeria* spp. in turkey meat at the supermarket and slaughterhouse level. J. Food Prot. 53:282–288.

George, S. M., L. C. C. Richardson, I. E. Pol, and M. W. Peck. 1998. Effect of oxygen concentration and redox potential on recovery of sublethally heat-damaged cells of *Escherichia coli* O157:H7, *Salmonella enteritidis* and *Listeria monocytogenes*. J. Appl. Microbiol. 84:903–909.

Gilbert, J., J. McLauchlin, and S. K. Velani. 1993. The contamination of pate by *Listeria monocytogenes* in England and Wales in 1989 and 1990. Epidemiol. Infect. 110:543–551.

Grant, I. R., and M. F. Patterson. 1995. Combined effect of gamma radiation and heating on the destruction of *Listeria monocytogenes* and *Salmonella typhimurium* in cook-chill roast beef and gravy. Int. J. Food Microbiol. 27:117–128.

Graves, L. M., S. B. Hunter, A. R. Ong, D. Schoonmaker-Bopp, K. Hise, L. Kornstein, W. E. DeWitt, P. S. Hayes, E. Dunne, P. Mead, and B. Swaminathan. 2005. Microbiological aspects of the investigation that traced the 1998 outbreak of listeriosis in the United States to contaminated hot dogs and establishment of molecular subtyping-based surveillance for *Listeria monocytogenes* in the PulseNet Network. J. Clin. Microbiol. 43:2350–2355.

Grønstøl, H. 1979. Listeriosis in sheep: *Listeria monocytogenes* from grass silage. Acta. Vet. Scand. 20:492–497.

Hassan, L., H. O. Mohammed, P. L. McDonough, and R. N. Gonzalez. 2000. A cross-sectional study on the prevalence of *Listeria monocytogenes* and *Salmonella* in New York dairy herds. J. Dairy Sci. 83:2441–2447.

Hatkin, J. M., and W. E. Philips Jr. 1986. Isolation of *Listeria monocytogenes* from an eastern wild turkey. J. Wildlife Dis. 22:110–112.

Hayman, M. M., I. Baxter, P. J. O'Riordan, and C. M. Stewart. 2004. Effects of high-pressure processing on the safety, quality and shelf life of ready-to-eat meats. J. Food Prot. 67:1709–1718.

Herbert, K. C., and S. Foster. 2001. Starvation survival in *Listeria monocytogenes*: characterization of the response and the role of known and novel components. Microbiol. 147:2275–2284.

Hofer, E. 1983. Bacteriologic and epidemiologic studies on the occurrence of *Listeria monocytogenes* in the feces of dairy cattle. J. Vet. Med. B. 37:276–282.

Holah, J. T., J. H. Taylor, D. J. Dawson, and K. E. Hall. 2002. Biocide use in the food industry and the disinfectant resistance of persistent strains of *Listeria monocytogenes* and *Escherichia coli*. J. Appl. Microbiol. 92:111S–120S.

Howard, L. R., and T. Dewi. 1995. Sensory, microbiological and chemical quality of mini-peeled carrots as affected by edible coating treatment. J. Food Sci. 60:142–144.

Howard, L. R., and T. Dewi. 1996. Minimal processing and edible coating effects on composition and sensory quality of mini-peeled carrots. J. Food Sci. 61:643–645.

Hudson, J. A. 1992. Efficacy of high sodium chloride concentrations for the destruction of *Listeria monocytogenes*. Lett. Appl. Microbiol. 14:178–180.

Huff, G. R., W. E. Huff, J. N. Beasley, N. C. Rath, M. G. Johnson, and R. Nannapaneni. 2005. Respiratory infection of turkeys with *Listeria monocytogenes* Scott A. Avian Dis. 49:551–557.

Husu, J. R. 1990. Epidemiological studies on the occurrence of *Listeria monocytogenes* in the feces of dairy cattle. Zentralbl Veterinarmed B. 37:276–282.

Ingham, S. C., G. Searls, S. Mohanan, and D. R. Buege. 2006. Survival of *Staphylococcus aureus* and *Listeria monocytogenes* on vacuum-packaged beef jerky and related products stored at 21°C. J. Food Prot. 69:2263–2267.

Jacquet, C., B. Catimel, R. Brosch, C. Buchreiser, P. Dehaumont, V. Goulet, A. Lepoutre, P. Veit, and J. Rocourt. 1995. Investigations related to the epidemic strain involved in the French listeriosis outbreak in 1992. Appl. Environ. Microbiol. 61:2242–2246.

Janes, M. E., S. Kooshesh, and M. G. Johnson. 2002. Control of *Listeria monocytogenes* on the surface of refrigerated, ready-to-eat chicken coated with edible zein film coatings containing nisin and/or calcium propionate. J. Food Sci. 67:2754–2757.

Jayarao, B. M., and D. R. Henning. 2001. Prevalence of foodborne pathogens in bulk tank milk. J. Dairy Sci. 84:2157–2162.

Jensen, A. 1993. Excretion of *Listeria monocytogenes* in feces after listeriosis: rate, quantity and duration. Med. Microbiol. Lett. 2:176–182.

Jiang, X., M. Islam, J. Morgan, and M. P. Doyle. 2004. Fate of *Listeria monocytogenes* in bovine manure-amended soil. J. Food Prot. 67:1676–1681.

Jofre, A., M. Garriga, and T. Aymerich. 2007. Inhibition of *Listeria monocytogenes* in cooked ham through active packaging with natural antimicrobials and high-pressure processing. J. Food Prot. 70:2498–2502.

Kalender, H. 2003. Detection of *Listeria monocytogenes* in feces from chickens, sheep and cattle in Elazig province. Tubitak 27:449–451.

Knabel, S. J., H. W. Walker, P. A. Hartman, and A. F. Mendonca. 1990. Effects of temperature and strictly anaerobic recovery on the survival of *Listeria monocytogenes* during pasteurization. Appl. Environ. Microbiol. 56:370–376.

Ko, S., M. E. Janes, N. S. Hettiarachchy, and M. G. Johnson. 2001. Physical and chemical properties of edible films containing nisin and their action against *Listeria monocytogenes*. J. Food Sci. 66:1006–1011.

Lawrence, L. M., and A. Gilmour. 1994. Incidence of *Listeria* spp. and *Listeria monocytogenes* in a poultry processing environment and in poultry products and their rapid confirmation by multiplex PCR. Appl. Environ. Microbiol. 60:4600–4604.

Levine, P., B. Rose, S. Green, G. Ransom, and W. Hill. 2001. Pathogen testing of ready-to-eat meat and poultry products collected at federally inspected establishment in the United States, 1990–1999. J. Food Prot. 64:1188–1193.

Li, Q., J. S. Sherwood, and C. M. Logue. 2004. The prevalence of *Listeria, Salmonella, Escherichia coli* and *E. coli* O157:H7 on bison carcasses during processing. Food Microbiol. 21:791–799.

Linnan, M. J., L. Mascola, X. D. Lou, V. Goulet, S. May, C. Salminen, D. W. Hird, M. L.

Yonekura, P. Hayes, and R. Weaver. 1988. Epidemic listeriosis associated with Mexican-style cheese. N. Engl J. Med. 319:823–828.

Liu, W., and N. Hansen. 1990. Some chemical and physical properties of nisin, a small protein antibiotic produced by *Lactococcus lactis*. Appl. Environ. Microbiol. 56:2551–2558.

Løken, T., E. Aspøy, and H. Grønstøl. 1982. *Listeria monocytogenes* excretion and humoral immunity in goats in a herd with outbreaks of listeriosis and in a dairy herd. Acta. Vet. Scand. 23:392–399.

Lou, Y., and A. E. Yousef. 1996. Resistance of *Listeria monocytogenes* to heat after adaptation to environmental stresses. J. Food Prot. 59:465–471.

Lou, Y., and A. E. Yousef. 1997. Adaption to sublethal environmental stresses protects *Listeria monocytogenes* against lethal preservation factors. Appl. Environ. Microbiol. 63:1252–1255.

Lu, Z., J. G. Sebranek, J. S. Dickson, A. F. Mendonca, and T. B. Bailey. 2005. Inhibitory effects of organic acid salts for control of *Listeria monocytogenes* on frankfurters. J. Food Prot. 68:499–506.

Lunden, J., T. Autio, A. Markkula, S. Hellstrom, and H. Korkeala. 2003. Adaptive and cross-adaptive responses of persistent and non-persistent *Listeria monocytogenes* strains to disinfectants. Int. J. Food Microbiol. 82:265–272.

Lungu, B., and M. G. Johnson. 2005a. Fate of *Listeria monocytogenes* inoculated onto the surface of model turkey frankfurter pieces treated with zein coatings containing nisin, sodium diacetate and sodium lactate at 4°C. J. Food Prot. 68:855–859.

Lungu, B., and M. G. Johnson. 2005b. Potassium sorbate does not increase control of *Listeria monocytogenes* when added to zein coatings with nisin on the surface of full fat turkey frankfurter pieces in a model system at 4°C. J. Food Sci. 70:M95–M99.

Lungu, B., S. C. Ricke, and M. G. Johnson. 2009. Growth, survival, proliferation and pathogenesis of *Listeria monocytogenes* under low oxygen and anaerobic conditions: a review. Anaerobe 15:7–17.

Marriott, N. G. 1999. Meat and poultry plant sanitation. Pages 258–282 in Principals of food sanitation. 4th ed. Aspen Publishers, Gaithersburg, MD.

Mascola, L., F. Sorvillo, V. Goulet, B. Hall, R. Weaver, and M. Linnan. 1992. Fecal carriage of *Listeria monocytogenes*-observations during a community wide, common source outbreak. Clin. Infect. Dis. 15:557–558.

McMahon, C. M. M., A. M. Doherty, J. J. Sheridan, I. S. Blair, D. A. McDowell, and T. Hegarty. 1999. Synergistic effect of heat and sodium lactate on the thermal resistance of *Yersinia enterocolitica* and *Listeria monocytogenes* in minced beef. Lett. Appl. Microbiol. 28:340–344.

Mead, P. S. 1999. Multistate outbreak of listeriosis traced to processed meats, August 1998–March 1999. Epidemiologic Investigation Report. Centers for Disease Control and Prevention, Atlanta, GA.

Mead, P. S., L. Slutsker, V. Dietz, L. F. McCaig, J. S. Bresee, C. Shapiro, P. M. Griffin, and R. V. Tauxe. 1999. Food-related illness and death in the United States. Emerg. Infect. Dis. At www.cdc.gov/ncidod/eid/vol5no5/mead.htm. Accessed August 2007.

Mendonca, A. F., and S. J. Knabel. 1994. A novel strictly anaerobic recovery and enrichment system incorporating lithium for the detection of heat-injured *Listeria monocytogenes* in pasteurized milk containing background microflora. Appl. Environ. Microbiol. 60:4001–4008.

Mendonca, A. F., M. G. Romero, M. A. Lihono, R. Nannapaneni, and M. G. Johnson. 2004. Radiation resistance and virulence of *Listeria monocytogenes* Scott A following starvation in physiological saline. J. Food Prot. 67:470–474.

Miettinen, M. K., A. Siitonen, P. Heiskanen, H. Haajanen, K. J. Bjorkroth, and H. J. Korkeala. 1999. Molecular epidemiology of an outbreak of febrile gastroenteritis caused by *Listeria monocytogenes* in cold-smoked rainbow trout. J. Clin. Microbiol. 37:2358–2360.

Ming, X., G. H. Weber, J. W. Ayres, and W. E. Sandine. 1997. Bacteriocins applied to food packaging materials to inhibit *Listeria monocytogenes* on meats. J. Food Sci. 62:413–415.

Montero, P., J. Gomez-Estaca, and M. C. Gomez-Guillen. 2007. Influence of salt, smoke and high pressure on growth of *Listeria monocytogenes* and spoilage microflora in cold-smoked dolphinfish (*Coryphaena hippurus*). J. Food Prot. 70:399–404.

Montville, T. J., A. M. Rogers, and A. Okereke. 1992. Differential sensitivity of *Clostridium botulinum* strains to nisin. J. Food Prot. 56:444–448.

Murphy, R. Y., L. K. Duncan, K. H. Driscoll, J. A. Marcy, and B. L. Beard. 2003. Thermal inactivation of *Listeria monocytogenes* on ready-to-eat turkey breast meat products during postcook in-package pasteurization with hot water. J. Food Prot. 66:1618–1622.

Murphy, R. Y., R. E. Hanson, N. R. Johnson, K. Chappa, and M. E. Berrang. 2006. Combining organic acid treatment with steam pasteurization to eliminate *Listeria monocytogenes* on fully cooked frankfurters. J. Food Prot. 69:47–52.

Murphy, R. Y., E. R. Johnson, L. K. Duncan, M. G. Johnson, and J. A. Marcy. 2001. Thermal inactivation of *Salmonella* spp. and *Listeria innocua* in the chicken breast patties processed in a pilot-scale air-convection oven. J. Food Sci. 66:734–741.

Murphy, R. Y., T. Osaili, L. K. Duncan, and J. A. Marcy. 2004. Thermal inactivation of *Listeria monocytogenes* and *Salmonella* in ground chicken thigh and leg meat. J. Food Prot. 67:1403–1407.

Natrajan, N., and B. W. Sheldon. 2000. Efficacy of nisin coated polymer films to inactivate *Salmonella typhimurium* on fresh broiler skin. J. Food Prot. 63:1189–1196.

Nesbakken, T., E. Nerbrink, O. J. Rotterud, and E. Borch. 1994. Reduction of *Yersinia enterocolitica* and *Listeria* spp. on pig carcasses by enclosure of the rectum during slaughter. Int. J. Food Microbiol. 23:197–208.

Nightingale, K. K., Y. H. Schukken, C. R. Nightingale, E. D. Fortes, A. J. Ho, Z. Her, Y. T. Grohn, P. L. McDonough, and M. Wiedmann. 2004. Ecology and transmission of *Listeria monocytogenes* infecting ruminants and in the farm environment. Appl. Environ. Microbiol. 70:4458–4467.

O'Bryan, C. A., P. G. Crandall, E. M. Martin, C. L. Griffis, and M. G. Johnson. 2006. Heat resistance of *Salmonella* spp., *Listeria monocytogenes*, *Escherichia coli* O157:H7 and *Listeria* innocua M1, a potential surrogate for *Listeria monocytogenes*, in meat and poultry: a review. J. Food Sci. 71:R23–R30.

O'Driscoll, B., C. G. M. Gahan, and C. Hill. 1996. Adaptive acid tolerance response in *Listeria monocytogenes*: isolation of an acid-tolerant mutant which demonstrates increased virulence. Appl. Environ. Microbiol. 62:1693–1698.

O'Driscoll, B., C. G. M. Gahan, and C. Hill. 1997. Two-dimensional polyacrylamide gel electrophoresis analysis of the acid tolerance response in *Listeria monocytogenes* L028. Appl. Environ. Microbiol. 63:2679–2685.

Oh, D., and D. L. Marshall. 1995. Monolaurin and acetic acid inactivation of *Listeria monocytogenes* attached to stainless steel. J. Food Prot. 59:249–252.

Padgett, T., I. Y. Han, and P. L. Dawson. 1998. Incorporation of food grade antimicrobial compounds into biodegradable packaging films. J. Food Prot. 61:1330–1335.

Painter, J., and L. Slutsker. 2007. Listeriosis in humans. Pages 85–110 in *Listeria*, listeriosis and food safety. 3rd ed. E. T. Ryser and E. H. Marth, eds. CRC Press, Taylor and Francis Group, New York, NY.

Podolak, R. K., J. F. Zayas, C. L. Kastner, and D. Y. C. Fung. 1996. Inhibition of *Listeria monocytogenes* and *Escherichia coli* O157:H7 on beef by application of organic acids. J. Food Prot. 59:370–373.

Pradhan, A. K., Y. Li, J. A. Marcy, M. G. Johnson, and M. L. Tamplin. 2007. Pathogen kinetics and heat and mass transfer-based predictive model for *Listeria innocua* in irregular-shaped poultry products during thermal processing. J. Food Prot. 70:607–615.

Ricke, S. C., M. M. Kundinger, D. R. Miller, and J. T. Keeton. 2005. Alternatives to antibiotics: chemical and physical antimicrobial interventions and foodborne pathogen response. Poult. Sci. 84:667–675.

Rivera-Betancourt, M., S. D. Shackelford, T. M. Arthur, K. E. Westmoreland, G. Bellinger, M. Rossman, J. O. Reagan, and M. Koohmaraie. 2004. Prevalence of *Escherichia coli* O157:H7, *Listeria monocytogenes* and *Salmonella* in two geographically distant commercial beef processing plants in the United States. J. Food Prot. 67:295–302.

Rogers, A. M., and T. J. Montville. 1994. Quantification of factors influencing nisin's inhibition of *Clostridium botulinum* 56A in a model food system. J. Food Sci. 59:663–668.

Rorvik, L. M., B. Aase, T. Alvestad, and D. A. Caugant. 2003. Molecular epidemiological survey of *Listeria monocytogenes* in broilers and poultry products. J. Appl. Microbiol. 94:633–640.

Salvat, G., M. T. Toquin, Y. Michel, and P. Colin. 1995. Control of *Listeria monocytogenes* in the delicatessen industries: the lessons of a listeriosis outbreak in France. Int. J. Food Microbiol. 25:75–81.

Samadpour, M., M. W. Barbour, T. Nguyen, T. M. Cao, F. Buck, G. A. Depavia, E. Mazengia, P. Yang, D. Alfi, M. Lopes, and J. D. Stopforth. 2006. Incidence of Enterohemorrhagic *Escherichia coli, Escherichia coli* O157, *Salmonella* and *Listeria monocytogenes* in retail fresh ground beef, sprouts and mushrooms. J. Food Prot. 69:441–443.

Samelis, J., and J. Metaxopoulos. 1999. Incidence and principal sources of *Listeria* spp. and *Listeria monocytogenes* contamination in processed meats and a meat processing plant. Food Microbiol. 16:465–477.

Samelis, J., J. N. Sofos, M. L. Kain, J. A. Scanga, K. E. Belk, and G. C. Smith. 2001. Organic acids and their salts as dipping solutions to control *Listeria monocytogenes* inoculated following processing of sliced pork bologna stored at 4°C in vacuum packages. J. Food Prot. 64:1722–1729.

Sanaa, M., B. Poutrel, J. L. Menard, and F. Serieys. 1993. Risk factors associated with contamination of raw milk by *Listeria monocytogenes* in dairy farms. J. Dairy Sci. 76:2891–2898.

Schlech, W. F., P. M. Lavigne, R. A. Bortolussi, A. C. Allen, C. V. Haldane, A. J. Wort, A. W. Hightower, S. E. Johnson, S. H. King, E. S. Nicholls, and C. V. Broome. 1993.

Epidemic listeriosis-evidence for transmission by food. N. Engl. J. Med. 308:203–206.

Schuchat, A., K. Deaver, P. Hayes, L. Graves, L. Mascola, and J. Wenger. 1993. Gastrointestinal carriage of *Listeria monocytogenes* in household contacts of patients with listeriosis. J. Infect. Dis. 167:1261–1262.

Sivarooban, T., N. S. Hettiarachchy, and M. G. Johnson. 2007. Inhibition of *L. monocytogenes* using with nisin and grape seed extract turkey frankfurters stored at 4 and 10°C. J. Food Protect. 70:1017–1020.

Skovgaard, N., and C. A. Morgen. 1988. Detection of *Listeria* spp. in feces from animals, in feeds and in raw foods of animal origin. Int. J. Food Microbiol. 6:229–242.

Skovgaard, N., and B. Norrung. 1989. The incidence of *Listeria* spp. in feces of Danish pigs and in minced pork meat. Int. J. Food Microbial 8:59–63.

Somers, E. B., and A. C. L. Wong. 2004. Efficacy of two cleaning and sanitizing combinations on *Listeria monocytogenes* biofilms formed at low temperature on a variety of materials in the presence of ready-to-eat meat residue. J. Food Prot. 67:2218–2229.

Sommers, C., and X. Fan. 2003. Gamma irradiation of fine emulsified sausage containing sodium diacetate. J. Food Prot. 66:819–824.

Stevens, K. A., B. W. Sheldon, N. A. Klapes, and T. R. Klaenhammer. 1991. Nisin treatment for inactivation of *Salmonella* species and other Gram negative bacteria, Appl. Environ. Microbiol. 57:3613–3615.

Thimothe, J., J. Walker, V. Suvanich, K. L. Gall, M. W. Moody, and M. Wiedmann. 2002. Detection of *Listeria* in crawfish processing plants and in raw, whole crawfish and processed crawfish (*Procambarus* spp). J. Food Prot. 65:1735–1739.

Van Kessel, J. S., J. S. Karns, L. Gorski, B. J. McClusky, and M. L. Perdue. 2004. Prevalence of *Salmonella*, *Listeria monocytogenes* and fecal coliforms in bulk tank milk on US dairies. J. Dairy Sci. 87:2822–2830.

Van Renterghem, B., F. Huysman, R. Rygole, and W. Verstraete. 1991. Detection and prevalence of *Listeria monocytogenes* in the agricultural ecosystem. J. Appl. Bacteriol. 71:211–217.

Wallace, F. M., J. E. Call, A. C. S. Porto, G. J. Cocoma, the ERRC special projects team, and J. B. Luchansky. 2003. Recovery rate of *Listeria monocytogenes* from commercially prepared frankfurters during extended refrigerated storage. J. Food Prot. 66:584–591.

Weis, J., and H. P. R. Seeliger. 1975. Incidence of *Listeria monocytogenes* in nature. Appl. Microbiol. 30:29–32.

Welshimer, H. J., and J. Donker-Voet. 1971. *Listeria monocytogenes* in nature. Appl. Microbiol. 21: 516–519.

Wilson, I. G. 1995. Occurrence of *Listeria* species in ready to eat foods. Epidem. Infect. 115:519–526.

Wimpfheimer, L., N. S. Altman, and J. H. Hotchkiss. 1990. Growth of *Listeria monocytogenes* Scott A serotype 4 and competitive spoilage organisms in raw chicken packaged under modified atmospheres and in air. Int. J. Food Microbiol. 11:205–214.

Yoshida, T., T. Sugimoto, M. Sato, and K. Hirai. 2000. Incidence of *Listeria monocytogenes* in wild animals in Japan. J. Vet. Med. Sci. 62:673–675.

Zhuang, R. Y., L. R. Beuchat, and F. J. Angulo. 1996. *Salmonella montevideo* on and in raw tomatoes as affected by temperature and treatment with chlorine. Appl. Environ. Microbiol. 61:2127–2131.

■ ■ ■

Rapid Methods and Detection Strategies for Foodborne Pathogens

■ ■ ■

▌12▐

Rapid Methods and Automation in Microbial Food Safety

Daniel Y. C. Fung

Rapid methods and automation in microbiology is a dynamic area in applied microbiology dealing with the study of improved methods in the isolation, early detection, characterization, and enumeration of microorganisms and their products in clinical, pharmaceutical, food, industrial, and environmental samples. In the past 25 years this field has emerged as an important subdivision of the general field of applied microbiology and is gaining momentum nationally and internationally as an area of research and application to monitor the numbers, kinds, and metabolites of microorganisms related to food spoilage, food preservation, food fermentation, food safety, and foodborne pathogens. Comprehensive reviews on the subject were made by Fung (2002, 2007), who summarized topics related to food safety.

Medical microbiologists became interested in rapid methods in the mid-1960s, work accelerated in the 1970s, and developments continued into the 1980s, 1990s, and up to the present day. Other disciplines such as pharmaceutical, environmental, industrial, and food microbiology were lagging about 10 years behind. Many symposia and conferences were held nationally and internationally to discuss the developments in this important applied microbiology topics. In 1980 the author developed a comprehensive week-long hands-on workshop at Kansas State University. The twenty-ninth workshop occurred in June 19–26, 2009.

Advances in Viable Cell Counts and Sample Preparation

Knowing the number of living organisms in the product, on the surface of manufacturing environment, and in the air of processing plants is very important for the food industry. Colony forming units is the standard way to express microbial loads. In the past 25 years ingenious systems in "massaging" solid and liquid food with a Stomacher, Smasher, or BioMixer, or "pulsifying" it with the Pulsifier were developed so that the samples could be plated in suitable agars after 1 to 2 minutes of operation. Typically one milliliter (after dilutions) is placed into melted agar or spread onto a solidified agar to encourage microorganisms to grow into discrete colonies for counting (Colony Forming Units, CFU/ml or g). These colonies can be isolated and further identified as pathogenic or nonpathogenic organisms. In the past 25 years, convenient systems such as nutrients housed in films (3M Petri-film systems), mechanical instruments to spread a sample over the surface of a pre-formed agar plate (Spiral Plater), trapping microorganisms on bacteriological membrane (IsoGrid System), and looking for growth on selective and nonselective culture media, agar-less solidifying system (Redigel), and others to reduce labor time in performing cell counts so important in the food industry.

The 3- or 5- tube most probable number (MPN) system has been in use for more than 100 years in food and public health water microbiology but it is very time consuming and labor intensive and uses large quantities of glassware and expensive culture media. In 2007 a completely mechanized, automated, and hands-off TEMPO (bioMerieux) 16-tube MPN system was developed and marketed for ease of operation. The operator only needs to make a 1:10 dilution of the food or water sample and then place 1 ml into a vial containing 3 ml of rehydrated nutri-ent medium for specific organisms (e.g., total count, coliform, etc.). A small deliv-ery tube is then placed into the vial with sample and nutrient. The delivery tube is attached to a sterile plastic card (4 x 3 x ⅛") that has three series of interconnected 16 chambers. The top 16 are very small holes, the second 16 are 10 times larger, and the third 16 are again 10 times larger in volume. Several of these cards (e.g., 6 cards) each containing one sample in the three 16-tube series (48 chambers in total) are placed in a rack and the rack is then inserted into a chamber. In the chamber the sample for each separate card is then "pressurized" into all 48 cham-bers to distribute exactly 1 ml of sample into the 16-tube MPN system. The filled cards, with the delivery tube cut and sealed, are incubated at 35°C overnight. The cards with growth results are placed in an instrument to read the presence or absence of fluorescence (medium dependent) in each of the 48 chambers to pro-vide MPN/ml or g of target organism in the sample. The instrument then auto-matically calculates the 16-tube 3-dilution series MPN of the sample. The entire procedure is highly automated. The savings of time of operation and materials are truly impressive and it is likely that the system will receive excellent reception by food and water microbiologists.

Air and Surface Sampling

The food industry also needs to ascertain the air quality as well as the cleanliness of surfaces for product manufacturing, storage, and transportation. An active air-sampling instrument such as the SAS system of pbi (Milan, Italy) can suction a known volume of air and deposit microorganisms on an agar surface (impaction). After incubation the number of organisms in per cubic meter of air can be obtained. Another method involves passage of a known volume of air through a tube of liquid (impingement) to capture the microorganisms and obtain CFU/ m^3 or ml. A variety of swabs, tapes, sponges, and contact agar methods have been developed to obtain surface count of food-manufacturing environments.

Aerobic, Anaerobic, and "Real-Time" Viable Cell Counts

By use of the correct gaseous environment or suitable reducing compounds one can obtain aerobic, anaerobic, or facultative anaerobic microbial counts of products. Typically, microbial counts were obtained in 24 to 48 hours. Several methods have been developed and tested in recent years that can provide real-time viable cell counts such as the use of "Vital" stains (Acridine Orange) to count fluorescing viable cells under the microscope (Direct Epifluorescent Filter Technique), scanning of samples for organism that have Fluorassure (a vital dye by the Chemunex Scan RDI system), or measuring ATP of micro-colonies trapped in special membranes (the MicroStar System of Millipore). These real-time tests can give viable cell counts in about four hours. A simple double-tube system using appropriate agar and incubation conditions has been developed by the author that can provide a *Clostridium perfringens* count for water testing in about six hours. Hawaii is the only U.S. state that uses *C. perfringens* as an indicator of fecal contamination for recreational waters. This information can be used to assess the suitability of swimming water in various beaches in the world.

Advances in Miniaturization and Diagnostic Kits

Identification of normal microflora, spoilage organisms, clinical and foodborne pathogens, starter cultures, etc., in many specimens is an important part of microbiology. In the past 25 years many miniaturized diagnostic kits have been developed and widely used to conveniently introduce the pure cultures into the system and obtain reliable identification in as short as two to four hours. Some systems could handle several or even hundreds of isolates at the same time. These diagnostic kits no doubt saved many lives by rapidly, accurately, and conveniently identifying pathogens so that treatments can be made correctly and rapidly. Some of the more common miniaturized systems on the market to identify enterics (e.g.,

Salmonella, Shigella, Proteus, Enterobacter), nonfermentors, anaerobes, gram positives, and even yeasts and molds are API, MicroID, Enterotube, Crystal ID systems, Biolog, and Vitek.

Advances in Immunological Testings

The antigen and antibody reaction has been used for decades to detect and characterize microorganisms and their components in medical and diagnostic microbiology. This is the basis for serotyping bacteria such as *Salmonella, Escherichia coli* O157:H7, and *Listeria monocytogenes*. Both polyclonal antibodies and monoclonal antibodies have been used extensively in applied food microbiology. The most popular format is the "Sandwiched" ELISA procedure. Recently some companies have completely automated the entire ELISA procedure and can complete an assay from 45 minutes to 2 hours after overnight incubation of the sample with suspect target organisms. VIDAS system (bioMerieux) is a totally automated system for identification of pathogens such as *Listeria, Listeria monocytogenes, Salmonella, E. coli* O157:H7, staphylococcal enterotoxins, and *Campylobacter*. There are more than 15,000 units in use worldwide. BioControl markets an Assurance EIA system that can be adapted to automation for high-volume testings.

Lateral Flow Technology (similar to a pregnancy test with a sample port, detection region, and control port on a small unit) offers a simple and rapid test for target pathogens (e.g., *E. coli* O157) after overnight incubation of food, or allergens (e.g., wheat gluten). The entire procedure takes only about 10 minutes with very little training necessary. BioControl VIP system, Neogen Reveal system, and Merck KgaA are some of the commercial systems available on the market.

A truly innovative development in applied microbiology is the immuno-magnetic separation (IMS) system. Homogeneous paramagnetic beads have been developed which can be coated with a variety of molecules such as antibodies, antigens, and DNA. to capture target cells such as *E. coli* O157:H7, *Listeria, Cryptosporidium,* and *Giardia.* These beads can then be immobilized, captured, and concentrated by a magnet stationed outside a test tube. After clean-up, the beads with the captured target molecules or organisms can be plated on agar for cultivation or used in ELISA, Polymerase Chain Reaction (PCR), microarray technologies, biochips, and so forth, for detection of target organisms. The Dynal company produces homogeneous paramagnetic beads that can be used for coating antigens, antibodies, DNA, and even cells for use in IMS system. A new and very efficient IMS system is the Pathatrix system which can circulate the entire liquid sample (e.g., 25 g of food in 225 ml of broth) over surfaces containing immobilized beads coated with specific antibodies for specific pathogens such as *E. coli* O157:H7. After capturing the target pathogens from the entire broth the liquid is discarded and the beads with pathogens can be released and then the detection of the pathogens can be made using conventional agar plates, ELISA, PCR, etc. This system increases detection efficiency and time greatly. In the author's laboratory as little as 1 to 10 *E. coli* O157:H7 CFU/25 g of ground beef can be detected in 5.25 hours from the time of placing the 25 gram of beef into 225 ml of broth for growth (4.5 hrs), circulating the broth in the Pathatrix system (0.5 hr) to completing the ELISA test (0.25 hr). This system was used to detect *E. coli* O157:H7 in the spinach outbreak in the United States in 2006–2007.

Advances in Instrumentation and Biomass Measurements

Instruments can be used to automatically monitor microbial growth kinetics and dynamics changes (such as ATP levels, specific enzymes, pH, electrical impedance, conductance, capacitance, turbidity, color, heat, and radioactive carbon dioxide) of a population (pathogens or nonpathogens) in a liquid and semi-solid sample. It is important to note that for the information to be useful, these parameters must correlate to a viable cell count of the same sample series. In general, the larger the number of viable cells in the sample, the shorter the detection time of these systems. A scatter gram is then plotted and used for further comparison of unknown samples. The assumption is that as the number of microorganisms increases in the sample, these physical, biophysical, and biochemical events will also increase accordingly. The detection time is inversely proportional to the initial population in the sample. When a sample has 5 log or 6 log organisms/ml, detection time can be achieved in about four hours. Some instruments can handle hundreds of samples at the same time. Instruments to detect ATP in a culture fluid include Lumac, BioTrace, Ligthning, Hy-Lite, Charm, Celsis, and Zyluz Profile 1. The Bactometer and Rapid Automated Bacterial Impedance Technique can monitor impedance changes in the test sample, and the Malthus system monitors conductance changes and relate the changes to microbial populations. Basically, any type of instrument that can continuously and automatically monitor turbidity and color changes of a liquid in the presence of microbial growth can be used for rapid detection of the presence of microorganisms.

Advances in Genetic Testings

Phenotypic expression of cells are subject to growth conditions such as temperature, pH, nutrient availability, and oxidation-reduction potentials. Genotypic characteristics of a cell are far more stable. Hybridization of DNA and RNA to known probes has been used for more than 30 years, but the polymerase chain reaction (PCR) is now an accepted method to detect viruses, bacteria, and even yeast and molds by amplification of the target DNA and detecting the PCR products. By use of reverse transcriptase, target RNA can also be amplified and detected. After a DNA (double-stranded) molecule is denatured by heat (e.g., 95°C), proper primers will anneal to target sequences of the single stranded DNA of the target organism (e.g., *Salmonella*) at a lower temperature (e.g., 50°C). A polymerase (e.g., *Taq* enzyme) will extend the primer at a higher temperature (e.g., 70°C) and complete the addition of complement bases to the single-stranded denatured DNA. After one thermal cycle one piece of DNA will become two pieces. After 21 and 31 cycles one piece will become 1 million and 1 billion copies, respectively. Early on, PCR products were detected by gel electrophoresis, which is very time consuming and a source of contamination of laboratory with amplified DNA. Now the occurrence of the PCR can be monitored in the PCR tube in real time with fluorescent probes or dyes. Some systems can monitor four different targets in the same sample (multiplexing) by using different reporting dyes. This method is now standardized and easy to use and interpret.

To distinguish closely related organisms, their genetic material is cleaved with restriction enzymes and the resulting DNA fragments are separated by pulsed field gel electrophoresis (PFGE) to obtain a fragment pattern (fingerprint). Alternatively, the ribosomal gene regions of these organisms can be analyzed by restriction enzyme-mediated cleavage of the DNA and probing for DNA fragments that hybridize to a ribosomal DNA probe. Generally, different species and even subspecies or strains will exhibit different hybridization patterns (riboprints) or PFGE patterns and can therefore be distinguished. For example, *Listeria monocytogenes* and *E. coli* has 49 and 60 distinct patterns, respectively. The discriminatory power of this technique permits "source tracking" by determining if, for example, a *L. monocytogenes* isolated from a food source is the same as an isolate obtained from an infected person.

Advances in Biosensor, Microchips, and Biochips

Biosensors are an exciting new field of study in applied microbiology. The basic idea is simple but the actual operation is quite complex and involves much instrumentation. Basically, a biosensor is a molecule or a group of molecules of biological origin attached to a signal recognition material. When an analyte comes in contact with the biosensor the interaction will initiate a recognition signal which can be reported in an instrument. Many types of biosensors have been developed. Whole cells can sometimes be used as biosensors. Analytes detected include toxins, specific pathogens, carbohydrates, insecticides and herbicides, ATP, antibiotics, etc. The recognition signals used include electrochemical (e.g., potenti - ometry, voltage changes, conductance and impedance, light addressable), optical (such as UV, bioluminescence, chemiluminescence, fluorescence, laser scattering, reflection and refraction of light, surface phasmon resonance, polarized light), and miscellaneous transducers (such as piezoelectric crystals, thermistor, acoustic waves, quartz crystal).

Recently, much attention has been directed to biochips and microchips developments to detect a great variety of molecules including foodborne pathogens. Due to the advancement in miniaturization technology as many as 50,000 individual spots (e.g., DNA microarrays) with each spot containing millions of copies of a specific DNA probe can be immobilized on a specialized microscope slide. Fluorescently labeled targets can be hybridized to these spots and be detected. Biochips can also be designed to detect all kinds of foodborne pathogens by imprinting a variety of antibodies or DNA molecules against specific pathogens on the chip for the simultaneous detection of pathogens such as *Salmonella*, *Listeria*, *E. coli*, and *Staphylococcus aureus*. The market value of biochips in general is estimated to be as high as $5 billion at this moment. This technology is especially important in the rapidly developing field of proteomics and genomics which require massive amount of data to generate valuable information. Advanced bioinformatics is necessary to interpret these data.

Certainly, the developments of these biochips and microchips are impressive for obtaining a large amount of information for biological sciences. As for foodborne pathogen detection there are several important issues to be considered. These biochips and related microfluidic systems are designed to detect minute quantities of target molecules. The target molecules must be free from contaminants before being applied to the biochips. In food microbiology, the minimum requirement for pathogen detection is 1 viable target cell in 25 grams of a food. Biochips will not be able to seek out such a cell from the food matrix without extensive cell amplification or sample preparation, separation, filtration, centrifugation, absorption, etc. Another concern is viability of the pathogens. Monitoring the presence of cell components in food does not necessarily indicate that the cells are alive or dead. Certainly human beings regularly consume killed pathogens such as *Salmonella* in cooked chicken without becoming sick. To ensure a system is detecting living cells some form of culture enrichment or vital dyes are necessary to indicate that the pathogens to be detected are alive.

Testing Trends and Predictions

There is no question that many microbiological tests are being conducted nationally and internationally in pharmaceutical and food products, environmental samples, medical specimens, and water samples. The most popular food microbiology tests are grouped as "indicator tests" (total viable cell count, coliform/*E. coli* count, and yeast and mold counts) and "specific pathogen tests" (*Salmonella*,

Listeria and *Listeria monocytogenes, E. coli* O157:H7, *Staphylococcus aureus, Campylobacter,* etc.). According to Strategic Consulting Inc. (Woodstock, VT, 2005) in 1998, the worldwide microbiological tests (including food, 56%; pharmaceutical, 30%; beverages, 10%; and environmental water, 4%) were estimated to be 755 million tests with a market value of $1 billion. Of the 420 million tests related to food laboratories, 360 million tests were for "indicator tests" and 60 million for "specific pathogens tests." The projection at this moment is that by 2008 the number of tests will be about 1.5 billion with a market value of $5 billion.

In 2007 about one-third of the tests were being performed in North America (the United States and Canada), another third in Europe, and the last third in the Rest of the World. This author predicts in 20 years the Rest of the World will perform 50% of the tests with North America and Europe performing 25% each. This is due to rapid economic development and food- and health-safety concerns of the world in the years ahead.

Prediction of the Future

The following are the ten predictions made by the author in 1995 at the ASM National Convention in Washington, D.C. Many predictions have been correct in 2008. (+) is a good prediction. (?) is an uncertain prediction.

1. Viable cell counts will still be used in the next 25 years. (+)
2. Real-time monitoring of hygiene will be in place. (+)
3. PCR, Ribotyping, and genetic tests will become reality in food laboratories. (+)
4. ELISA and immunological tests will be completely automated and widely used. (+)
5. Dipstick technology will provide rapid answers. (+)
6. Biosensors will be in place for Hazard Analysis Critical Control Point Programs. (?)
7. Biochips, microchips, and microarrays will greatly advance in the field. (+)
8. Effective separation, concentration of target cells will assist rapid identification. (+)
9. Microbiological alert systems will be in food and pharmaceutical packages. (?/+)
10. Consumers will have rapid alert kits for pathogens at home. (?)

In conclusion, it is safe to say that the field of rapid method and automation in microbiology will continue to grow in numbers and kinds of tests to be done in the future due to the increased concern on food safety and public health. The future looks very bright for the field of rapid methods and automation in microbiology. The potential is great and many exciting developments will certainly unfold in the near and far future in food safety and security areas.

References

Fung, D. Y. C. 2000. Rapid methods and automation in microbiology: comprehensive reviews in food science and food safety. Inaugural Issue 1(1):3–22. At www.ift.com. Accessed September 15, 2007.

Fung, D. Y. C. 2007. Rapid methods and automation in microbiology in pharmaceutical samples. Am. Pharma. Rev. 10(3):82–86.

Strategic Consulting, Inc. 2005. Global review of microbiology testing in the food processing market. Strategic Consulting, Inc., Woodstock, VT. At weschler@stragetic-consult.com. Accessed September 15, 2007.

13

Genomic Approaches to Bacterial Pathogens Using Transposon Mutagenesis: Food-Safety Applications

Young Min Kwon, Dhruva Bhattacharya, and Steven C. Ricke

Introduction

One main characteristic feature of bacterial pathogens is their ability to cope with the adverse conditions in the environment and host. Bacterial pathogens can reach and survive in host microenvironments that are usually not accessible to nonpathogenic bacteria. Accordingly, pathogens typically possess special mechanisms to overcome the threatening conditions in the host and survive in these niches (Foster and Spector 1995; Slauch et al. 1997). The survival strategies depend upon the complex yet coordinated expression of bacterial factors in a timely manner to evade various host immune mechanisms encountered in different niches and temporal stages of infection. The identification and characterization of genetic factors responsible for the expression of these phenotypes have been of major interest in the study of bacterial pathogens (Chiang et al. 1999). Increased understanding of the virulence factors has provided the basis for development of effective control strategies such as vaccines or antibiotics. Particularly, bacterial genes essential for survival *in vitro* or *in vivo* provide promising targets for development of novel antimicrobials (Rosamond and Allsop 2000).

The virulence strategies of numerous bacterial pathogens depends upon their ability to infect humans by contaminating foods that are subsequently consumed. These foodborne bacterial pathogens should have additional capability to persist in the stressful conditions during food processing and storage. For instance, foodborne pathogens are expected to encounter a variety of stressors at this stage of their life cycle, including osmotic stress, temperature fluctuation, desiccation, starvation, and high pressure. Considering overlapping stress signals that exist in both host environments and food-production systems, there are likely common pathways in bacterial cells that are essential for survival in both conditions, food environments and human hosts. Based upon this hypothesis, there are factors uniquely associated with survival in food environments. Understanding that there are specific genetic pathways essential for survival of foodborne pathogenic bacteria in food systems should provide the basis for the development of target-specific and more effective preventative strategies. However, as compared to the pathways or genetic factors essential for *in vitro* or *in vivo* survival, those responsible for persistent survival within the food-production system have been poorly characterized. The reason for this discrepancy could be partially explained by dominant use of traditional approaches in the area of food microbiology and a lack of sufficient resources and expertise in food microbiology laboratories to conduct genetic or genomic analyses to study these factors. However, the powerful methods that are already established and widely used to dissect genetic factors important during host infection could be directly and readily applicable to understand the survival mechanisms of foodborne bacterial pathogens in the context of food-production environments.

In this review we give an overview of the various transposon-based functional genomic approaches that have been developed to identify *in vitro* and *in vivo* survival factors, along with their significance in current efforts to develop effective antibacterial strategies. We will extend the discussion to possible application of those approaches to the study of foodborne bacterial pathogens in the context of food safety to promote effective control of the pathogens in food-production environments and final food products.

Discovery of Target Pathways for Development of Novel Antibiotics

Infectious diseases are currently the third-leading cause of human death in the United States and the leading cause worldwide (Cohen 2000). In the United States alone, over 1,000,000 people die from infectious diseases every year. Traditionally, the prevention and control of bacterial pathogens have been largely dependent on the use of vaccines and antibiotics. Particularly, antibiotics have proved to be highly effective in treating and controlling bacterial infections. However, antibiotic resistance has steadily increased in the last three decades, and it is now well accepted that the current arsenal of antibiotics will not be sufficient in the combat against pathogenic bacteria. Almost every antibiotic placed into clinical use has engendered resistance irrespective of its chemical class or molecular target (Miesel et al.

2003). In response to this seemingly serious situation in public health, the pharmaceutical companies have a renewed commitment to develop new classes of antibiotics to treat these resistant bacterial pathogens. Such endeavors are best appreciated in the light of the enormous budgets spent on antibiotic research in the United States, which is estimated to be $3.6 billion every year (Cheng et al. 2003). Furthermore, the possible use of such multidrug-resistant pathogenic bacteria in bioterrorism requires urgent development of a new class of antibiotics to fight against such attacks. Traditionally, the development of antibacterial agents has been mainly based on screening for natural compounds exhibiting antibacterial properties. However, such broad cell-based screening usually does not provide information on the biochemical targets, which hinders efforts to optimize the lead compounds on the basis of structure-activity relationships (Miesel et al. 2003). To overcome this obstacle, the pharmaceutical industry redirected its antibacterial discovery approach in the early 1980s from screening for antibacterial activity to inhibiting specific biochemical targets (Rosamond and Allsop 2000). Most antibiotics that have been used clinically target a limited number of macromolecules involved in essential cellular functions. This suggests that many novel targets remain to be identified for the development of novel classes of antibiotics (Miesel et al. 2003).

The molecular targets of most antibiotics developed by whole cell-based screening are the essential gene products, which still remain important targets for development of novel antibiotics. Inhibition of those essential functions usually leads to eradication of the bacteria in the infected hosts. However, the bactericidal activity in those antibiotics is a major factor in the emergence of antibiotic resistance bacteria in recent years (Rosamond and Allsop 2000). In an effort to reduce the rate of resistance development, the bacterial genes that are not essential for *in vitro* growth but required for survival in infected host tissues have been suggested as alternative targets. Such *in vivo* essential gene products are more likely to provide relatively narrow therapeutic spectrums, thereby decreasing the likelihood of selection for broad antibiotic resistance. However, the difficulty in establishing a minimum inhibitory concentration (MIC) assay for inhibitors of *in vivo* essential processes remains a practical problem. Recently, targeting nonmultiplying bacteria has been suggested as a promising strategy for antibacterial drug development (Coates et al. 2002). Clinically latent bacteria, which are not killed efficiently by existing antibiotics, usually prolong the duration of chemotherapy. Therefore, killing such nonmultiplying bacteria may reduce the duration of chemotherapy, thus lowering the rate of emergence of resistance.

Increasing Demands for Comprehensive Identification of *in vivo* Survival Factors

The growing concern about antibiotic resistance has coincided with revolutionary progress in the availability of genome sequences of more than 250 microorganisms in recent years (Raskin et al. 2006). This has generally changed the way bacterial pathogens are studied and concomitantly revealed a large number of genes in bacterial pathogens with unknown biological functions. Traditionally, virulence genes have been discovered by individually assessing defined mutant strains for virulence-associated phenotypes or by screening a mutant library for those with attenuated virulence. When a gene product is suspected to be important in the host infection by homologies to known virulence factors or deduced biochemical properties implicated in virulence, a mutant strain with defined deletion in the target gene of interest can be constructed and tested to determine its virulence potential as compared to that of wild type. However, mutant libraries such as transposon mutant libraries of various pathogens have been screened for the mutants that exhibit decreased virulence-associated phenotype or attenuation during an animal model of infection (Bowe et al. 1998; Turner et al. 1998).

Transposons are short genetic elements that can be randomly inserted into the genome. Since transposon insertion usually inactivates the gene where the insertion is made, transposon mutagenesis has been widely used for efficient random mutagenesis of bacteria (Maloy 2007). Although these conventional approaches have made major contributions in shaping our current understanding of bacterial pathogenesis, it is obvious that such methods are unable to provide comprehensive information for thousands of genes waiting to be assigned functions. Regarding virulence factors required for *in vivo* survival of bacterial pathogens during host infection, the limitation of the conventional approaches is more obvious due to the large numbers of animals required for individually assessing thousands of mutants as well as the need for suitable animal models (Turner et al. 1998).

To overcome the limitation of assessing individual mutants for attenuation during infection of a live host, during the last decade several clever genetic strategies have been devised and used successfully for comprehensive identification of bacterial virulence genes required for *in vivo* survival. Methods are being modified to allow rapid analysis of microbial genomes on a genome-wide scale, compelling the pharmaceutical companies to take advantage of such advances in search for new antibiotics. Development of the new technologies and strategies combined with genome sequences and bioinformatic tools have changed the way researchers approach antibacterial drug discovery. Great progress has already been made in the development of several experimental approaches to identify potential antibiotic targets on a genome-wide scale; namely, bacterial essential genes or virulence genes for *in vivo* survival (Sassetti et al. 2001; Hughes 2003).

Essential genes that are indispensable for *in vitro* growth and virulence genes that are required for *in vivo* survival can be collectively regarded as essential genes, defined operationally according to selective conditions of interest (*in vitro* vs. *in vivo*). Accordingly, there are some common features in the experimental schemes for comprehensive identification of both classes of genes in spite of their different biological functions. Most of the experimental schemes consist of three common steps as shown in Figure 13.1: first, generation of a mutant library either by a random or systematic method (input pool); second, selection of the mutant library under different conditions (output pool); and third, comparison of the original library with that after selection to identify the mutants that are underrepresented or lost from the pool. A transposon is often the choice for the mutagenesis, due to the ease with which mutational sites can be identified and the ability to specifically disrupt a single gene in each mutant. Mutants that are not recovered after selection are assumed to harbor insertions in genes that are required for survival under a specific growth condition. The third step is where different genomic methods vary

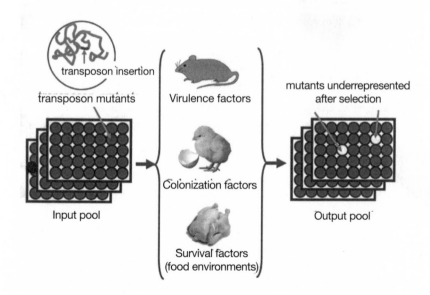

Figure 13.1. Experimental scheme for comprehensive negative selection of transposon mutants.

substantially, because the task for this step is technically most challenging. The issue is how each transposon mutant in a complex pool of mutants can be distinguished from other mutants in the same pool, and how the relative level of each mutant can be reliably quantified to allow comprehensive comparison of the mutant library with those recovered after selection in the animal infection model. All of the mutants in a library will have the exact same genomic contents except for the site of transposon insertion and that is the only characteristic that distinguishes each from the others at DNA sequence level. Therefore, all of the approaches make use of the different transposon insertion sites or the transposon tagged with a unique marker as a means to identify unique insertions.

Comprehensive Functional Genomic Approaches

Signature-tagged mutagenesis (STM) is one of the first negative selection methods that was devised to screen large number of transposon mutants of bacterial pathogens simultaneously to identify mutants that have reduced virulence in a live host (Hensel et al. 1995). STM employs the insertion of random sequence tags into the transposon as a means to differentiate the individual mutants within a complex pool of transposon mutants and uses a comparative hybridization strategy to identify transposon mutants that are missing in recovered pools from host tissues. Since its first application to murine infections with *Salmonella enterica* serovar Typhimurium, the use of STM, and its variations, has increased continuously for comprehensive identification of virulence genes in diverse pathogenic microorganisms (Mei et al. 1997; Chiang and Mekalanos 1998; Tsolis et al. 1999). Two similar strategies have been developed to eliminate rather time-consuming steps required in STM such as probe labeling and hybridization. In these methods, the DNA fragments of varying sizes have been used for tagging to allow differentiation of each transposon mutant within the pool of mutants. Hildago-Grass et al. (2002) cloned a DNA fragment without homology to the bacterial pathogen under study into a restriction enzyme site inside an IS256 element located on a suicide delivery plasmid. The IS256 variants containing different sizes of tag fragments were constructed by bidirectional deletion of the tag fragment from the internal restriction enzyme site. Input pools were constructed by pooling 21 mutants each harboring varying sizes of tag fragments and used for infection. For identification of attenuated mutants, the input and output pools were used to amplify the tags by multiplex PCR using universal primers flanking the tag fragments, one of them being labeled with fluorescence and the resulting fragments being separated on a sequencing gel. This method, termed polymorphic-tag-lengths-transposon-mutagenesis (PTTM), was applied to the mouse infection model of Group A *Streptococcus* (GAS) and subsequently led to identification of a mutant that failed to survive in the spleen. In a similar method called DNA size marker identification technology (SMIT), varying lengths of unrelated DNA fragments were cloned in a shuttle vector, which then was integrated into the same locus on the genome of *Staphylococcus aureus* (Benton et al. 2004). Twenty-five strains harboring different size tags were then subjected to transposon mutagenesis, and used to form each pool consisting of 25 random mutants of *S. aureus*. From the input and output pools, the size markers were simultaneously amplified and separated on a polyacrylamide gel. In the screening of a total of 6,300 mutants, 23 unique insertions leading to attenuation were identified. The strength of these methods includes simplicity in experimental protocol and feasibility in processing multiple samples but pool size is limited by the electrophoretic resolution.

Genetic footprinting is a PCR-based strategy that was designed to determine the functions of sequenced genes in microorganisms (Smith et al. 1995; 1996; Hare et al. 2001; Akerley et al. 2002). This method uses PCR to determine transposon insertion sites in the genome with a gene-specific primer and a transposon-specific primer. When conducted with input and output pools, the footprinting patterns generated by gel electrophoresis allow identification of the genes required for growth or survival under selective conditions. Although selection can be conducted only once initially with a mutant pool, the contribution of each gene to cell survival during selection must be assessed individually using position-specific PCR primers followed by electrophoretic resolution. Since the resolution of the method relies on the density of insertions in the target genes, a very efficient transposon system is required to ensure random insertions saturate the genome. Although not demonstrated yet, genetic footprinting can be used for screening of virulence genes in animal infection models. We previously devised a method termed "transposon footprinting" (Kwon et al. 2002) in which the transposon-flanking sequences are simultaneously amplified from a pool of mutants such that different mutants generally produce PCR products of different sizes. The amplified sequences generate a "footprinting" pattern for each pool upon separation by electrophoresis. Bands present in the input pool but not in the output pool are indicative of mutants that have not survived the selection in host tissue. We used this method to identify Tn5 mutants of S. Typhimurium defective in survival in mouse tissues (Kwon et al. 2002). We have also applied this approach to identify Tn5 mutants of S. Typhimurium with increased sensitivity to desiccation (Park et al. 2002) and short-chain fatty acids (Kwon et al. 2003a) and defective for survival on egg-shell surface (Kwon et al. 2003b). Like other gel-based methods for separation of fragments, the limited pool size due to limited electrophoretic resolution is the major disadvantage.

Genome-wide Functional Genomic Approaches

The methods described above allow parallel comparison of a pair of mutant pools, input and output pools. However, all of the methods have limitations in the size of the pools that can be simultaneously screened. In STM, the pool size is limited by the hybridization kinetics and the sensitivity of the membrane-based blotting system. In all other methods, which use gel electrophoresis as means to create the profile for a mutant pool, the major factor limiting the pool size is the resolution that the gel allows to separate the multiplex-type PCR products. In mutant screening using an animal infection model, it is often the case that there are biological bottlenecks associated with the animal model or the route of infection adopted. When a significant bottleneck exists, mutants in a complex pool can be lost stochastically even though they retain virulence potential similar to that of wild type. In this situation, the pool size must be reduced to the level where loss of any mutants in the output pool faithfully represents attenuation even though the method of choice for negative selection theoretically allows a larger pool size for mutant screening.

However, when there is no significant biological bottleneck existing in the infection model, it would be desirable if the pool size could be increased to such an extent that a genome-saturating mutant library can be used for simultaneous screening. The number of mutants to saturate the genome depends upon the size of the bacterial genome. It would be in the range of 10^4–10^5 mutants in a library for genomes between 1.0 and 6.0 Mbp in size. Achieving such a goal was thought to be unrealistic in the early stages of the post-genomic era. However, it has become quite possible in recent years with the aid of microarray technology. Different experimental variations have been developed to accomplish this goal, based on the same underlying principle of using microarray for simultaneous mapping of thousands of sequences derived from transposon-flanking sequences.

Sassetti et al. (2001) developed a microarray-based method called transposon site hybridization (TraSH), for genome-wide identification of conditionally essential genes. TraSH utilizes transposon harboring outward-facing T7 RNA polymerase promoters. This allows preparation of labeled RNA complementary to the transposon-flanking sequences in a complex pool of mutants. By using labeled RNA probes prepared from a pool of mutants that have been selected under different conditions for a microarray, differential gene requirements of *Mycobacterium tuberculosis* were determined (Sassetti et al. 2001; 2003). As a natural transition, they successfully expanded the use of TraSH to genome-wide screening for virulence genes in *M. tuberculosis* required for *in vivo* survival (Sassetti and Rubin 2003). Since the development of TraSH, methods based on this same principle have been developed and tested with different bacterial species. Badarinarayana et al. (2001) developed a microarray-based method for simultaneous mapping of genome-saturating mutant library and validated the approach by identifying conditionally essential genes in *Escherichia coli* that are required for growth in minimal medium. The well-characterized capability of a microarray in detecting target sequences supports the accuracy of the mapping of transposon insertions. Similar methods but based on a different experimental protocol to retrieve signals corresponding to numerous insertion sites were developed by Salama et al. (2004) and used to identify essential genes in *Helicobacter pylori*. In the previous methods, a Y-shaped linker was used to amplify transposon-flanking sequences from which mRNA sequences were generated by T7 RNA polymerase to generate labeled probes for microarray hybridization. However, the method developed by Salama et al. (2004), termed microarray tracking of transposon mutants (MATT), uses random primers to amplify transposon-flanking sequences, which then are directly labeled to prepare microarray probes. The microarray-based methods for simultaneous mapping of genome-saturating mutants are powerful methods for categorizing gene function on a genomic scale that should be applicable to a variety of microorganisms. More recently this experimental approach has been adopted and modified to analyze genetic requirements of other bacterial species such as *Bacillus anthracis* and *Francisella tularensis* (Weiss et al. 2007; Day et al. 2007).

Conclusions

The comprehensive functional screening approaches described in the previous sections have been developed with the aim of identifying genes in bacterial pathogens required for survival *in vitro* or *in vivo*. However, these approaches could be readily extended to accomplish experimental goals to understand bacterial pathways important in other processes. The main two requirements for application of those functional genomics approaches are a transposon mutagenesis system and that mutants can be recovered after selection in sufficient numbers to represent all surviving mutants. In the case of foodborne bacterial pathogens, our understanding of the genetic mechanisms used by foodborne bacterial pathogens to allow prolonged survival during food processing or storage would be of high interest. This information will be crucial in designing more effective strategies to control the pathogens in food environments. Most strategies to reduce bacterial pathogens in the postharvest stages of food production are based on the general bacteriostatic or bactericidal treatments of foods and food-processing environments. Although effective in reducing bacterial loads in contaminated food, it is obvious that more effective control measures are in great demand to ensure supply of more safe foods. These strategies are not based on the knowledge of the specific pathways or mechanisms in a bacterial pathogen that are used for survival against various stresses during the process. However, if we know what genetic factors are essential for counteracting stresses at different stages of production, it would be possible to develop control measures targeting particular pathways or sensitizing the pathogens to the stresses. Therefore, with appropriate experimental design, essential genetic factors could be easily identified using real food materials or a series of selective processing conditions. We see this approach as analogous to the recent strategies to develop novel antibiotics, in which efforts are focused on targeting narrow and specific pathways for efficient control while minimizing the probability of development of antibiotic resistance. We envision that this line of efforts will eventually create new directions of research to promote enhanced food safety targeting specific pathways essential for survival of foodborne pathogens in food environments.

Acknowledgments

This study was supported by the funding from a Food Safety Consortium grant and a National Institute of Health grant. We thank Julia Sonka (Center for Food Safety, IFSE, University of Arkansas) for her help on editing.

References

Akerley, B. J., E. J. Rubin, V. L. Novick, K. Amaya, N. Judson, and J. J. Mekalanos. 2002. A genome-scale analysis for identification of genes required for growth or survival of *Haemophilus influenzae*. Proc. Natl. Acad. Sci. USA 99:966–971.

Badarinarayana, V., P. W. EstepIII, J. Shendure, J. Edwards, S. Tavazoie, F. Lam, and G. M. Church. 2001. Selection analyses of insertional mutants using subgenic-resolution arrays. Nature Biotechnol. 19:1060–1065.

Benton, B. M., J. P. Zhang, S. Bond, C. Pope, T. Christian, L. Lee, K. M. Winterberg, M. B. Schmid, and J. M. Buysse. 2004. Large-scale identification of genes required for full virulence of *Staphylococcus aureus*. J. Bacteriol. 186:8478–8489.

Bowe, F., C. J. Lipps, R. M. Tsolis, E. Groisman, F. Heffron, and J. G. Kusters. 1998. At least four percent of the *Salmonella typhimurium* genome is required for fatal infection of mice. Infect. Immun. 66:3372–3377.

Cheng, Q., S. Wang, and A. A. Salyers. 2003. New approaches for anti-infective drug discovery: antibiotics, vaccines and beyond. Current Drug Targets–Infectious Disorders 3:65–76.

Chiang, S. L., and J. J. Mekalanos. 1998. Use of signature-tagged transposon mutagenesis to identify *Vibrio cholerae* genes critical for colonization. Mol. Microbiol. 27:797–805.

Chiang, S.L., J. J. Mekalanos, and D. W. Holden. 1999. *In vivo* genetic analysis of bacterial virulence. Annu. Rev. Microbiol. 53:129–154.

Coates, A., Y. Hu, R. Bax, and C. Page. 2002. The future challenges facing the development of new antimicrobial drugs. Nature Reviews Drug Discovery 1:895–910.

Cohen, M. L. 2000. Changing patterns of infectious disease. Nature 406:762–767.

Day, W. A., Jr., S. L. Rasmussen, B. L. Carpenter, S. N. Peterson, and A. M. Friedlander. 2007. Microarray analysis of transposon insertion mutations in *Bacillus anthracis:* global identification of genes required for sporulation and germination. J. Bacteriol. 189:3296–3301.

Foster, J. W., and M. P. Spector. 1995. How *Salmonella* survive against the odds. Ann. Rev. Microbiol. 49:145–174.

Hare, R. S., S. S. Walker, T. E. Dorman, J. R. Greene, L. M. Guzman, T. J. Kenney, M. C. Sulavik, K. Baradaran, C. Houseweart, H. Yu, Z. Foldes, A. Motzer, M. Walbridge, G. H. Shimer Jr., and K. J. Shaw. 2001. Genetic footprinting in bacteria. J. Bacteriol. 183:1694–1706.

Hensel, M., J. E. Shea, C. Gleeson, M. D. Jones, E. Dalton, and D. W. Holden. 1995. Simultaneous identification of bacterial virulence genes by negative selection. Science 269:400–403.

Hidalgo-Grass, C., M. Ravins, M. Dan-Goor, J. Jaffer, A. E. Moses, and E. Hanski. 2002. A locus of group A Streptococcus involved in invasive disease and DNA transfer. Mol. Microbiol. 46:87–99.

Hughes, D. 2003. Exploiting genomics, genetics and chemistry to combat antibiotic resistance. Nature Reviews in Genetics 4:432–441.

Kwon, Y. M., L. F. Kubena, D. J. Nisbet, and S. C. Ricke. 2002. Functional screening of bacterial genome for virulence genes by transposon footprinting. Methods in Enzymol. 358:141–152.

Kwon, Y. M., L. F. Kubena, D. J. Nisbet, and S. C. Ricke. 2003a. Genetic screening for identification of *Salmonella typhimurium* Tn5 mutants with potential hypersensitivity to short-chain fatty acids. J. Rapid Method. Auto. Microbiol. 11:89–95.

Kwon, Y. M., L. F. Kubena, D. J. Nisbet, and S. C. Ricke. 2003b. Isolation of *Salmonella typhimurium* Tn5 mutants defective for survival on egg shell surface using transposon footprinting. J. Environ. Sci. Health Part B. B38:103–109.

Maloy, S. R. 2007. Use of antibiotic-resistant transposons for mutagenesis. Methods in Enzymol. 421:11–17.

Mei, J. M., F. Nourbakhsh, C. W. Ford, and D. W. Holden. 1997. Identification of *Staphylococcus aureus* virulence genes in a murine model of bacteremia using signature-tagged mutagenesis. Mol. Microbiol. 26:399–407.

Miesel, L., J. Greene, and T. A. Black. 2003. Genetic strategies for antibacterial drug discovery. Nature Reviews in Genetics 4:442–456.

Park, S. Y., Y. M. Kwon, S. G. Birkhold, L. F. Kubena, D. J. Nisbet, and S. C. Ricke. 2002. Application of a transposon footprinting technique for rapid identification of *Salmonella typhimurium* Tn5 mutants required for survival under desiccation stress conditions. J. Rapid Method. Auto. Microbiol. 10:197–206.

Raskin, D. M., R. Seshadri, S. U. Pukatzki, J. J. Mekalanos. 2006. Bacterial genomics and pathogen evolution. Cell 124:703–714.

Rosamond, J., and A. Allsop. 2000. Harnessing the power of the genome in the search for new antibiotics. Science 287:1973–1976.

Salama, N. R., B. Shepherd, and S. Falkow. 2004. Global transposon mutagenesis and essential gene analysis of *Helicobacter pylori*. J. Bacteriol. 186:7926–7935.

Sassetti, C. M., D. H. Boyd, and E. J. Rubin. 2001. Comprehensive identification of conditionally essential genes in mycobacteria. Proc. Natl. Acad. Sci. USA 98:12712–12717.

Sassetti, C. M., D. H. Boyd, and E. J. Rubin. 2003. Genes required for mycobacterial growth defined by high-density mutagenesis. Mol. Microbiol. 48:77–84.

Sassetti, C. M., and E. J. Rubin. 2003. Genetic requirements for mycobacterial survival during infection. Proc. Natl. Acad. Sci. USA 100:12989–12994.

Slauch, J., R. Taylor, and S. Maloy. 1997. Survival in a cruel world: how *Vibrio cholerae* and *Salmonella* respond to an unwilling host. Genes Dev. 11:1761–1774.

Smith, V., D. Botstein, and P. O. Brown. 1995. Genetic footprinting: a genomic strategy for determining a gene's function given its sequence. Proc. Natl. Acad. Sci. USA 92:6479–6483.

Smith, V., K. N. Chou, D. Lashkari, D. Botstein, and P. O. Brown. 1996. Functional analysis of the genes of yeast chromosome V by genetic footprinting. Science 274:2069–2074.

Tsolis, R. M., S. M. Townsend, E. A. Miao, S. I. Miller, T. A. Ficht, L. G. Adams, and A. J. Bäumler. 1999. Identification of a putative *Salmonella enterica* serotype Typhimurium host range factor with homology to IpaH and YopM by signature-tagged mutagenesis. Infect. Immun. 67:6385–6393.

Turner, A. K., M. A. Lovell, S. D. Hulme, L. Zhang-Barber, and P. A. Barrow. 1998. Identification of *Salmonella typhimurium* genes required for colonization of the chicken alimentary tract and for virulence in newly hatched chicks. Infect. Immun. 66:2099–2106.

Weiss, D. S., A. Brotcke, T. Henry, J. J. Margolis, K. Chan, and D. M. Monack. 2007. In vivo negative selection screen identifies genes required for Francisella virulence. Proc. Natl. Acad. Sci. USA 104:6037–6042.

▌14▐

Advances in Antibody-Based Technologies for *Listeria monocytogenes*

Ramakrishna Nannapaneni and Michael G. Johnson

Introduction

In the United States each year, foodborne illnesses affect 6 million to 80 million persons, cause 9,000 deaths, and cost an estimated $5 billion. *Listeria monocytogenes* is one of the major foodborne bacterial pathogens and causes over 2,518 infections, 2,322 hospitalizations, and 504 deaths annually in the United States (Mead et al. 1999). The 25% fatality rate of *L. monocytogenes* is the second-highest among human bacterial foodborne pathogens. The Centers for Disease Control and Prevention have established a future target of 2.5 cases of human listeriosis per million population by 2010, a significant reduction from 7 cases of human listeriosis per million population reported by the 1987 baseline study (U.S.DHHS 2007). In line with this future target, FoodNet data from 1996 show a gradual reduction in human listeriosis cases, attributed to food-safety measures implemented by food industries and regulatory agencies (MMWR 2006).

    L. monocytogenes is a dangerous intracellular pathogen that can cause meningitis and abortion or still-born death of fetuses if ingested in low numbers in foods by pregnant women. The populations that are most prone to listeriosis are newborn babies, the elderly, and immunocompromised individuals. Epidemiological studies show that the consumption of contaminated foods containing as little as 100 to 1,000 cells of *L. monocytogenes* were associated with listeriosis infections. Over the years, *L. monocytogenes* was implicated in several serious outbreaks (Mead et al. 1999, 2006), which include pasteurized milk, Mexican-style cheese, pate, and coleslaw. This led to intensification of efforts to develop rapid and sensitive detection and enumeration techniques that are applicable to a wide range of food products. Of all the potential food products, dairy products and cold smoked seafood products are implicated in numerous cases of sporadic listeriosis outbreaks in the United States that have led to many Class I (voluntary) recalls attributed to *L. monocytogenes* since 2000. Undercooked poultry has been reportedly implicated as a risk factor for *L. monocytogenes* foodborne infections (Pinner et al. 1992; Schuchat et al. 1991). In late 1998 through early 1999, there were two massive USDA mandated recalls of cooked meat and poultry products due to *L. monocytogenes* contamination; 20 million pounds of product were recalled by Thorn Apple Valley and Sara Lee Food Companies (USDA-FSIS 1998, 1999a, 1999b). In 2002, there was the largest recall in history involving 27.4 million pounds of turkey products under the Wampler Foods brand of Pilgrims Pride Corporation due to *L. monocytogenes* contamination (USDA-FSIS 2002). Controlling the persistence and survival of *L. monocytogenes* is a critical issue for all food-processing industries. USDA-FSIS has a zero tolerance for *L. monocytogenes* in ready-to-eat processed meat and poultry products (Schuchat et al. 1991; Swaminathan 2001).

L. monocytogenes is highly versatile and can be found in a wide range of environments. Persistence of low numbers of *L. monocytogenes* in food-processing environments is of great concern for the food industry. This pathogen acquires enhanced resistance during stress through multiple mechanisms (Ferreira et al. 2003; Miller et al. 2000; Taormina et al. 2001) enabling it to survive food-processing treatments. Unlike other human foodborne pathogenic bacteria, *L. monocytogenes* is capable of growth at refrigeration temperatures, thus leading to contamination of cold-stored foods. It grows slowly at refrigerated temperatures and contaminates ready-to-eat cooked meats and poultry during processing stages. It is likely that *L. monocytogenes* commonly encounters conditions of sublethal injury and stress in the food-processing environments that induce a stress-responsive alternative sigma factor or starvation survival response (SSR) (Herbert and Foster 2001). In 2001, USDA-FSIS enforced new performance standards for processed meat and poultry products. All establishments that produce RTE meat and poultry products are required to conduct (1) environmental testing of food-contact surfaces for the presence of *L. monocytogenes;* and (2) finished product testing. This is intended to reduce the incidence of *L. monocytogenes* in RTE meat and poultry products (USDA-FSIS 2001).

The food industry strives to cut down pathogen detection times while maintaining the safety of the food-processing equipment and environment. Some of the current microbiological monitoring techniques involve very long delay times between the time that samples are assayed and the times when the results are known. To reduce the risk of exposure and avoid expensive product recalls, many processors put their products "on hold" and do not ship until the results of key microbiological analyses are known. Such policies involve maintaining very large inventories of refrigerated/frozen product with significant investments in storage costs. All industries associated with processing of animal, vegetable, and fruit products—where there is a stringent need for the zero tolerance of processed foods for pathogenic bacteria—will need new highly sensitive pathogen-detection technologies. Currently, these industries spend millions of dollars per year testing their products to assure an absence of pathogens in these foods prior to release for sale. Culture methods traditionally used for screening for the presence or absence of *L. monocytogenes* in food-processing environments and in food products can take up to a week to produce confirmed results. To decrease this assay time, rapid methods employing antibody- and/or nucleic acid-based technologies combined with standard enrichment methods are continuously evolving toward accurate, reliable, and faster detection of *L. monocytogenes*. Compared to nucleic acid-based technologies, antibody-based technologies are more user friendly and less expensive, and can be routinely implemented in diagnostic laboratories without expensive equipment. To effectively prevent or control *L. monocytogenes* contamination in food-processing environments and in finished products, there is a need for accurate diagnostics that are specific for *L. monocytogenes* and do not react with the other major *Listeria* species, especially *L. innocua,* which generally occurs in the same foods in which *L. monocytogenes* is found. This chapter focuses on antibody-based technologies for *L. monocytogenes*—the diversity of antibody probes, their strengths and limitations, and emerging rapid antibody-based technologies for this pathogen.

Diversity of Polyclonal and Monoclonal Antibodies Developed against *L. monocytogenes*

Antibodies are the most critical reagents of any immunoassay. Antibody probes capable of distinguishing human pathogenic *L. monocytogenes* from nonpathogenic *Listeria* spp. are in great demand. There are eight species within the genus *Listeria*—*L. monocytogenes, L. innocua, L. seeligeri, L. welshimeri, L. ivanovii, L. grayi, L. murrayi,* and *L. denitrificans* (Farber and Peterkin 1991)—out of which only *L. monocytogenes* is pathogenic for humans and other animals and *L. ivanovii* is pathogenic for animals, and the remaining six are nonpathogenic species. In addition to this, detecting all serotypes of *L. monocytogenes* is also important since it occurs in 13 distinct serotypes, which are 1/2a, 1/2b, 1/2c, 3a, 3b, 3c, 4a, 4b, 4c,

4d, 4e, 4ab, and 7 (Bhunia 1997; Farber and Peterkin 1991). Of these 13 serotypes, three main serotypes, 1/2a, 1/2b, and 4b, are involved as the predominant causes (90%) of human and animal cases of listeriosis (Swaminathan 2001; Swaminathan et al. 2001). Among the three epidemiologically relevant *L. monocytogenes* serotypes, 50% of the human listeriosis-associated cases worldwide are caused by serotype 4b, but by contrast the majority of the foodborne isolates belongs to serotype 1/2 (1/2a, 1/2b, 1/2c) (Swaminathan 2001). Since only a few serotypes are predominant in human listeriosis outbreaks, there is also a growing demand for specific antibodies capable of distinguishing highly virulent serotypes of *L. monocytogenes* in food products.

Polyclonal antibodies are a mixture of different antibody types binding to multiple sites on the antigen that was used for immunization. Conversely, monoclonal antibodies consist of a single antibody species binding to a single specific site on the antigen (Harlow and Lane 1988). Table 14.1 shows the diversity of polyclonal antibodies developed against *Listeria* and *L. monocytogenes*. Except for the high specificity of those recognizing p60 protein (Bubert et al. 1992; 1994), all other polyclonal antibodies raised against flagella (Peel et al. 1988ab), internalin (Gaillard et al. 1994), and listeriolysin (Traub et al. 1995) were found to be suitable only for genus-specific detection of *Listeria* and not suitable for species-specific detection of *L. monocytogenes*. Table 14.2 shows the diversity of monoclonal antibodies raised against *Listeria* and *L. monocytogenes*. Monoclonal antibodies recognizing cell surface antigens (Bhunia 1997; Bhunia and Johnson 1992), flagellar antigens (Farber and Speirs 1987), listerial toxins (Kuhn and Goebel 1989), other virulence determinants (Erdenlig et al. 1999), or for cytoplasmic antigens (Loiseau

Table 14.1. Diversity of Polyclonal Antibodies (PAb) to *Listeria* spp. and *L. monocytogenes*

Protein/antigen recognized	Specificity	Reference
29 kDa Flagella recognizing PAb	*Listeria* genus-specific	Peel et al. 1988a, 1998b
58–60 kDa Listeriolysin recognizing Pab	*Listeria* genus-specific	Traub and Bauer 1995
60 kDa protein specific PAb	*Listeria* genus-specific	Ruhland et al. 1993
60 kDa protein, peptide A specific PAb	*L. monocytogenes* species-specific, all serotypes except 4a, 4c	Bubert et al. 1994
60 kDa protein, peptide D specific PAb	*L. monocytogenes* species-specific, reacts with all serotypes	Bubert et al. 1994

Table 14.2. Diversity of Monoclonal Antibodies (MAb) to *Listeria* spp. and *L. monocytogenes*

Protein/antigen recognized	Specificity	Reference
30 to 38 kDa recognizing MAbs	*Listeria* genus-specific	Butman et al. 1988
38 to 41 kDa recognizing MAbs	*Listeria* genus-specific	Loiseau et al. 1995
18.5 kDa recognizing MAb P5C9	*L. monocytogenes, L. innocua, and L. welshieri*	Siragusa and Johnson 1990
55–58, 55–37, 55–23 kDa recognizing MAbs	*Listeria* genus-specific	Torensma et al. 1993
60 kDa protein recognizing MAb:p6017	*Listeria* genus-specific	Yu et al. 2004
60 kDa protein recognizing MAb:p6007	*L. monocytogenes* species-specific	Yu et al. 2004
66 kDa recognizing, MAb EM-7G1	*L. monocytogenes* species-specific, all serotypes except 3b, 4a, 4c	Bhunia and Johnson 1992a; Nannapaneni et al. 1998a
43 kDa and 94–97 kDa recognizing, MAb EM-7G1	*Listeria* genus-specific	Bhunia and Johnson 1992a; Nannapaneni et al. 1998b
76, 66, 56, and 52 kDa recognizing, MAb C11E9	*Listeria monocytogenes* and *L. innocua*	Bhunia et al. 1991
27, 29, or 60 kDa protein recognizing MAbs	*Listeria* genus-specific	Erdenlig et al. 1999
80 kDa Internalin A recognizing, MAb mAb2B3	*L. monocytogenes* species-specific, all serotypes with enhanced reactivity to ½a	Hearty et al. 2006
66 and 76 kDa recognizing, MAbs 22D10 and 24F6	*Listeria* genus-specific	Heo et al. 2007
MAbs Cc74.33, c74.180	*L. monocytogenes* high degree of specificity for 4b	Kathariou et al. 1994
MAbs M2365 and M2367	*L. monocytogenes* serotype 4b specific	Lin et al. 2006

et al. 1995) have been developed for their utilization in *Listeria* diagnostics (Table 14.2). The flagellum consists of a 29 kDa subunit flagellin protein that is encoded by the *fla*A gene (Dons et al. 1992). To date, none of the monoclonal antibodies raised to flagella (29 kDa), listeriolysin O (58–60 kDa), actA (90 kDa), or other *Listeria* cell antigens (18.5 kDa, 30–38 kDa, 58 kDa) has proved to be specific for the *L. monocytogenes* serotypes alone (Bhunia 1997). Siragusa and Johnson (1990) described and characterized the first monoclonal antibody with specificity for *L. monocytogenes, L. welshimeri,* and *L. innocua.* A flagellar antibody has been used for the immunomagnetic separation of genus-specific *Listeria* antigens (Peel et al. 1988a), but flagellar expression is temperature dependent and thus these flagella may not be present in bacteria at all times, particularly if grown at 37°C or exposed to shear forces (Farber and Speirs 1987; Farber and Peterkin 1991). The majority of antibodies raised against extracellular p60 antigen of *Listeria* have proved to be genus-specific (Kuhn and Goebel 1989), but a unique sequence variable region of p60 was shown to be specific for *L. monocytogenes* (Bubert et al. 1994).

Panels of species-specific monoclonal antibodies exhibiting different reactivity patterns against different *L. monocytogenes* serotypes are yet to be developed for meeting the strong need for the rapid detection methods. Production of specific monoclonal antibodies against *L. monocytogenes* has been difficult to achieve; there is no general consensus in the literature regarding the type of immunogen (whole cells, cell walls, soluble antigen, extracellular components, etc.) likely to result in the most specific polyclonal or monoclonal antibody. Continuing research is needed for the identification of specific monoclonal antibodies capable of recognizing different cell surface epitopes, species-specific or serotype-specific for *L. monocytogenes* strains (Nannapaneni et al. 1998a; Clark et al. 2000; Lei et al. 2001).

A species-specific antibody probe, EM-7G1, developed by Bhunia and Johnson (1992) recognizes a 66 kDa protein that is present as a cell surface protein. No direct role for this protein was implicated in pathogenesis, but it may be essential for the metabolism of certain peptides in the growth medium (Bhunia 1997). Generally, cell surface antigens are believed to be more stable irrespective of growth conditions. Conversely, flagellar antigens or listeriolysin may only be produced under certain favorable environmental conditions (Peel et al. 1988a, 1998b). MAb EM-7G1 has proved to be useful for detecting heat-injured cells (Patel and Beuchat 1995) or live cells of selected serotypes of *L. monocytogenes* from food samples (Carroll et al. 2000). Presently there is a lack of species-specific monoclonal antibody-probes for *L. monocytogenes* in the public domain—except for MAb EM-7G1. This monoclonal antibody has certain limitations such as inability to recognize all serotypes of *L. monocytogenes* or loss of detection if cells are heated or grown in the USDA recommended Modified University of Vermont Broth

(UVM1) (Nannapaneni et al. 1998a). SDS-PAGE and Western blotting tests proved that the MAb EM-7G1 detected a specific protein epitope present on the 66 kDa cell surface antigen of only *L. monocytogenes* and not that of the other *Listeria* species (Bhunia and Johnson 1992). However, subsequent characterization tests determined that (a) MAb EM-7G1, while effectively detecting 10 of the 13 major serotypes of live cells of *L. monocytogenes* both by microcolony immunoblot technique and ELISA, did not however detect the three remaining serotypes, 3b, 4a, and 4c (Nannapaneni et al. 1998a); (b) MAb EM-7G1 detected live bacterial cells of *L. monocytogenes* but failed to detect heat-killed cells of *L. monocytogenes* (Nannapaneni et al. 1998a). Further, MAb EM-7G1 exhibited greatly decreased strength of ELISA reactions against *L. monocytogenes* cell serotypes when cells were grown in certain enrichment broth media, notably UVM1 as recommended by the U.S. Department of Agriculture for poultry products and Fraser broth (FRB) as recommended for dairy products by the Food and Drug Administration (Nannapaneni et al. 1998a). Another MAb, EM-6E11, recognizing *Listeria* genus-specific epitopes of 43 and 94 to 97 kDa also lost reactivity to *Listeria* cells in ELISA when heat killed and had lower reactivity to *Listeria* cells grown in UVM1 or FRB (Nannapaneni et al. 1998b). Anti-LLO MAbs or polyclonal antibodies (PAbs) (Nato et al. 1991) may be specific of *L. monocytogenes,* but unfortunately these antigens are structurally homologous to the hemolysins also found in many other gram-positive bacteria, thus resulting in false positive cross-reactions. Also, production of hemolysin can be suppressed by the growth environment such as temperature, pH, salt, and preservatives present in foods, thereby causing false negative results (Datta and Kothary 1993). Two recently developed monoclonal antibodies, 22D10 and 24F6, reacted with heat-killed cells of *L. monocytogenes* but were found to be non-species-specific (Heo et al. 2007). In contrast with MAb EM-7G1, which is broadly reactive to 11 of the 13 serotypes, the other species-specific MAbs developed were narrowly reactive with only a few serotypes of *L. monocytogenes.* Similarly, Kathariou et al. (1994) developed two other MAbs c74.33 and c74.180, which showed a high degree of specificity against serotype 4b but also recognized 4d and 4e and no other serotypes of *L. monocytogenes.* Recently another species-specific MAb probe, mAb2B3, developed by Hearty et al. (2006) reacted with a 80 kDa protein that was identified as internalin A and distinguished *L. monocytogenes* from other *Listeria* spp. but reacted strongly with *L. monocytogenes* serotype ½a and weakly with serotypes ½c, 4a, 4b, and 4c. Of the 35 MAbs recently developed by Lin et al. (2006), two monoclonal antibodies, M2365 and M2367, were found to be highly specific for *L. monocytogenes* of serotype 4b without any reactions to other serotypes; interestingly, these two serotype 4b-specific MAb probes, M2365 and M2367, did not recognize any *L. monocytogenes* protein bands in Western blot assays. Conversely, the other MAbs that recognized 20 kDa, 35 kDa, 36 kDa, 62 kDa, 75 kDa, 77 kDa, and 87–88 kDa in Western blot assays reported by Lin et al. (2006) were not specific for *L. monocytogenes*. More recently, MAb:p6607 by Yu et al. (2004) was reported to be species-specific for an *L. monocytogenes* p60 protein and differed from other anti-p60 protein species-specific polyclonal antibodies developed by Bubert et al. (1994). MAb probes that can selectively recognize the two most important *L. monocytognes* serotypes, 4b and ½ a, have yet not been developed but this could be solved by using a mixture of two species-specific MAbs selective for each of these two serotypes summarized in Table 14.2. Therefore, for detecting all *L. monocytogenes* serotypes, continuing research work must focus on the development of other species-specific monoclonal antibodies recognizing different epitopes. Such MAbs must target (a) epitopes present on the surface of affinity-purified 66 kDa cell surface antigen; (b) epitopes occurring on the 60 kDa protein or 80 kDa internalin A or 90 kDa cell surface Acta protein; or (c) against other specific predominant protein fractions common among *L. monocytogenes* serotypes produced by cells when grown in different preenrichment media. The cell surface 90 kDa ActA protein proved to be crucial for the induction of actin assembly in most serotypes of *L. monocytogenes*. Application of anti-internalin, anti-actA, or anti-phospholipase antibodies for detecting *L. monocytogenes* in food and clinical samples is yet to be fully investigated. In parallel with the reports of Nannapaneni et al. (1998a, 1998b), differential expression of *L. monocytogenes* proteins, for example, Internalin B and ActA, may differ significantly when cells are grown in selective versus nonselective enrichment broth media (Lathrop et al. 2008). Such differential expression will adversely affect antibody-based detection. More discoveries are yet to be made for other species-specific or serotype-specific cell surface, intra- or extracellular antigens of *L. monocytogenes* and their expression. *L. monocytogenes* is subjected to various stresses including cold, heat, salt, or acid during food processing and storage but there is little information on how these stresses affect modification of epitopes in immunodiagnostic assays (Geng et al. 2003, 2006; Hahm and Bhunia 2006; Solve et al. 2000). Also, MAbs recognizing different species-specific epitopes of 66, 80, and 90 kDa protein or those against other specific antigens may need to be combined in mixtures for the development of routine diagnostics for major serotypes of *L. monocytogenes* cells grown in different enrichment broth environments. The unknown cocktail of monoclonal and/or polyclonal antibodies used in the commercial Vidas test kit for this pathogen is claimed to be specific for all serotypes of *L. monocytogenes* but no refereed publications are available describing these antibodies, to the best of our knowledge.

Emerging Antibody-Based Technologies for Rapid Detection of *L. monocytogenes*

Immunodiagnostic assays are extremely useful as management tools for faster detection. A number of solid-phase support systems are available for detection of foodborne pathogens using antibodies. Antibodies that exhibit extremely high specificity to the target antigen are highly valuable for rapidly detecting the presence or absence of specific antigens and also for purifying antigens of interest. The Enzyme-Linked Immunosorbent Assay (ELISA) has become a widely accepted detection method owing to its high sensitivity, ease of use, and rapidity; however, it does not readily lend itself to quantitation of *L. monocytogenes* cells (Feldsine et al. 1997; Mattingly et al. 1988). Recent developments in antibody-based qualitative detection technologies will result in more convenient, effective, and specific immunodiagnostic assays for qualitative detection of foodborne *L. monocytogenes*. In one device, *L. monocytogenes* antigen-antibody complex is detected in a lateral flow immunoassay by Neogen Reveal that is routinely used in food-processing industries and this assay can be completed in 10–15 minutes but requires a large number of cells in the order of 10^7 to 10^9 cfu/ml. Immunomagnetic beads are also being used to speed up the bacterial capture in antibody assays. Antibody-coated magnetic beads are used to selectively concentrate target *L. monocytogenes* cells directly from samples or preenriched samples and resulting antigen-antibody complexes are captured magnetically. Immunomagnetic separation (IMS) protocols are simple and take one hour to complete. Sensitivity and specificity were increased when IMS was used for *L. monocytogenes* detection, achieving detection limits in the range of 10^4 to 10^6 cfu/ml (Duffy et al. 1997; Mitchell et al. 1994; Skjerve et al. 1990). Recently, Paoli et al. (2007) coupled single-chain antibody fragments (scFvs) to immunomagnetic beads for the detection of *L. monocytogenes* and these beads showed a higher efficiency of capture as compared to commercially available anti-*Listeria* immunomagnetic beads. Hibi et al. (2006) developed an assay using immunomagnetic separation combined with flow cytometry for the rapid detection of *L. monocytogenes* with a claimed detection range of 10^2 to 10^8 cfu/ml.

Antibodies directed against *L. monocytogenes* have also been used for the construction of biosensors that will allow rapid detection of this pathogen in the target matrices. Antibody-based biosensors are simple to operate, have fewer steps, and may require shorter preenrichment times (10–18 hours). The biosensor assay per se is capable of decreasing the final assay time to about one hour once sufficient target pathogen cells are present. The use of in-line, real-time biosensors promises to further reduce detection times and provide faster automation of procedures. The construction of a biosensor requires stable attachment of an antibody probe to the sensor surface; for example, a gold electrode of the piezo-electric quartz crystal (Nannapaneni and Johnson 1997; Gullbault and Luong 1994). Upon attachment of an antibody probe, the quartz crystal acts as a sensor surface for the capturing of bacterial cells. Another approach to the biosensor construction for foodborne pathogens is the use of surface plasmon resonance (SPR) phenomenon for the development of a fiber-optic sensor. SPR is an optical phenomenon that occurs as a result of a change in total internal reflection of light on the surface of metal film–liquid interface. When the light is absorbed, SPR is created. When SPR occurs, energy from the incident light is lost to the metal film, resulting in a drop of the reflected light intensity. The optical phenomenon of SPR causes changes in refractive index on the surface of a sensor chip coated with antibody as a result of binding of target bacterial cells (Leonard et al. 2005). To date, optically based biosensors have been developed in several research labs (Geng et al. 2004; Wang et al. 2007) and by BIAcore International and Texas Instruments. Texas Instruments introduced the first miniature integrated surface plasmon resonance liquid sensor. A compact SPR sensor using antibody phage system Lm P4:A8, expressing the scFv antibody against ActA protein was used for detection of whole cells of *L. monocytogenes* with a minimum detection limit of about 2 x 10^6 cfu/ml (Nanduri et al. 2007). A species-specific MAb mAb2B3 against Internalin A developed by Hearty et al. (2006) showed a sensitivity of 1 x 10^7 cfu/ml in both ELISA and SPR assay formats. The minimum detection of *L. monocytogenes* using polyclonal antibodies to Internalin B in the SPR sensor system was 2 x 10^5 cfu/ml (Leonard et al. 2005). Taylor et al. (2006) developed an SPR sensor for the detection of *L. monocytogenes*, with limits of detection in the range of 3.4 x 10^3 to 1.2 x 10^5 cfu/ml. An electrochemical immunoassay system utilizing highly dispersed carbon particles as a flow through matrix for the immobilization of antibodies that act as a working electrode was developed and demonstrated to be sensitive and highly specific for detection of *L. monocytogenes* (Chemburu et al. 2005). Geng et al. (2004) developed an antibody-based optical sensor using cyanine labeled murine monoclonal antibody C11E9, followed by an enrichment step to detect low levels of *L. monocytogenes*. Based on monoclonal antibody MAb-C11E9, a resonant mirror biosensor was used for the detection of soluble protein extracts of *L. monocytogenes* but showed cross-reaction with *L. innocula* starins (Lathrop et al. 2003). An electrochemical immunosensor capable of reacting with 23 kDa *L. monocytogenes* surface Internalin B protein at picogram levels was developed by Tully et al. (2006). The new fluorescent quantum-dot biosensor technologies developed by Li and associates (Biodetection Technologies, LLC) detected *L. monocytogenes* within 1½ hours using a biotin-labeled anti-*Listeria* polyclonal antibody in a complex with streptavidin-coated magnetic nanobeads and yielded a sensitivity of detection as low as 30 cfu/ml in pure culture experiments (Wang et al. 2007).

In contrast to rapid detection on the basis of presence-absence of cells, enumerating the numbers of *L. monocytogenes* present on the surfaces of raw or in finished food products is a continuing challenge. The universally applicable DNA-based pulse field gel electrophoresis method yields accurate fingerprinting of suspect isolates for epidemiological comparisons, but does not provide enumeration of bacteria in a food product. Widely applicable PCR or biosensor or ELISA or electrochemiluminescence methods are capable of accurately detecting but not enumerating the number of pathogen cells present. Among the so-called rapid-detection methods, only the microcolony immunoblotting (or microcolony immunoprinting) technique has the capability to simultaneously detect and enumerate pathogen levels in a food product (Hitchins 2002). Several laboratories are currently working on adapting this technique for quantitative enumeration of pathogens from food products. However, the full potential of this established technique to integrate the quantitative data needs for both HACCP and risk assessment and to develop new USDA-FSIS regulations for the safety of poultry and meat products are yet to be exploited. Since any given single antibody or DNA probe will not be able to detect all serotypes of *L. monocytogenes*, it is necessary to use a panel of several antibody probes simultaneously in an assay for the concurrent specific detection of all *L. monocytogenes* cells that may be present in a given food sample. Bhunia et al. (1992) first demonstrated the use of the microcolony immunoblot technique using MAb C11E9 to quantitate *L. monocytogenes* in food products but this antibody also reacted with *L. innocua* and therefore this assay was not specific for *L. monocytogenes* enumeration. Patel and Buechat (1995) demonstrated the application of species-specific MAb EM-7G1 for detecting heat-injured cells of *L. monocytognes* using microimmunoblot techniques in 34 hours. By evaluating a large collection of *L. monocytogenes* and other *Listeria* spp., a high degree of species-specificity of EM-7G1 was demonstrated in a microcolony immunoblot assay (Carroll et al. 2000). To the best of these reviewers' knowledge, none of the other species-specific MAbs against *L. monocytogenes* were tested in microcolony immunoblot techniques for the enumeration of live or stressed cells of *L. monocytogenes*. Exposure of *L. monocytogenes* cells to environmental stresses, including heating, freezing, and chemical agents may result in structural and/or metabolic damage without causing cell death. Improving the detectability of such injured *L. monocytogenes* cells occurring in food-processing environments needs to be attempted by using *L. monocytogenes* species-specific microcolony immunoblot assays employing cell surface recognizing antibodies (Geng et al. 2006; Kang and Siragusa 1999).

Conclusions and Future Perspectives

L. monocytogenes causes severe illness and even death in immune-compromised individuals such as young children. Early detection of pathogenic *L. monocytogenes* in food-processing environments and in contaminated foods is essential to prevent food-poisoning outbreaks. The prevention of foodborne diseases depends on careful food production, handling of raw products, and preparation of finished products. Hazards can be introduced at any stage from farm to table. The HACCP programs require food industries to identify points in food production where contamination may occur and target the reduction or elimination of these by systematic application of monitoring and control strategies. There is a need for development of rapid, user-friendly, time- and cost-effective pathogen-detection techniques that do not require expensive instrumentation and that can be implemented in routine QA laboratories. The conventional standard microbiological methods for detection of *L. monocytogenes* in food are time consuming and require a minimum of one week. Nucleic-acid-based detection methods (such as polymerase chain reaction) may require less time *if* suitable numbers of the target pathogen are present and the nucleic acid of these cells can be separated from the food matrix. Generally, antibody-based detection methods are more user friendly than nucleic-acid-based methods in a regular food microbiology quality control laboratory. Antibody-based assays have the potential for rapid detection with high sensitivity, accuracy, and minimal sample preparation time, and are easy to use and inexpensive when compared to PCR-based assays. A wide range of polyclonal and monoclonal antibodies that are genus-specific and species-specific have been developed against *L. monocytogenes* and tested in different biosensor assay formats using pure cultures. However, none of these have been applied for the detection of *L. monocytognes* in actual food sample extracts. By further improvement of these new technologies, real-time immunobiosensor assays promise to further reduce detection times and speed up automation procedures for *L. monocytogenes*. Future antibody-based technologies must also be able to rapidly detect the presence of low numbers of live, stress-adapted, or stress-injured cells of *L. monocytogenes* surviving in food-processing environments and in finished food products.

References

Bhunia, A. K., P. H. Ball, A. T. Fuad, B. W. Kurz, J. W. Emerson, and M. G. Johnson. 1991. Development and characterization of a monoclonal antibody specific for *Listeria monocytogenes* and *Listeria innocua*. Infect. Immun. 59:3176–3184.

Bhunia, A. K., and M. G. Johnson. 1992. Monoclonal antibody specific for *Listeria monocytogenes* associated with a 66-kilodalton cell surface antigen. Appl. Environ. Microbiol. 58:1924–1929.

Bhunia, A. K., P. H. Ball, and M. G. Johnson. 1992. A 20–24 h microcolony-immunoblot technique to detect and enumerate *Listeria monocytogenes* inoculated into foods. J. Rapid Methods and Automation in Microbiol. 1:67–82.

Bhunia, A. K. 1997. Antibodies to *Listeria monocytogenes*. Crit. Rev. Microbiol. 23:77–107.

Bubert, A., M. Kuhn, M. Goebel, and S. Kohler. 1992. Structural and functional properties of the p60 proteins from different *Listeria* species. J. Bacteriol. 174:8166.

Bubert, A., P. Schubert, S. Köhler, R. Frank, and W. Goebel. 1994. Synthetic peptides derived from the *Listeria monocytogenes* p60 protein as antigens for the generation of polyclonal antibodies specific for secreted cell-free *L. monocytogenes* p60 proteins. Appl. Environ. Microbiol. 60:3120–3127.

Butman, B. T., M. C. Plank, R. J. Durham, and J. A. Mattingly. 1988. Monoclonal antibodies which identify a genus-specific *Listeria* antigen. Appl. Environ. Microbiol. 54:1564–1569.

Carroll, S. A., L. E. Carr, E. T. Mallinson, C. Lamichhanne, B. E. Rice, D. M. Rollins, and S. W. Joseph. 2000. A colony lift immunoassay for the specific identification and quantification of *Listeria monocytogenes*. J. Microbiol. Methods. 41:145–153.

Chemburu, S., E. Wilkins, and I. Abdel-Hamid. 2005. Detection of pathogenic bacteria in food samples using highly-dispersed carbon particles. Biosens. Bioelectron. 21:491–499.

Clark, E. E., I. Wesley, F. Fiedler, N. Promadej, and S. Kathariou. 2000. Absence of serotype-specific surface antigen and altered teichoic acid glycosylation among epidemic-associated strains of *Listeria monocytogenes*. J. Clin. Microbiol. 38:3856–3859.

Datta, A. R., and M. H. Kothary. 1993. Effects of glucose, growth temperature, and pH on listeriolysin O production in *Listeria monocytogenes*. Appl. Environ. Microbiol. 59:3495–3497.

Dons, L., O. F. Rasmussen, and J. E. Olsen. 1992. Cloning and characterization of a gene encoding flagellin of *Listeria monocytogenes*. Mol. Microbiol. 6:2919–2929.

Duffy, G., J. J. Sheridan, H. Hofstra, D. A. McDowell, and I. S. Blair. 1997. A comparison of immunomagnetic and surface adhesion immunofluorescent techniques for the rapid detection of *Listeria monocytogenes* and *Listeria innocua* in meat. Lett. Appl. Microbiol. 24:445–450.

Erdenlig, S., A. J. Ainsworth, and F. W. Austin. 1999. Production of monoclonal antibodies to *Listeria monocytogenes* and their application to determine the virulence of isolates from channel catfish. Appl. Environ. Microbiol. 65:2827–2832.

Farber, J. M., and P. I. Peterkin. 1991. *Listeria monocytogenes*, a food-borne pathogen. Microbiol. Rev. 55:476–511.

Farber, J. M., and J. I. Speirs. 1987. Monoclonal antibodies directed against the flagellar antigens of *Listeria* species and their potential use in EIA-based methods. J. Food Prot. 50:479–484.

Feldsine, P. T., A. H. Lienau, R. L. Forgey, and R. D. Calhoon. 1997. Assurance polyclonal enzyme immunoassay for detection of *Listeria monocytogenes* and related *Listeria* species in selected foods: collaborative study. J. AOAC Int. 80:775–790.

Ferreira, A., D. Sue, C. P. O'Byrne, and K. J. Boor. 2003. Role of *Listeria monocytogenes* sigB in survival of lethal acidic conditions and in the acquired acid tolerance response. Appl. Environ. Microbiol. 69:2692–2698.

Gaillard, J. L., S. Dramsi, P. Berche, and P. Cossart. 1994. Molecular cloning and expression of internalin in *Listeria*. Methods in Enzymol. 236:551–565.

Geng, T., B. K. Hahm, and A. K. Bhunia. 2006. Selective enrichment media affect the antibody-based detection of stress-exposed *Listeria monocytogenes* due to differential expression of antibody-reactive antigens identified by protein sequencing. J. Food Prot. 69:1879–1886.

Geng, T., K. P. Kim, R. Gomez, D. M. Sherman, R. Bashir, M. R. Ladisch, and A. K. Bhunia. 2003. Expression of cellular antigens of *Listeria monocytogenes* that react with monoclonal antibodies C11E9 and EM-7G1 under acid-, salt- or temperature-induced stress environments. J. Appl. Microbiol. 95:762–772.

Geng, T., M. T. Morgan, and A. K. Bhunia. 2004. Detection of low levels of *Listeria monocytogenes* cells by using a fiber-optic immunosensor. Appl. Environ. Microbiol. 70:6138–6146.

Gullbault, G. G., and J. H. T. Luong. 1994. Piezoelectric immunosensors and their applications in food analysis. Pages 151–172 in Food biosensor analysis. G. Wagner and G. G. Gullbault, eds. Marcel Dekkar, New York, NY.

Hahm, B. K., and A. K. Bhunia. 2006. Effect of environmental stresses on antibody-based detection of *Escherichia coli* O157:H7, *Salmonella enterica* serotype Enteritidis and *Listeria monocytogenes*. J. Appl. Microbiol. 100:1017–1027.

Harlow, E. D., and D. Lane. 1998. Antibodies: a laboratory manual. Cold Spring Harbor Laboratory Press, Cold Spring Harbor, New York.

Hearty, S., P. Leonard, J. Quinn, and R. O'Kennedy. 2006. Production, characterization and potential application of a novel monoclonal antibody for rapid identification of virulent *Listeria monocytogenes*. J. Microbiol. Methods. 66:294–312.

Heo, S. A., R. Nannapaneni, R. P. Story, and M. G. Johnson. 2007. Characterization of new hybridoma clones producing monoclonal antibodies reactive against both live and heat-killed *Listeria monocytogenes*. J. Food Sci. 72:M008–M015.

Herbert, K. C., and S. J. Foster. 2001. Starvation survival in *Listeria monocytogenes:* characterization of the response and the role of known and novel components. Microbiology 147:2275–2284.

Hibi, K., A. Abe, E. Ohashi, K. Mitsubayashi, H. Ushio, T. Hayashi, H. Ren, and H. Endo. 2006. Combination of immunomagnetic separation with flow cytometry for detection of *Listeria monocytogenes*. Anal. Chim. Acta. 573–574:158–163.

Hitchins, A. D. 2002. Detection and enumeration of *Listeria monocytogenes* in foods. Chapter 10 in Bacteriological analytical manual. 8th ed. At http://www.cfsan.fda.gov/~ebam/bam-10.html [online]. Accessed March 30, 2008.

Kang, D. H., and G. R. Siragusa. 1999. Agar underlay method for recovery of sublethally heat-injured bacteria. Appl. Environ. Microbiol. 65:5334–5337.

Kathariou, S., C. Mizumoto, R. D. Allen, A. K. Fok, and A. A. Benedict. 1994. Monoclonal antibodies with a high degree of specificity for *Listeria monocytogenes* serotype 4b. Appl. Environ. Microbiol. 60:3548–3552.

Kuhn, M., and W. Goebel. 1989. Identification of an extracellular protein of *Listeria monocytogenes* possibly involved in intracellular uptake by mammalian cells. Infect. Immun. 57:355–358.

Lathrop, A. A., P. P. Banada, and A. K. Bhunia. 2008. Differential expression of InlB and

ActA in *Listeria monocytogenes* in selective and nonselective enrichment broths. J. Appl. Microbiol. 104:627–639.

Lathrop, A. A., Z. W. Jaradat, T. Haley, and A. K. Bhunia. 2003. Characterization and application of a *Listeria monocytogenes* reactive monoclonal antibody C11E9 in a resonant mirror biosensor. J. Immunol. Methods 281:119–128.

Lei, X. H., F. Fiedler, Z. Lan, and S. Kathariou. 2001. A novel serotype-specific gene cassette (*gltA-gltB*) is required for expression of teichoic acid-associated surface antigens in *Listeria monocytogenes* of serotype 4b. J. Bacteriol.183:1133–1139.

Leonard, P., S. Hearty, G. Wyatt, J. Quinn, and R. O'Kennedy. 2005. Development of a surface plasmon resonance-based immunoassay for *Listeria monocytogenes*. J. Food Prot. 68:7287–7235.

Lin, M., D. Todoric, M. Mallory, B. S. Luo, E. Trottier, and H. Dan. 2006. Monoclonal antibodies binding to the cell surface of *Listeria monocytogenes* serotype 4b. J. Med. Microbiol. 55:291–299.

Loiseau, O., J. Cottin, R. Robert, G. Tronchin, C. Mahaza, and J. M. Senet. 1995. Development and characterization of monoclonal antibodies specific for the genus *Listeria*. FEMS Immunol. Med. Microbiol. 11:219–230.

Mattingly, J. A., B. T. Butman, M. C. Plank, R. J. Durham, and B. J. Robison. 1988. Rapid monoclonal antibody-based enzyme-linked immunosorbent assay for detection of *Listeria* in food products. J. Assoc. Off. Anal. Chem. 71:679–681.

Mead, P. S., E. F. Dunne, L. Graves, M. Wiedmann, M. Patrick, S. Hunter, E. Salehi, F. Mostashari, A. Craig, P. Mshar, T. Bannerman, B. D. Sauders, P. Hayes, W. Dewitt, P. Sparling, P. Griffin, D. Morse, L. Slutsker, B. Swaminathan, and *Listeria* Outbreak Working Group. 2006. Nationwide outbreak of listeriosis due to contaminated meat. Epidemiol. Infect. 134:744–751.

Mead, P. S., L. Slutsker, V. Dietz, L. F. McCaig, J. S. Bressee, C. Shapiro, P. M. Griffin, R. B. Tauxe. 1999. Food-related illness and death in the United States. Emerg. Infect. Dis. 5:607–634.

Miller, A. J., D. O. Bayles, and B. S. Eblen. 2000. Cold shock induction of thermal sensitivity in *Listeria monocytogenes*. Appl. Environ. Microbiol. 66:4345–4350.

Mitchell, B. A., J. A. Milbury, A. M. Brookins, and B. J. Jackson. 1994. Use of immuno-magnetic capture on beads to recover *Listeria* from environmental samples. J. Food Prot. 57:743–745.

MMWR. 2006. Preliminary FoodNet data on the incidence of infection with pathogens transmitted commonly through food—10 states, United States, 2005. MMWR 55(14):392–395. At www. cdc.gov/mmwr/preview/mmwrhtml/mm5514a2.htm. Accessed April 14, 2006.

Nanduri, V., A. K. Bhunia, S. I. Tu, G. C. Paoli, and J. D. Brewster. 2007. SPR biosensor for the detection of *L. monocytogenes* using phage-displayed antibody. Biosens Bioelectron. 23:248–252.

Nannapaneni, R., and M. G. Johnson. 1997. Piezoelectric- and surface plasmon resonance (SPR)-based biosensors for detecting foodborne pathogens. Pages 43–45 in NSF/FDA/USDA Workshop on Enhancing Food Safety Through the Use of Sensors. September 24, 1997, Washington, DC.

Nannapaneni, R., R. Story, A. K. Bhunia, and M. G. Johnson. 1998a. Unstable expression and thermal instability of a species-specific cell surface epitope associated with a 66-kilodalton antigen recognized by monoclonal antibody EM-7G1 within serotypes of *Listeria monocytogenes* grown in nonselective and selective broths. Appl. Environ. Microbiol. 64:3070–3074.

Nannapaneni, R., R. Story, A. K. Bhunia, and M. G. Johnson. 1998b. Reactivities of genus-specific monoclonal antibody EM-6E11 against *Listeria* species and serotypes of *Listeria monocytogenes* grown in nonselective and selective enrichment broth media. J. Food Prot. 61:1195–1198.

Nato, F., K. Reich, S. Lhopital, S. Rouyre, C. Geoffroy, J. C. Mazie, and P. Cossart. 1991. Production and characterization of neutralizing and nonneutralizing monoclonal antibodies against listeriolysin O. Infect. Immun. 59:4641–4646.

Patel, J. R., and L. R. Beuchat. 1995. Evaluation of enrichment broths for their ability to recover heat-injured *Listeria monocytogenes*. J. Appl. Bacteriol. 78:366–372.

Peel, M., W. Donachie, and A. Shaw. 1988a. Temperature-dependent expression of flagella of *Listeria monocytogenes* studied by electron microscopy, SDS-PAGE and western blotting. J. Gen. Microbiol. 134:2171–2178.

Peel, M., W. Donachie, and A. Shaw. 1988b. Physical and antigenic heterogeneity in the flagellins of *Listeria monocytogenes* and *Listeria ivanovii*. J. Gen. Microbiol. 134:2593–2598.

Pinner, R. W., A. Schuchat, B. Swaminathan, P. S. Hayes, K. A. Deaver, R. E. Weaver, B. D. Plikaytis, M. Reeves, C. V. Bonnme, and J. D. Wenger. 1992. Role of foods in sporadic listeriosis. II. JAMA 267:2046–2050.

Paoli, G. C., L. G. Kleina, and J. D. Brewster. 2007. Development of *Listeria monocytogenes*-specific immunomagnetic beads using a single-chain antibody fragment. Foodborne Pathog. Dis. 4:74–83.

Ruhland, G. J., M. Hellwig, G. Wanner, and F. Fiedler. 1993. Cell-surface location of *Listeria*-specific protein p60—detection of *Listeria* cells by indirect immunofluorescence. J. Gen. Microbiol. 139:609–616.

Schuchat, A., B. Swaminathan, and C. V. Broome. 1991. Epidemiology of human listeriosis. Clin. Microbiol. Rev. 4:169–183.

Siragusa, G. R., and M. G. Johnson. 1990. Monoclonal antibody specific for *Listeria monocytogenes*, *Listeria innocua*, and *Listeria welshimeri*. Appl. Environ. Microbiol. 56:1897–1904.

Skjerve, E., L. M. Rorvik, and O. Olsvik. 1990. Detection of *Listeria monocytogenes* in foods by immunomagnetic separation. Appl. Environ. Microbiol. 56:3478–3481.

Solve, M., J. Boel, and B. Norrung. 2000. Evaluation of a monoclonal antibody able to detect live *Listeria monocytogenes* and *Listeria innocua*. Int. J. Food Microbiol. 57:219–224.

Swaminathan, B. 2001. *Listeria monocytogenes*. Chapter 18, pages 383–409, in Food microbiology: fundamentals and frontiers. 2nd ed. American Society for Microbiology Press, Washington, DC.

Swaminathan, B., T. J. Barrett, S. B. Hunter, R. V. Tauxe, and CDC PulseNet Task Force. 2001. PulseNet: the molecular subtyping network for foodborne bacterial disease surveillance, United States. Emerg. Infect. Dis. 7:382–389.

Taormina, P. J., and L. R. Beuchat. 2001. Survival and heat resistance of *Listeria monocytogenes* after exposure to alkali and chlorine. Appl. Environ. Microbiol. 67:2555–2563.

Taylor, A. D., J. Ladd, Q. Yu, S. Chen, J. Homola, and S. Jiang. 2006. Quantitative and simultaneous detection of four foodborne bacterial pathogens with a multi-channel SPR sensor. Biosens. Bioelectron. 22:752–758.

Torensma, R., M. J. Visser, C. J. Aarsman, M. J. Poppelier, A. C. Fluit, and J. Verhoef. 1993. Monoclonal antibodies that react with live *Listeria* spp. Appl. Environ. Microbiol. 59:2713–2716.

Traub, W. H., and D. Bauer. 1995. Simplified purification of *Listeria monocytogenes* listeriolysin O and preliminary application in the enzyme-linked immunosorbent assay (ELISA). Zentralbl. Bakteriol. 283:29–42.

Tully, E., S. Hearty, P. Leonard, and R. O'Kennedy. 2006. The development of rapid fluorescence-based immunoassays, using quantum dot-labelled antibodies for the detection of *Listeria monocytogenes* cell surface proteins. Int. J. Biol. Macromol. 39:127–134.

Yu, K.Y., Y. Noh, M. Chung, H. J. Park, N. Lee, M. Youn, B. Y. Jung, and B. S. Youn. 2004. Use of monoclonal antibodies that recognize p60 for identification of *Listeria monocytogenes*. Clin. Diagn. Lab. Immunol. 11:446–451.

USDA-FSIS. 1998. Florida firm expands recall of franks for *Listeria*. At http://www.fsis. usda.gov/OA/recalls/prelease/pr035–98b.htm. Accessed July 7, 2009.

USDA-FSIS. 1999a. BIL MAR *Listeria* recall—Additional brands sold at retail. At http://www.fsis.usda.gov/OA/recalls/prelease/pr044–98a.htm. Accessed July 7, 2009.

USDA-FSIS. 1999b. Thorn Apple Valley frankfurters and lunch combination products recalled for potential *Listeria* contamination. At http://www.fsis.usda.gov/OA/ recalls/prelease/pr005–99.htm. Accessed July 7, 2009.

USDA-FSIS. 2001. Performance standards for the production of processed meat and poultry products; proposed rule. 9 CFR Parts 301, 303, et al. Volume 66, Number 39, 48p. At http://www.fsis.usda.gov/OPPDE/rdad/FRPubs/97–013P.pdf. Accessed February 27, 2001.

USDA-FSIS. 2002. Pennsylvania firm expands recall of turkey and chicken products for possible *Listeria* contamination. At http://www.fsis.usda.gov/OA/recalls/prelease/ pr090–2002.htm. Accessed July 7, 2009.

U.S.DHHS. 2007. Healthy people 2010 midcourse review. Focus area, chapter 10—food safety. At http://www.healthypeople.gov/data/midcourse/pdf/fa10.pdf [online]. Accessed March 30, 2008.

Wang, H., Y. Li, and M. Slavik. 2007. Rapid detection of *Listeria monocytogenes* using quantum dots and nanobeads based optical biosensor. J. Rapid Methods Automation Microbiol. 15:67–76.

▋ 15 ▋

The Potential for Application of Foodborne *Salmonella* Gene Expression Profiling Assays in Postharvest Poultry Processing

Sujata A. Sirsat, Arunachalam Muthaiyan, Scot E. Dowd,
Young Min Kwon, and Steven C. Ricke

Introduction

Pathogenic contamination of foods is a threat to human health and has the potential for causing fatalities. Foodborne pathogens have been estimated to cause approximately 76 million illnesses, 325,000 hospitalizations, and 5,000 deaths in the United States each year (Mead et al. 1999). Based on reports by the Food Safety and Inspection Services, *Salmonella* is the most common cause of foodborne enteric illness (USDA-FSIS, 1996) and has been estimated to cause 26% of hospitalizations and more than 30% of deaths related to foodborne pathogens (Mead et al. 1999). The World Health Organization (WHO) has reported that salmonellosis is reemerging as an important infectious disease worldwide (Nakaya et al. 2003).

Transmission of *Salmonella* in humans is via the oral-fecal route and is usually due to consumption of raw or undercooked food products (Darwin and Miller 1999). Several food items including peanut butter, tomatoes, grain puffs, milk, ground beef, eggs, and poultry have been implicated in salmonellosis outbreaks. Salmonellosis is most often associated with foods of animal origin such as poultry or poultry products (Bryan and Doyle 1995). According to the USDA-FSIS reports, *Salmonella* contamination due to fecal contamination of carcasses is a major issue for the poultry industry. Hence, the USDA-FSIS compliance guide asserts the need for multistage intervention strategies for poultry carcasses that test positive for *Salmonella*.

Salmonella enterica serovar Typhimurium is a common cause of gastroenteritis, a self-limiting disease marked by diarrhea and abdominal cramps (Miller and Pegues 2000). Fever is the common symptom of *Salmonella* infection. However, in rare cases infections have the potential to be severe, resulting in fatalities. For instance, the peanut butter *Salmonella* outbreak was responsible for at least four deaths according to press reports (Enoch 2007). In 2005, a *Salmonella* outbreak that caused at least one death took place in a restaurant in South Carolina. The cause was reported to be undercooked turkey (Jordan 2005). Recently, reports have addressed the possibility of complications several years after the victims recover from foodborne diseases. For instance, *E. coli* victims may suffer from kidney failure; *Salmonella* and *Shigella* victims may complain of arthritis; and patients who underwent a mild *Campylobacter* infection may develop paralysis (Neergaard 2008).

In order to limit contamination on raw poultry products, processing plants are expected to have control systems to ensure that birds are processed by methods that minimize contamination as much as possible. This involves minimizing any external contamination that may come in contact with the carcass, along with the application of various biological, chemical, and physical treatments. Biological treatments can consist of bacteriophages and bacteriocins (Campbell 2003; Joerger 2003; Weld et al. 2004). Examples of common chemical treatments used in the processing industry include organic acids, cetylpyridinium chloride, trisodium phosphate, chlorine, salts, and spices (Ricke et al. 2005; Ricke 2003). Organic acids at various concentrations and temperatures, steam, and hot water rinses are most commonly used in the poultry industry to decontaminate the carcass (Castillo et al. 1998; Ellebracht et al. 1999; Ricke et al. 2005; Ricke 2003). Physical intervention treatments involve the use of heat, cold, radiation, and newer techniques such as high hydrostatic pressure (HPP), ultrasonication, and electroporation (Farkas 2007; Jay 2000). These intervention methods pose physical obstacles to microorganisms that may be present on the surface of the carcass. Studies have shown that when two intervention steps are applied to an organism, the application of the first intervention treatment may cause the surviving pathogens to be resistant to the second intervention (Kwon and Ricke 1998; Kwon et al. 2000; Leyer and Johnson 1993). Hence the first treatment may elicit protective mechanisms in the pathogen so that it may survive further stressors.

This review focuses on a concept to potentially use microarrays for gene expression profiling of *Salmonella* exposed to a particular antimicrobial treatment or hurdle to identify the best possible combinations of antimicrobial treatments and the sequence in which they may be applied. This in turn would result in combinations that could be synergistic and most efficient in limiting contamination in poultry processing. The next section discusses *Salmonella* genetics and the effect of environmental factors on the transcriptome of the pathogen.

Genetics of *Salmonella* Pathogenesis

An increase in gastric pH leads to a decrease in dose of *Salmonella* required to cause an infection (Giannella et al. 1973). However, when exposed to low pH, *Salmonella* exhibits an acid-tolerance response, which may explain why it can survive extreme acidic conditions of the stomach (Garcia-del Portillo et al. 1993; Kwon and Ricke 1998; Kwon et al. 2000). *Salmonella* expresses several fimbriae that are responsible for adherence of the pathogen to the intestinal epithelial cells (Bäumler et al. 1996). The ability of *Salmonella* to attach and penetrate the intestinal epithelia of the host is crucial for its pathogenesis. Once *Salmonella* adheres to the host epithelium, its subsequent invasion is followed by cytoskeletal rearrangements in the host cell leading to disruption and membrane ruffling of the host cells (Francis et al. 1992; 1993). It is subsequently internalized by a process called pinocytosis (Chen et al. 1996; Galán 1996). Once *Salmonella* invades the host epithelium it encounters the macrophages of the host. This leads to activation of virulence mechanisms of the pathogen in order to survive and replicate (Alpuche-Aranda et al. 1994). It is believed that *Salmonella* virulence has evolved as a result of horizontal gene transfer (Bäumler et al. 1998; Bäumler 1997). This concept is supported by the fact that large numbers of virulence genes are clustered within the chromosome. The five *Salmonella* pathogenicity islands (SPI) identified in *Salmonella* are located in various regions on the chromosomes containing sets of virulence genes (Marcus et al. 2000) and are listed in Table 15.1.

Invasion is believed to be controlled by genetic and environmental regulators. SPI-1 encodes a type III secretion system (TTSS), which is essential for the process of cell invasion. The protein forms a secretory "needle complex" that spans the inner and outer bacterial membrane (Kubori et al. 1998). Control of invasion genes leads to the formation of the type III secretion apparatus at the point of infection (Altier 2005). This secretion system is used by microorganisms to translocate virulence-associated effector proteins into the cytoplasm of the host cells resulting in a cross-talk that leads to downstream responses such as membrane ruffling and bacterial internalization (Suárez and Rüssmann 1998; Wood et al. 1998). SPI-1 specifically codes for transcriptional regulators such as *hilA* (Bajaj et al. 1995), *hilC* (Johnston et al. 1996), and *hilD* (Schechter et al. 1999). Environ mental signals such as high pH (Bajaj et al. 1996), low osmolarity (Galán and Curtiss 1990), and low oxygen (Jones et al. 1994) are believed to increase invasion of *Salmonella* in the host. This has been proven by testing the effects of these factors on *Salmonella* strains containing *lacZ* fusions in the invasion genes and

Table 15.1. Pathogenicity Islands of *Salmonella*

Salmonella Pathogenicity Island (SPI)	Function	Size	Key genes	References
SPI-1	Invasion, cross-talk, host membrane ruffling	40 kb	*hilA, hilC, hilD, invF, sspC*	Collazo and Galán 1997
SPI-2	Intracellular replication	40 kb	*ssa, ssr*	Cirillo et al. 1998; Hensel et al. 1998
SPI-3	Intramacrophage survival, virulence in mice	17 kb	*mgtB, mgtC*	Blanc-Potard and Groisman 1997
SPI-4	TTSS mediates toxin production	25 kb	*ssb, soxSR*	Wong et al. 1998
SPI-5	Enteropathogenesis phenotype	9 kb	*pipA, pipD, sopB*	Wood et al. 1998

performing betagalactosidase assays (Bajaj et al. 1995, 1996). The *hilA* gene has been shown to be required for the expression of three other invasion genes: *invF, sspC,* and *orgA* (Bajaj et al. 1995). The gene *invF* is a transcriptional regulator (Eichelberg and Galán 1999); *sspC* codes for an invasion protein (Hueck et al. 1995); and *orgA* product is a component of an export machinery system (Galán 1996). PhoPQ is a regulatory two-component system, which is not contained within SPI-1 but is crucial for invasion of *Salmonella* (Behlau and Miller 1993; Pegues et al. 1995). This two-component system responds to extracellular cation levels (Garcia Vescovi et al. 1996). In conditions of low cation concentration, sensor kinase PhoQ phosphohorylates the regulator PhoP, which activates *pag* transcription. Induction and expression of *pag* genes are required for the survival of the bacteria in the macrophage (Alpuche-Aranda et al. 1994). Hence, gene expression of *Salmonella* is a complex system and environmental impacts including postprocessing intervention treatments used in the food industry may be potential stressors, which could trigger *Salmonella* virulence gene expression.

Post-Processing Hurdle Methods

Most antimicrobial treatments have been used for 50 to 100 years in the food industry (Davidson and Harrison 2002). However, it is only more recently that concern regarding resistance to these treatments has been raised. This concern is because microorganisms previously exposed to different stressors have been shown to exhibit resistance to antimicrobials and sanitizer treatments (Davidson and

Harrison 2002; Davis et al. 2005). These treatments are used on the food product to inactivate or inhibit growth of pathogenic bacteria. For instance, organic acids are used on beef carcass, nisin and lysozyme are used to inhibit *Clostridium botulinum* in cheese, and nitrite is used to prevent growth of *C. botulinum* in cured meats (Davidson and Harrison 2002). These hurdle techniques may be classified as biological, physical, and chemical hurdles (Ricke et al. 2005) and examples are listed in Tables 15.2 and 15.3.

The treatments listed in Tables 15.2 and 15.3 are typically used in combination with one another or in combination with other inhibitory agents. The combined use of different antimicrobial treatments is usually referred to as "multiple

Table 15.2. Mode of Action of the Most Commonly Used Chemical Hurdles in Food Preservation

Chemical hurdle	Mode of Action	References
Chlorine (used in combined form such as calcium chloride, chlorine dioxide, cetylpyridium chloride (CPC), acidified sodium chlorite (ACS), among others)	Targets amino groups in proteins, inhibits metabolism of the organism	Denyer and Stewart 1998; Denyer 1995; Jay 2000; Kemp et al. 2000
Nitrite	Inhibits microorganisms, prevents synthesis of ATP from pyruvate	Woods et al. 1981
Organic acids (Lactic, acetic, propionic, etc.)	Undissociated molecule enter cell membranes, disturb the PMF, decrease in pH, affects amino acid transportation, inhibits metabolism	Brul and Coote 1999; Ronning and Frank 1987
Salts	Dries microorganism, growth inhibition, and death	Jay 2000
Trisodium phosphate (TSP)	Chelates divalent cations in the outer membrane, leads to increased permeability and cell death	Sampathkumar et al. 2003

Table 15.3. Mode of Action of the Most Commonly Used Physical Hurdles in Food Preservation

Physical hurdle	Mode of Action	References
Cold	Causes changes in pH, denatures cellular proteins, metabolic injury, leads to sudden or slow death	Jay 2000
Heat	Affects cell membrane proteins and lipids, leads to cell damage and death	Jay 2000
High hydrostatic pressure (HPP)	Pressure leads to solubilization and leakage of key ions from membrane, leads to cell damage and death	Farkas 2007; San Martin et al. 2002
Radiation	Wavelength absorbed by proteins and nucleic acids, leads to cell death	Jay 2000

interventions" or "hurdle technology" (Leistner and Gorris 1995). Hurdle technology involves combining food-preservation methods in order to create "hurdles" for a foodborne pathogen, limit its survival, and improve the stability and quality of food. Employing a combination of hurdles can also result in synergistic activity among hurdles enabling the use of lower concentrations of each of these treatments in order to decontaminate the carcass and fresh produce.

The most common hurdles used for food preservation are temperature (high or low), redox potential, salts, lactic acid bacteria (LAB), acids, and water activity. These hurdles act as food preservatives in order to inhibit growth, decrease survival, disturb homeostasis, and even cause microbial death (Leistner 2000). However, synergistic antimicrobial activity does not always occur when multiple intervention steps are applied either simultaneously or in a sequential manner if these pathogens are inherently resistant to a given treatment or if the intervention treatment targets the same genetic pathway. Although the hurdles target different genetic pathways to exert antimicrobial activities, the exposure to one hurdle may confer resistance to the second hurdle. In this possible scenario, applying multiple hurdles not only fails to cause a synergistic effect but also can produce a bacterial subpopulation that has enhanced resistance to general stressors. This phenomenon, termed crossprotection, may be defined as the ability of one stressor to provide protection to the pathogen against other stressors (Bearson et al. 1997). For instance, it has been demonstrated that acid-adapted bacteria are resistant to stressors such as heat, salt, crystal violet, and polymyxin B (Leyer and Johnson 1993). In similar studies, Kwon and Ricke (1998) and Kwon et al. (2000) reported that *S.* Typhimurium, which had adapted to short-chain fatty acids at neutral pH also became more resistant to extreme conditions such as high pH, high osmolarity, and reactive oxygen. Singh et al. (2006) examined the effect of thermal treatment on a cocktail of acid-adapted *Salmonella* and *E. coli* O157:H7. Acid-adapted and nonacid adapted cultures were inoculated on the irradiated ground beef and exposed to 62°C and 65°C for 10 minutes. It was noted that acid-adapted microorganisms were more resistant to thermal treatments when compared to nonacid resistant bacteria. This study shows that application of the first treatment may render the surviving pathogen more resistant to the next treatment. These studies demonstrate that hurdles, which may seem mechanistically independent at the cellular level, may render the pathogen resistant to other hurdles. Hence, it is important to understand the genetic pathways of *Salmonella* affected by each hurdle on a genome-wide scale to select not only the best possible combination of treatments, but also the sequence in which such hurdles should be applied in order to better ensure elimination of pathogenic microorganisms from food.

Quantifying Gene Expression Patterns: Concepts and Methods

The transcriptome refers to all of the genes that are expressed by a given organism. Transcriptomics can in turn be defined as the study of gene expression. However, only a fraction of genes in the bacterial chromosome are turned on and expressed at any given point in time (Jacob et al. 2005). The collection of mRNA that is transcribed from the genes at a particular time is often referred to as the gene expression profile. The transcriptome, unlike the genome, is extremely dynamic and changes rapidly in response to the environmental conditions (Lockhart and Winzeler 2000). Gene expression analysis or transcriptional profiling is essentially quantifying the abundance of gene-specific mRNA or transcriptomes of an organism, respectively, under specific conditions at a specified time. The analysis of gene expression provides evidence on which regulatory pathways and biochemical mechanisms of the organism are active at any given time, and different gene expression quantification methods are shown in Figure 15.1. Techniques such as real-time quantitative PCR that use mRNA as starting material are applicable in the food industry as they can differentiate between the viable and nonviable cells since mRNA can only be extracted from viable cells for identification of the pathogen (Vaitilingom et al. 1998).

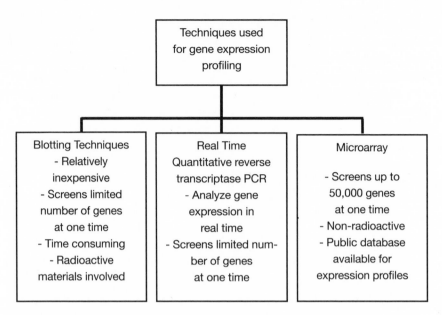

Figure 15.1. Techniques used for gene expression profiling.

Blotting Techniques: Northern Blot

The fundamental principle of the Northern blot technique is that RNA is separated on a agarose gel on the basis of its size, blotted on a membrane by capillary action, hybridized with a complementary probe, and quantified by various detection techniques such as radioactivity or fluorescence (Sambrook and Russell 2001; Trayhurn 1996). Northern blots have been used in several studies to analyze the gene expression of the *S.* Typhimurium genome to quantify gene responses and characterize the physiological state of the cell. Piddock et al. (2000) tested for multiple-drug-resistant *Salmonella* in patients treated with ciprofloxacin to study the role of efflux in the resistance phenotype by analyzing and quantifying seven genes. Holmstrøm et al. (1999) tested the levels of *groEL* and *tsf* mRNAs per cell of *S.* Typhimurium in various culture conditions such as fast, slow, or no growth, and heat shock. They observed that number of *groEL* mRNA was maximized in heat shock cultures and *tsf* mRNA numbers were at their highest in fast-growing cultures. The technique has also been used to test the effect of mutations in one gene on the expression levels of other genes. After disrupting the flagellin *fliC* gene of *S.* Enteritidis, Van Asten et al. (2000) determined the effect of this mutation on the gene expression level of the downstream *fliU* gene, which is one of the genes responsible for secretion of flagellin protein and expression of motility. Similarly, Wood et al. (1998) assessed the effect of SPI-5 mutations on the gene expression level of SopB, a novel secreted effector protein. An independent study was performed to test the expression of pathogenicity-related protein SEp22 from *S.* Enteritidis. The investigators studied the expression of the gene in various stages of *Salmonella* growth using Northern blot analysis and concluded that expression of SEp22 decreases after the culture enters stationary phase (Terai et al. 2005).

In light of these applications it should be noted that Northern blots can analyze only a limited number of genes simultaneously and may take up to 24 hours to complete. Hence, it is extremely labor intensive and time consuming. In addition, the application of this technique in the food industry is probably limited as a large amount of mRNA is required. This implies that there needs to be high level of pathogenic contamination in order to facilitate effective extraction of the nucleic acid. Also, mRNA is highly unstable and requires special facilities and handling skills. Depending on the extent of bacterial contamination, it may be challenging to apply this technique to a complex food matrix as it may be difficult to isolate the microorganisms in sufficient numbers to carry out further analysis. This would in turn lead to problems in isolating detectable quantity of mRNA required for Northern blot hybridization.

Real-Time Quantitative Reverse Transcription PCR (qRT-PCR)

Polymerase chain reaction assays essentially involve logarithmic amplification of short DNA sequences within a longer double stranded DNA molecule (Mullis 1990). Real-time PCR is a variation in this conventional PCR methodology, and its advantage is the ability to detect a target fluorescent signal as amplification takes place, which means that the amplification process is monitored in "real time" (Wawrik et al. 2002). Hence, any problems with the process can be detected easily and quantification of mRNA copy numbers tends to be very accurate. The first step to analyze gene expression by this method is converting mRNA to cDNA using reverse transcriptase to synthesize a double-stranded DNA molecule followed by monitoring of gene amplification in real time with appropriate gene-specific primers and/or probes (Hanna et al. 2005; Wawrik et al. 2002). The principle of qRT-PCR is based on measuring the accumulation of product during the exponential phase of the PCR cycles which can be observed on an amplification plot (Giulietti et al. 2001); qRT-PCR leaves behind the more laborious and time-consuming agarose gel electrophoresis used for detection. Instead, using various detection techniques described in the following sections, qRT-PCR methodology involves detection of fluorescence accumulation at the end of every cycle via target amplification. The cycle at which threshold fluorescence reaches a critical detection limit is called C_T, which is inversely proportional to the amount of target present in the sample at any given time (Giulietti et al. 2001).

There are three popular technologies used for qRT-PCR. SYBR Green, arguably the most popular and least expensive method, does not require a target-specific fluorescent dye attached probe for the reaction. SYBR Green uses a non-specific nucleic acid dye that intercalates DNA and hence its fluorescence increases along with the amplification product (Zipper et al. 2004). However, this dye can also bind to any double-stranded DNA molecule that may include primer dimers or any nonspecific PCR product leading to false-positive signals. Detection of DNA amplification is made more reliable by using either molecular beacons or TaqMan assays. Molecular beacons are single stranded DNA molecules that may be 25 to 35 nucleotides in length and are designed such that the 3-foot and 5-foot ends are complementary to each other and form a hairpin loop structure. The 3-foot end contains a quencher and the 5-foot end contains a fluorescent dye. Hence, when the hairpin loop structure is formed, no signal is detected. However, when the beacon binds the target DNA in single-strand form, the quencher and fluorophore are physically separated and the signal may be detected (Hanna et al. 2005).

A third development that improved quantitative detection using real-time PCR is the TaqMan probe. The TaqMan assay involves the use of primers and a fluorophore attached single-stranded probe (also called a double-dye oligonucleotide), which is specific to one of the strands of amplicon. The fluorescent dye is present on the 5-foot end and a quenching dye on the 3-foot end of the TaqMan probe. Once the probe is bound to the target DNA, 5-foot exonuclease activity cleaves the probe and frees the 5-foot reporter dye so the fluorescence signal can be recorded (Smith et al. 2001). This principle makes the TaqMan probe more specific compared to the SYBR Green detection component.

Dunkley et al. (2007) measured *Salmonella* Enteritidis *hilA* expression in fecal and cecal contents of hens that were fed different diets and concluded that changes in the gastrointestinal environment induced *hilA* that mirrored colonization and organ infection by this organism. Bader et al. (2003) demonstrated induction of *S.* Typhimurium PhoP/PhoQ and RpoS regulons and repression of genes related to flagella synthesis and invasion after exposure to antimicrobial peptides. They also demonstrated that growth of *Salmonella* in low concentrations of the peptides leads to development of resistance mechanisms to them that are dependent on the PhoP/PhoQ regulon.

Although qRT-PCR has been widely used to quantify and detect viable pathogens in foods, problems often associated with this technique are inconsistent data and the lack of strong supporting *in vivo* studies. A combination of different primers, probes, and detection methods makes this technique less reproducible from lab to lab (Bustin 2002). Another problem associated with qRT-PCR is choosing a reliable gene that always has the same expression level as the controls (Klein 2002). Fey et al. (2004) developed a quantitative PCR method to quantify not only the number of bacterial cells but also the number of RNA copies. They tested gene expression levels of *invA* and 16S rRNA genes in *Salmonella* cultures and water samples inoculated with *Salmonella* using qRT-PCR and concluded that both these approaches led to successful, sensitive, and accurate quantification of microorganisms and target RNA molecules. When Kundinger et al. (2007) simulated starvation conditions by incubating *S.* Typhimurium in spent media, they confirmed with qRT-PCR that the *rsmC* gene was a stable baseline gene and could be used successfully as a reference gene.

In addition, PCR is often associated with nonspecific amplification problems. Often, due to low reaction temperatures, primers may anneal to nonspecific regions on the bacterial genome, resulting in unwanted PCR products (Bustin 2002). For instance, Rahn et al. (1992) reported nonspecific amplification of non-*Salmonella* genes while using PCR technology to amplify the *invA* gene of *Sal-monella*. They confirmed the specificity of amplification by hybridizing the PCR products to radio-labeled *invA* gene fragments. Therefore, the amplification of specific PCR products requires careful design of primers. Ideally a primer should be designed with GC content of 40–60% and a melting temperature of 52–58°C. Finally, the primer sequences must be designed so they are not complementary and hence do not self anneal, which can lead to the formation of primer dimers (Burpo 2001).

Microarrays

Microarrays are used for various applications on the gene and protein level. The flow chart in Figure 15.2 shows the various types of microarrays used in transcriptional and translational research. A microarray usually consists of a glass slide on which probes (DNA or protein spots) representing the organism are spotted using methods such as high-speed robotics or photolithography. These robotics are made up of a computer-controlled three-axis robot and a pen-tip assembly that can spot the probes on the glass slides in a precise manner (Cheung et al. 1999). Labeled target DNA or protein molecules are subsequently allowed to hybridize with these probes leading to quantification of gene or protein expression levels (Ramsay 1998). After a brief description of proteomic arrays, the next sections will mainly focus on DNA transcriptional arrays.

A protein microarray consists of different affinity reagents such as antibodies spotted on a solid surface. These antibodies are designed to capture their target

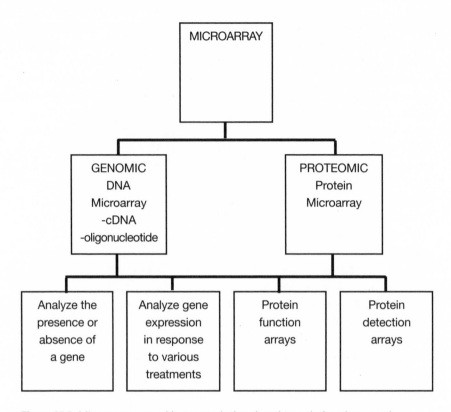

Figure 15.2. Microarrays used in transcriptional and translational research.

proteins from a complex mixture of proteins. This captured protein may then be detected and/or quantified (MacBeath 2002). There are two types of protein arrays: protein function arrays and protein detection arrays. Protein function arrays are constructed by obtaining protein samples from cDNA by an *in vitro* transcription/translation technique and spotting on a slide. Two different markers are used to label the test and control proteins, which are then allowed to bind to the capture proteins (e.g., antibodies) on the array. Protein function arrays allow the detection of a physiological effect on a cell and hence the function of the protein (Kodadek 2001). The protein detection array is made of arrayed protein-binding agents and allows for protein expression profiling (Kodadek 2001).

DNA microarrays provide global scale analysis at the level of transcription and enable gene expression to be examined in response to specific conditions where some genes may be upregulated and others downregulated (Goldsmith and Dhanasekaran 2004). This in turn provides a "gene expression profile" or "transcriptome" for the organism under specific conditions. This type of global analysis may also lead to the identification of novel genes that may play a crucial role in the physiology of the organism under a particular environmental condition.

The approach of DNA microarrays are essentially the opposite of that used in Northern blot technology (Schulze and Downward 2001). While Northern blots are based on immobilizing the target on a nitrocellulose paper and exposing it to the probe, microarrays are based on the principle of immobilizing oligonucleotide or PCR probes on a glass surface and exposing this to the target mRNA, cDNA, or DNA sample.

DNA microarrays also have other applications involving the identification and genotyping of mutations and polymorphisms. A sequence variation in the gene sequence will alter the hybridization pattern that enables the identification of mutations in the genome (Lander 1999). Additionally, DNA microarrays are also applied for the purposes of drug discovery and development. This is done by monitoring the change in gene expression levels in response to various drugs (Debouck and Goodfellow 1999; Muthaiyan et al. 2008).

A microarray detects the abundance or presence of fluorescently labeled nucleic acids in a given sample. Microarrays also have advantages when compared to qRT-PCR because the latter allows only a few genes to be characterized at one time while microarrays enable the analysis of tens of thousands of genes on a single glass slide at one time (Lucchini et al. 2001). An individual DNA microarray chip may contain from a few hundred to several thousand nucleic acid probes. The use of microarrays is especially advantageous when the question regarding effect on gene expression needs to be asked on a global scale.

DNA Microarrays for Transcriptional Profiling of Microbial Responses

DNA microarray technologies are of two types—cDNA arrays and oligonucleotide arrays. DNA microarray technologies involve array fabrication, experimental methodology, and data analysis. Microarrays are created by spotting PCR products or oligonucleotides to a glass slide or created by *in situ* synthesis of specific oligo nucleic acid sequence corresponding to each gene in the genome using photolithography (for example, Affymetrix high-density oligo arrays) (Lockhart et al. 1996; Schena et al. 1995). Before printing the spots on the glass slides, the slides are chemically treated to improve the efficiency of the procedure. UV crosslinking is performed so that the spots are immobilized on the glass slide (Nguyen et al. 2002). *In situ* synthesis of sequences has an advantage over the spotted microarrays as the sequences may be generated from the sequence database directly, eliminating the need to produce PCR products corresponding to thousands of genes. It is crucial to incorporate negative control DNA that is known to be absent in the test strain and hence not hybridize with the target DNA sample (Liu-Stratton et al.

2004). The spots on the microarray may be described as capturing probes that react with the labeled nucleic acid in the samples (Bedná 2000).

The experimental use of a typical transcriptional profiling study is briefly explained in Figure 15.3. For any gene expression profile analyses it is crucial to use a control where the microorganism is not exposed to treatment conditions (Conway and Schoolnik 2003; Cummings and Relman 2000). Following exposure to the specific conditions, mRNA from the test and control samples can be extracted and stabilized using commercial reagents to prevent degradation of mRNA and confirm that no factors other than the defined conditions under study influence the genomic expression of the pathogen (Ambion Inc. 2007). The mRNA is subsequently subjected to reverse transcription to synthesize labeled cDNA probes with different fluorescecent dyes for each control and test sample. Fluorescent dyes of the cyanine dye family Cy3 and Cy5 that attach with an aminoacyl moiety of modified nucleotide are used for visual and quantification purposes (Ernst et al. 1989). Most commonly the differentially labeled control and test samples are cohybridized to measure the relative abundance of sample mRNAs as compared to that of the control mRNA. This approach, called dual channel microarrays, involves mixing the control and test samples labeled with different fluorescent dyes and hybridizing the mixture on a single microarray (Xiang et al. 2003). One method to perform comparative microarray analysis would be to use one slide as a control slide and the other as a test slide and hybridize with cDNA labeled with different dyes. This is often referred to as single-channel microarrays (Butte 2002; Li et al. 2002). Following hybridization, emission of two different wavelengths of the cyanine dyes are quantified with the resulting test/control sample's fluorescence ratios calculated with a specific software program. Roughly, if the test sample is labeled with red fluorescence and the control is labeled with green, a yellow signal on a particular spot on the chip means that the gene is not affected by the treatment. A red signal signifies the gene is upregulated and a green signal signifies downregulation (Bedná 2000).

In the data analysis step, microarrays must be normalized to eliminate variation that may be introduced during preparation of the sample or spotting the sample on the array. Hence, this procedure is crucial for further steps of microarray experiments and data analysis. The most common normalization procedures used include total intensity, regression, and ratio statistics (Dopazo et al. 2001; Eisen et al. 1998; Tavazoie et al. 1999). Total intensity normalization assumes that both the test and the control contain the same amount of RNA and hence the total RNA hybridized on the slide from each sample is the same (Quackenbush 2001). The regression method is based on the assumption that a significant number of genes are expressed to the same extent in both samples. These genes usually cluster

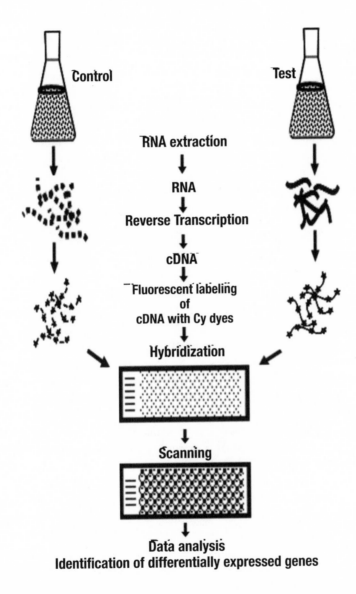

Figure 15.3. Diagrammatic representation of genes affected by three different hurdles. Each of the circles represents genes of the pathogen affected after treatment with either hurdle A, B, or C.

together in a scatter plot and best-fit slope is calculated using regression analysis (Saeed et al. 2003). Lastly, the ratio statistics normalization method assumes that there exists some subset of genes with the same expression levels in both samples and hence calculate confidence limits in order to identify differentially expressed genes (Quackenbush 2001).

Although microarrays can be used to screen thousands of genes in a single experiment, it is crucial that resulting array data be supported by other molecular technologies and physiological analysis (Stekel 2003). Techniques such as qRT-PCR or Northern blotting must be considered to provide independent experimental evidence to compare with the microarray responses. In addition, microarray experiments should be performed as replicates in order to maximize statistical significance and improve analysis quality (Conway and Schoolnik 2003).

Applications of Microarray Technology for Foodborne *Salmonella*

Detection and Phylogenetic Relationship

Goldschmidt (2006) has previously reviewed biosensor, microarray, and nanotechnology methods for detection of *Salmonella* in various systems, as well as technologies which have not yet been applied for detection of *Salmonella* serovars. DNA microarrays have proven to be a comparatively rapid technique to detect foodborne pathogens when compared to traditional culture-based methods involving pre - enrichment and selective isolation. A DNA microarray overcomes the challenge of analyzing these pathogens on complex food matrices by development of a molecular signature that is specific to certain forms of contamination (Liu-Stratton et al. 2004). Taitt et al. (2004) developed a multianalyte array biosensor (MAAB), which is a sandwich immunoassay to perform a rapid analysis for detection of *S*. Typhimurium in spiked foods. This approach possessed advantages over molecular techniques such as PCR since the need for enrichment processes and genomic analysis were eliminated.

However, microarray experiments are not just limited to detection of foodborne *Salmonella*. Some studies have employed DNA microarrays to analyze the presence or absence of a gene. Chan et al. (2003) used DNA arrays to study evolution and phylogenetic relationships between various serovars of *Salmonella*. When they used an *S*. Typhimurium DNA microarray to perform hybridizations on the genetic organization of *S. enterica* (subspecies I and IIIa) and *S. bongori*, they observed variability in the Arizona SPI-2 as compared to other *Salmonella*, indicating that it had evolved differently. They also found shared genetic features between *S. enterica* serovars Typhi, Paratyphi A, and Sendai that contributed to their ability to cause enteric fever in humans. In an independent study, Porwollik et al. (2002) constructed a microarray containing ORFs that represented 97% of the *S*. Typhimurium genome and were able to predict which genes were obtained at various stages of *Salmonella* evolution. This type of microarray allowed analyses of gain, loss, or divergence of genes during evolution. In similar studies, Reen et al. (2005) examined genomic variation between 12 environmental, veterinary, and clinical *Salmonella enterica* serovar Dublin, Agona, and Typhimurium strains isolated in Ireland between 2000 and 2003, along with two clinical isolates from Canada and four archival isolates, which belonged to serovars Dublin and Agona using DNA microarray methodology. Boyd et al. (2003) used a microarray of *S.* Typhimurium LT2 and Typhi CT18 to assess the genomic content of a diverse set of isolates of serovar Typhi. Using comparative genomic hybridization they observed that there were several gene content differences and concluded that the genomic reservoir is extremely unstable in spite of the fact that the bacterial population is clonal.

Alvarez et al. (2003) identified 32 different *Salmonella* serovars from raw plant-based feed in Spanish animal feed mills. They observed that *Salmonella enterica* serovar California was the most prominent serovar and used microarrays for comparative genomic hybridization of this serovar. Pelludat et al. (2005) developed a microarray with 83 probes for the surveillance and typing of epidemic *Salmonella* strains that would yield reproducible data for routine applications.

Gene Expression Quantification

If the functions of these genes are known, gene expression analysis has the potential to indicate the physiological state of the pathogen (Conway and Schoolnik 2003). By utilizing microarrays, genes can be screened that may not be necessarily directly linked to virulence but may affect upstream or downstream virulence gene expression. De Keersmaecker et al. (2005) analyzed the gene expression profile of *S.* Typhimurium in spent culture supernatant of a probiotic *Lactobacillus* strain. Their investigation revealed downregulation of a cluster of genes that were *hilA* regulated. These genes included *invF, prgH,* and *sicA,* among others. In addition, they performed site-directed mutagenesis, beta-galactosidase assays, and gel mobility shift experiments to further confirm the fact that inclusion of probiotics leads to decreased invasion efficiency of *Salmonella*. Similarly, expression analysis was used by Clements et al. (2002) to test the effect of polynucleotide phosphorylase (PNPase) on bacterial invasion and intracellular replication. They used microarray analysis and concluded that PNPase affects gene expression of SPI-1 and SPI-2. This study demonstrated that there is a relationship between PNPase and *Salmonella* virulence.

Huang et al. (2007) used microarrays to test the effect of hyperosmotic conditions on *Salmonella* generating gene expression profiles at 0, 30, and 120 minutes after exposure of the bacterium to hyperosmotic conditions. They concluded that invasion increases only after *Salmonella* is exposed to high salt conditions for 120 minutes and regulatory factors such as PhoP and sigma factors enable bacteria to adapt to hyperosmotic conditions. When examining the effect of bile on transcription and protein synthesis in *S.* Typhimurium, it was observed that bile repressed key invasion, flagellar, and motility genes of *Salmonella* (Prouty et al. 2004). In addition, they observed that the *marRAB* operon, which is involved in multiple antibiotic resistances, and an efflux pump, *acrAB,* is activated in the presence of bile. They concluded that bile was an environmental signal that regulated virulence gene expression in *Salmonella.*

Dowd et al. (2007) performed gene expression analysis using microarrays to analyze the effect of antibiotics on *S.* Typhimurium. When exposed to nalidixic acid, there were differential regulation of SPI-1 and -2 and induction of multidrug-resistance efflux pumps and outer membrane lipoproteins. They concluded that an export of antimicrobials from cells and limited diffusion of nalidixic acid into the cell occurred. They also noted that this antibiotic induced a transcriptional positive adaptive state (error-prone DNA repair was induced) that made *Salmonella* more likely to develop long-term resistance to this antibiotic. Gantois et al. (2006) compared gene expression microarray analysis of *S.* Enteritidis and *S.* Typhimurium in response to butyric acid by growing bacterial cultures in LB media supplemented with 10 mM butyrate and control cultures in LB media without butyrate. They observed that both the serovars exhibited downregulation of SPI-1 genes including *hilA* and *hilD* following exposure to butyric acid. Lawhon et al. (2003) investigated the role of *csrA,* which regulates invasion genes in *Salmonella,* by comparing gene expression of a *csrA* mutant with a wild-type *Salmonella.* They grew the *Salmonella* cultures in LB broth buffered to pH 8.0 with 100 mM HEPES. They found reduced gene expression levels of SPI-1 genes and flagellar genes in the *csrA* mutant. In similar studies, Monsieurs et al. (2005) compared gene expression of mutant *Salmonella* with knocked-out genes encoding the PhoPQ system with wild-type *Salmonella.* From the data it was seen that a mutation in these genes affects 2,855 other genes in *Salmonella.* Sirsat et al. (2007) tested the effect of hot water at 42°C and 48°C on *S.* Typhimurium ATCC 14028 and analyzed the gene expression using a 1152 virulence genes-targeted oligonucleotide microarray. They observed repression of several invasion genes including *hilA* that are located on SPI-1. However, they also found that genes on SPI-2 and SPI-5, which are responsible for bacterial survival and replication within the host, were induced. In addition, several fimbriae genes were upregulated in response to the hot-water treatments.

Conclusions

In this review we have described various hurdles that are applied on food products in the food industry as decontaminants. These hurdles may be applied in combination or alone. Studies have shown that if the hurdles are mechanistically independent, then combination treatments are more effective compared with single-hurdle applications. Hurdles, however, are stressors to pathogens on the food surface and may render surviving pathogens more virulent and/or resistant to other hurdles. It is crucial to apply the most effective combinations and sequence of hurdles to decontaminate the food surface in a way that they do not lead to crossprotection. This is important since hurdles used in the food industry may lead to crossprotection even though they are regarded as mechanistically independent on the cellular level. In this way bacterial cells have the ability to adapt to stressors such as acidic pH, which may result in increased virulence expression. This phenomenon has been demonstrated in both S. Typhimurium (Bearson et al. 1997; Gahan and Hill 1999) and *L. monocytogenes* (Gahan and Hill 1999). Hence, application of the first stressor or hurdle may result in the surviving microorganisms becoming resistant to incoming hurdles.

Microarrays are an effective tool to study on a global level the effect of one or several combinations of various hurdles on the transcriptome of a pathogen and to decide the best combination and sequence in which the hurdles may be applied on the food. Microarrays may be used for several applications, such as testing the effect of single or multiple processing treatments or of hurdles on the gene expression profiles of foodborne pathogens. As seen in Figure 15.4, the comparison could be informative of possible means of crossprotection, cross reaction, or synergism between two or more antimicrobial treatments. Hence, transcriptome analysis enables investigators to probe which treatments elicit similar gene expression responses in pathogens leading to crossprotection. In Figure 15.4a, the circles represent genes regulated by hurdles A, B, and C, respectively. The region that overlaps between these circles depicts the genes that these hurdles affect in common and consequently the combination of hurdles with the minimum overlapping surface area would be the most effective in eliminating the pathogen. It can be seen from Figure 15.4b that either hurdles D and E or D and F would be potentially better combinations than E and F. If the hurdles are proven to be mechanistically independent at the cellular level, a combination treatment is definitely more effective in reducing pathogen contamination. Therefore, global analysis of gene expression of S. Typhimurium in response to various post-processing treatments will lead to insightful data regarding the effect of these treatments on the transcriptome of the pathogen. This in turn will enable investigators to identify a "signature" of the organism in response to various hurdles and to better understand the best possible combination of treatments and the sequence in which they may be applied to maximize the synergistic effect of multiple-hurdle technologies.

Figure 15.4. Venn diagrams describing with an example the application of microarrays on hurdle interventions.

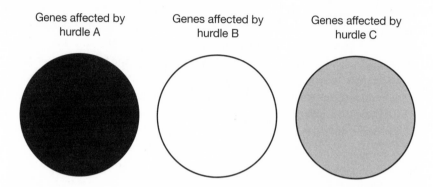

Figure 15.4.a. Diagrammatic representation of genes affected by three different hurdles. Each of the circles represent genes of the pathogen affected after treatment with either hurdle A, B, or C. No genes are mutually affected by any of the iinterventions.

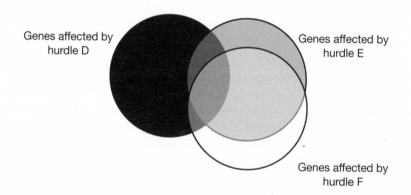

Figure 15.4.b. As observed from the Venn diagram, hurdle D and hurdle F affect several common genes of the pathogen. Similar observations can be made for hurdle F and hurdle E. However, combinations of either hurdle D and hurdle E or D and F have affected fewer genes in common. Hence, this combination may prove most effective and least crossprotective.

Acknowledgments

This review is supported by a USDA Food Safety Consortium grant.

References

Alpuche-Aranda, C. M., E. L. Racoosin, J. A. Swanson, and S. I. Miller. 1994. *Salmonella* stimulate macrophage macropinocytosis and persist within spacious phagosomes. J. Exp. Med. 179:601–608.

Altier, C. 2005. Genetic and environmental control of *Salmonella* invasion. J. Microbiol. 43:85–92.

Alvarez, J., S. Porwollik, I. Laconcha, V. Gisakis, A. B. Vivanco, I. Gonzalez, S. Echenagusia, N. Zabala, F. Blackmer, M. McClelland, A. Rementeria, and J. Garaizar. 2003. Detection of a *Salmonella enterica* serovar California strain spreading in Spanish feed mills and genetic characterization with DNA microarrays. Appl. Environ. Microbiol. 69:7531–7534.

Ambion Inc. 2007. RNA *later* tissue collection: RNA stabilization solution. At http://www.ambion.com/techlib/prot/bp_7020.pdf. 2007. Accessed October 1, 2007.

Bader, M. W., W. W. Navarre, W. Shiau, H. Nikaido, J. G. Frye, M. McClelland, F. C. Fang, and S. I. Miller. 2003. Regulation of *Salmonella typhimurium* virulence gene expression by cationic antimicrobial peptides. Mol. Microbiol. 50:219–230.

Bajaj, V., C. Hwang, and C. A. Lee. 1995. *hilA* is a novel *ompR/toxR* family member that activates the expression of *Salmonella typhimurium* invasion genes. Mol. Microbiol. 18:715–727.

Bajaj, V., R. L. Lucas, C. Hwang, and C. A. Lee. 1996. Co-ordinate regulation of *Salmonella typhimurium* invasion genes by environmental and regulatory factors is mediated by control of *hilA* expression. Mol. Microbiol. 22:703–714.

Bäumler, A. J. 1997. The record of horizontal gene transfer in *Salmonella*. Trends Microbiol. 5:318–322.

Bäumler, A. J., R. M. Tsolis, T. A. Ficht, and L. G. Adams. 1998. Evolution of host adaptation in *Salmonella enterica*. Infect. Immun. 66:4579–4587.

Bäumler, A. J., R. M. Tsolis, and F. Heffron. 1996. Contribution of fimbrial operons to attachment to and invasion of epithelial cell lines by *Salmonella typhimurium*. Infect. Immun. 64:1862–1865.

Bearson, S., B. Bearson, and J. W. Foster. 1997. Acid stress responses in enterobacteria. FEMS Microbiol. Lett. 147:173–180.

Bedná , M. 2000. DNA microarray technology and application. Med. Sci. Monit. 6:796–800.

Behlau, I., and S. I. Miller. 1993. A PhoP-repressed gene promotes *Salmonella typhimurium* invasion of epithelial cells. J. Bacteriol. 175:4475–4484.

Blanc-Potard, A. B., and E. A. Groisman. 1997. The *Salmonella selC* locus contains a pathogenicity island mediating intramacrophage survival. EMBO J. 16:5376–5385.

Boyd, E. F., S. Porwollik, F. Blackmer, and M. McClelland. 2003. Differences in gene content among *Salmonella enterica* serovar Typhi isolates. J. Clin. Microbiol. 41:3823–3828.

Brul, S., and P. Coote. 1999. Preservative agents in foods: mode of action and microbial resistance mechanisms. Int. J. Food Microbiol. 50:1–17.

Bryan, F. L., and M. P. Doyle. 1995. Health risks and consequences of *Salmonella* and *Campylobacter jejuni* in raw poultry. J. Food Prot. 58:326–344.

Burpo, F. J. 2001. A critical review of PCR primer design algorithms and cross-hybridization case study. At http://cmgm.stanford.edu/biochem218/Projects%202001/Burpo.pdf. Accessed October 9, 2007.

Bustin, S. A. 2002. Quantification of mRNA using real-time reverse transcription PCR (RT-PCR): trends and problems. J. Mol. Endocrinol. 29:23–39.

Butte, A. 2002. The use and analysis of microarray data. Nat. Rev. Drug Discov. 1:951–960.

Campbell, A. 2003. The future of bacteriophage biology. Nat. Rev. Genet. 4:471–477.

Castillo, A., L. M. Lucia, K. J. Goodson, J. W. Savell, and G. R. Acuff. 1998. Use of hot water for beef carcass decontamination. J. Food Prot. 61:19–25.

Chan, K., S. Baker, C. C. Kim, C. S. Detweiler, G. Dougan, and S. Falkow. 2003. Genomic comparison of *Salmonella enterica* serovars and *Salmonella bongori* by use of an *S. enterica* serovar Typhimurium DNA microarray. J. Bacteriol. 185:553–563.

Chen, L.-M., S. Hobbie, and J. E. Galán. 1996. Requirement of CDC42 for *Salmonella*-induced cytoskeletal and nuclear responses. Science 274:2115–2118.

Cheung, V. G., M. Morley, F. Aguilar, A. Massimi, R. Kucherlapati, and G. Childs. 1999. Making and reading microarrays. Nat. Genet. 21:15–19.

Cirillo, D. M., R. H. Valdivia, D. M. Monack, and S. Falkow. 1998. Macrophage-dependent induction of the *Salmonella* pathogenicity island 2 type III secretion system and its role in intracellular survival. Mol. Microbiol. 30:175–188.

Clements, M. O., S. Eriksson, A. Thompson, S. Lucchini, J. C. D. Hinton, S. Normark, and M. Rhen. 2002. Polynucleotide phosphorylase is a global regulator of virulence and persistency in *Salmonella enterica*. Proc. Natl. Acad. Sci. USA 99:8784–8789.

Collazo, C. M., and J. E. Galán. 1997. The invasion-associated type-III protein secretion system in *Salmonella*—a review. Gene 192:51–59.

Conway, T., and G. K. Schoolnik. 2003. Microarray expression profiling: capturing a genome-wide portrait of the transcriptome. Mol. Microbiol. 47:879–889.

Cummings, C. A., and D. A. Relman. 2000. Using DNA microarrays to study host-microbe interactions. Emerg. Infect. Dis. 6:513–525.

Darwin, K. H., and V. L. Miller. 1999. Molecular basis of the interaction of *Salmonella* with the intestinal mucosa. Clin. Microbiol. Rev. 12:405–428.

Davidson, P. M., and M. A. Harrison. 2002. Resistance and adaptation to food antimicrobials, sanitizers, and other process controls. Food Technology 56:69–78.

Davis, A. O., J. O. O'Leary, A. Muthaiyan, M. J. Langevin, A. Delgado, A. T. Abalos, A. R. Fajardo, J. Marek, B. J. Wilkinson, and J. E. Gustafson. 2005. Characterization of *Staphylococcus aureus* mutants expressing reduced susceptibility to common housecleaners. J. Appl. Microbiol. 98:364–372.

De Keersmaecker, S. C. D., K. Marchal, T. L. Verhoeven, K. Engelen, J. Vanderleyden, and C. S. Detweiler. 2005. Microarray analysis and motif detection reveal new targets of the *Salmonella enterica* Serovar Typhimurium HilA regulatory protein, including *hilA* itself. J. Bacteriol. 187:4381–4391.

Debouck, C., and P. N. Goodfellow. 1999. DNA microarrays in drug discovery and development. Nat. Genet. 21:48–50.

Denyer, S. P. 1995. Mechanisms of action of antibacterial biocides. International Biodeterioration & Biodegradation 36:227–245.

Denyer, S. P., and G. S. A. B. Stewart. 1998. Mechanisms of action of disinfectants. International Biodeterioration & Biodegradation 41:261–268.

Dopazo, J., E. Zanders, I. Dragoni, G. Amphlett, and F. Falciani. 2001. Methods and approaches in the analysis of gene expression data. J. Immunol. Methods 250:93–112.

Dowd, S. E., K. Killinger-Mann, J. Blanton, M. San Francisco, and M. Brashears. 2007. Positive adaptive state: microarray evaluation of gene expression in *Salmonella enterica* Typhimurium exposed to nalidixic acid. Foodborne Pathog. Dis. 4:187–200.

Dunkley, K. D., J. L. McReynolds, M. E. Hume, C. S. Dunkley, T. R. Callaway, L. F. Kubena, D. J. Nisbet, and S. C. Ricke. 2007. Molting in *Salmonella* Enteritidis-challenged laying hens fed alfalfa crumbles. I. *Salmonella* Enteritidis colonization and virulence gene *hilA* response. Poult. Sci. 86:1633–1639.

Eichelberg, K., and J. E. Galán. 1999. Differential regulation of *Salmonella typhimurium* type III secreted proteins by pathogenicity island 1 (SPI-1)-encoded transcriptional activators InvF and HilA. Infect. Immun. 67:4099–4105.

Eisen, M. B., P. T. Spellman, P. O. Brown, and D. Botstein. 1998. Cluster analysis and display of genome-wide expression patterns. Proc. Natl. Acad. Sci. U.S.A 95:14863–14868.

Ellebracht, E. A., A. Castillo, L. M. Lucia, R. K. Miller, and G. R. Acuff. 1999. Reduction of pathogens using hot water and lactic acid on beef trimmings. J. Food Sci. 64:1094–1099.

Enoch, J. S. 2007. Fourth peanut butter death reported. At http://www.consumeraffairs. com/news04/2007/03/peanut_butter_recall15.html. Accessed January 24, 2008.

Ernst, L. A., R. K. Gupta, R. B. Mujumdar, and A. S. Waggoner. 1989. Cyanine dye labeling reagents for sulfhydryl groups. Cytometry 10:3–10.

Farkas, J. 2007. Physical methods of food preservation. Pages 685–712 in Food microbiology: fundamentals and frontiers. 3rd ed. M. P. Doyle and L. R. Beuchat, eds. ASM Press, Washington, DC.

Fey, A., S. Eichler, S. Flavier, R. Christen, M. G. Höfle, and C. A. Guzmán. 2004. Establishment of a real-time PCR-based approach for accurate quantification of bacterial RNA targets in water, using *Salmonella* as a model organism. Appl. Environ. Microbiol. 70:3618–3623.

Francis, C. L., T. A. Ryan, B. D. Jones, S. J. Smith, and S. Falkow. 1993. Ruffles induced by *Salmonella* and other stimuli direct macropinocytosis of bacteria. Nature 364:639–642.

Francis, C. L., M. N. Starnbach, and S. Falkow. 1992. Morphological and cytoskeletal changes in epithelial cells occur immediately upon interaction with *Salmonella typhimurium* grown under low-oxygen conditions. Mol. Microbiol. 6:3077–3087.

Gahan, C. G. M., and C. Hill. 1999. The relationship between acid stress responses and virulence in *Salmonella typhimurium* and *Listeria monocytogenes*. Int. J. Food Microbiol. 50:93–100.

Galán, J. E. 1996. Molecular and cellular bases of *Salmonella* entry into host cells. Curr. Top. Microbiol. Immunol. 209:43–60.

Galán, J. E., and R. Curtiss III. 1990. Expression of *Salmonella typhimurium* genes required for invasion is regulated by changes in DNA supercoiling. Infect. Immun. 58:1879–1885.

Gantois, I., R. Ducatelle, F. Pasmans, F. Haesebrouck, I. Hautefort, A. Thompson, J. C.

Hinton, and F. Van Immerseel. 2006. Butyrate specifically down-regulates *Salmonella* pathogenicity island 1 gene expression. Appl. Environ. Microbiol. 72:946–949.

Garcia Vescovi, E., F. C. Soncini, and E. A. Groisman. 1996. Mg^{2+} as an extracellular signal: environmental regulation of Salmonella virulence. Cell 84:165–174.

Garcia-del Portillo, F., J. W. Foster, and B. B. Finlay. 1993. Role of acid tolerance response genes in *Salmonella typhimurium* virulence. Infect. Immun. 61:4489–4492.

Giannella, R. A., O. Washington, P. Gemski, and S. B. Formal. 1973. Invasion of HeLa cells by *Salmonella typhimurium:* a model for study of invasiveness of *Salmonella.* J. Infect. Dis. 128:69–75.

Giulietti, A., L. Overbergh, D. Valckx, B. Decallonne, R. Bouillon, and C. Mathieu. 2001. An overview of real-time quantitative PCR: applications to quantify cytokine gene expression. Methods 25:386–401.

Goldschmidt, M. C. 2006. The use of biosensor and microarray techniques in the rapid detection and identification of salmonellae. J. AOAC Int. 89:530–537.

Goldsmith, Z. G., and N. Dhanasekaran. 2004. The microrevolution: applications and impacts of microarray technology on molecular biology and medicine (review). Int. J. Mol. Med. 13:483–495.

Hanna, S. E., C. J. Connor, and H. H. Wang. 2005. Real-time polymerase chain reaction for the food microbiologists: technologies, applications, and limitations. J. Food Sci. 70:49–53.

Hensel, M., J. E. Shea, S. R. Waterman, R. Mundy, T. Nikolaus, G. Banks, A. Vazquez-Torres, C. Gleeson, F. C. Fang, and D. W. Holden. 1998. Genes encoding putative effector proteins of the type III secretion system of *Salmonella* pathogenicity island 2 are required for bacterial virulence and proliferation in macrophages. Mol. Microbiol. 30:163–174.

Holmstrøm, K., T. Tolker-Nielsen, and S. Molin. 1999. Physiological states of individual *Salmonella typhimurium* cells monitored by in situ reverse transcription-PCR. J. Bacteriol. 181:1733–1738.

Huang, X., H. Xu, X. Sun, K. Ohkusu, Y. Kawamura, and T. Ezaki. 2007. Genome-wide scan of the gene expression kinetics of *Salmonella enterica* serovar Typhi during hyperosmotic stress. Int. J. Mol. Sci. 8:116–135.

Hueck, C. J., M. J. Hantman, V. Bajaj, C. Johnston, C. A. Lee, and S. I. Miller. 1995. *Salmonella typhimurium* secreted invasion determinants are homologous to *Shigella* Ipa proteins. Mol. Microbiol. 18:479–490.

Jacob, F., D. Perrin, C. Sanchez, and J. Monod. 2005. The operon: a group of genes with expression coordinated by an operator. C.R. Acad. Sci. Paris 250 (1960):1727–1729. C. R. Biol. 328:514–520.

Jay, J. M. 2000. Modern food microbiology. 6th ed. Aspen Publishing, Gaithersburg, MD.

Joerger, R. D. 2003. Alternatives to antibiotics: bacteriocins, antimicrobial peptides and bacteriophages. Poult. Sci. 82:640–647.

Johnston, C., D. A. Pegues, C. J. Hueck, C. A. Lee, and S. I. Miller. 1996. Transcriptional activation of *Salmonella typhimurium* invasion genes by a member of the phosphorylated response-regulator superfamily. Mol. Microbiol. 22:715–727.

Jones, B. D., N. Ghori, and S. Falkow. 1994. *Salmonella typhimurium* initiates murine infection by penetrating and destroying the specialized epithelial M cells of the Peyer's patches. J. Exp. Med. 180:15–23.

Jordan, J. 2005. Undercooked turkey blamed for *Salmonella* outbreak. At http://www.marlerblog.com/2005/06/articles/case-news/undercooked-turkey-blamed-for-salmonella-outbreak/. Accessed January 24, 2008.

Kemp, G. K., M. L. Aldrich, and A. L. Waldroup. 2000. Acidified sodium chlorite antimicrobial treatment of broiler carcasses. J. Food Prot. 63:1087–1092.

Klein, D. 2002. Quantification using real-time PCR technology: applications and limitations. Trends Mol. Med. 8:257–260.

Kodadek, T. 2001. Protein microarrays: prospects and problems. Chem. Biol. 8: 105–115.

Kubori, T., Y. Matsushima, D. Nakamura, J. Uralil, M. Lara-Tejero, A. Sukhan, J. E. Galán, and S.-I. Aizawa. 1998. Supramolecular structure of the *Salmonella typhimurium* type III protein secretion system. Science 280:602–605.

Kundinger, M. M., I. B. Zabala-Díaz, V. I. Chalova, W.-K. Kim, R. W. Moore, and S. C. Ricke. 2007. Characterization of *rsmC* as a potential reference gene for *Salmonella* Typhimurium gene expression during growth in spent media. Sensing and Instrumentation for Food Quality and Safety 1:99–103.

Kwon, Y. M., S. Y. Park, S. G. Birkhold, and S. C. Ricke. 2000. Induction of resistance of *Salmonella typhimurium* to environmental stresses by exposure to short-chain fatty acids. J. Food Sci. 65:1037–1040.

Kwon, Y. M., and S. C. Ricke. 1998. Induction of acid resistance of *Salmonella typhimurium* by exposure to short-chain fatty acids. Appl. Environ. Microbiol. 64:3458–3463.

Lander, E. S. 1999. Array of hope. Nat. Genet. 21:3–4.

Lawhon, S. D., J. G. Frye, M. Suyemoto, S. Porwollik, M. McClelland, and C. Altier. 2003. Global regulation by CsrA in *Salmonella typhimurium*. Mol. Microbiol. 48:1633–1645.

Leistner, L. 2000. Basic aspects of food preservation by hurdle technology. Int. J. Food Microbiol. 55:181–186.

Leistner, L., and L. G. M. Gorris. 1995. Food preservation by hurdle technology. Trends Food Sci. Technol. 6:41–46.

Leyer, G. J., and E. A. Johnson. 1993. Acid adaptation induces cross-protection against environmental stresses in *Salmonella typhimurium*. Appl. Environ. Microbiol. 59:1842–1847.

Li, X., W. Gu, S. Mohan, and D. J. Baylink. 2002. DNA microarrays: their use and misuse. Microcirculation 9:13–22.

Liu-Stratton, Y., S. Roy, and C. K. Sen. 2004. DNA microarray technology in nutraceutical and food safety. Toxicol. Lett. 150:29–42.

Lockhart, D. J., H. Dong, M. C. Byrne, M. T. Follettie, M. V. Gallo, M. S. Chee, M. Mittmann, C. Wang, M. Kobayashi, H. Horton, and E. L. Brown. 1996. Expression monitoring by hybridization to high-density oligonucleotide arrays. Nat. Biotechnol. 14:1675–1680.

Lockhart, D. J., and E. A. Winzeler. 2000. Genomics, gene expression and DNA arrays. Nature 405:827–836.

Lucchini, S., A. Thompson, and J. C. D. Hinton. 2001. Microarrays for microbiologists. Microbiology 147:1403–1414.

MacBeath, G. 2002. Protein microarrays and proteomics. Nat. Genet. 32 Suppl:526–532.

Marcus, S. L., J. H. Brumell, C. G. Pfeifer, and B. B. Finlay. 2000. *Salmonella* pathogenicity islands: big virulence in small packages. Microbes Infect. 2:145–156.

Mead, P. S., L. Slutsker, V. Dietz, L. F. McCaig, J. S. Bresee, C. Shapiro, P. M. Griffin, and R. V. Tauxe. 1999. Food-related illness and death in the United States. Emerg. Infect. Dis. 5:607–625.

Miller, S. I., and D. A. Pegues. 2000. *Salmonella* species, including *Salmonella typhi*. Pages 2344–2363 in Principles and practice of infectious diseases. Churchill Livingstone, Philadelphia, PA.

Monsieurs, P., S. De Keersmaecker, W. W. Navarre, M. W. Bader, F. De Smet, M. McClelland, F. C. Fang, B. De Moor, J. Vanderleyden, and K. Marchal. 2005. Comparison of the PhoPQ regulon in *Escherichia coli* and *Salmonella typhimurium*. J. Mol. Evol. 60:462–474.

Mullis, K. B. 1990. Target amplification for DNA analysis by the polymerase chain reaction. Ann. Biol. Clin. Paris 48:579–582.

Muthaiyan, A., J. A. Silverman, R. K. Jayaswal, and B. J. Wilkinson. 2008. Transcriptional profiling reveals that daptomycin induces the *Staphylococcus aureus* cell wall stress stimulon and genes responsive to membrane depolarization. Antimicrob. Agents Chemother. 52:980–990.

Nakaya, H., A. Yasuhara, K. Yoshimura, Y. Oshihoi, H. Izumiya, and H. Watanabe. 2003. Life-threatening infantile diarrhea from fluoroquinolone-resistant *Salmonella enterica* Typhimurium with mutations in both *gyrA* and *parC*. Emerg. Infect. Dis. 9:255–257.

Neergaard, L. 2008. Food poisoning can be long-term problem. At http://www.uspharmd.com/health/2008_01_22.html#Food_poisoning_can_be_longterm_problem. Accessed February 6, 2008.

Nguyen, D. V., A. B. Arpat, N. Wang, and R. J. Carroll. 2002. DNA microarray experiments: biological and technological aspects. Biometrics 58:701–717.

Pegues, D. A., M. J. Hantman, I. Behlau, and S. I. Miller. 1995. PhoP/PhoQ transcriptional repression of *Salmonella typhimurium* invasion genes: evidence for a role in protein secretion. Mol. Microbiol. 17:169–181.

Pelludat, C., R. Prager, H. Tschäpe, W. Rabsch, J. Schuchhardt, and W.-D. Hardt. 2005. Pilot study to evaluate microarray hybridization as a tool for *Salmonella enterica* serovar Typhimurium strain differentiation. J. Clin. Microbiol. 43:4092–4106.

Piddock, L. J. V., D. G. White, K. Gensberg, L. Pumbwe, and D. J. Griggs. 2000. Evidence for an efflux pump mediating multiple antibiotic resistance in *Salmonella enterica* serovar Typhimurium. Antimicrob. Agents Chemother. 44:3118–3121.

Porwollik, S., R. M.-Y. Wong, and M. McClelland. 2002. Evolutionary genomics of *Salmonella*: gene acquisitions revealed by microarray analysis. Proc. Natl. Acad. Sci. USA 99:8956–8961.

Prouty, A. M., I. E. Brodsky, J. Manos, R. Belas, S. Falkow, and J. S. Gunn. 2004. Transcriptional regulation of *Salmonella enterica* serovar Typhimurium genes by bile. FEMS Immunol. Med. Microbiol. 41:177–185.

Quackenbush, J. 2001. Computational analysis of microarray data. Nat. Rev. Genet. 2:418–427.

Rahn, K., S. A. De Grandis, R. C. Clarke, S. A. McEwen, J. E. Galán , C. Ginocchio, R. Curtiss III, and C. L. Gyles. 1992. Amplification of an *invA* gene sequence of *Salmonella typhimurium* by polymerase chain reaction as a specific method of detection of *Salmonella*. Mol. Cell. Probes. 6:271–279.

Ramsay, G. 1998. DNA chips: state-of-the art. Nat. Biotechnol. 16:40–44.

Reen, F. J., E. F. Boyd, S. Porwollik, B. P. Murphy, D. Gilroy, S. Fanning, and M. McClelland. 2005. Genomic comparisons of *Salmonella enterica* serovar Dublin, Agona, and Typhimurium strains recently isolated from milk filters and bovine samples from Ireland, using a *Salmonella* microarray. Appl. Environ. Microbiol. 71:1616–1625.

Ricke, S. C. 2003. Perspectives on the use of organic acids and short chain fatty acids as antimicrobials. Poult. Sci. 82:632–639.

Ricke, S. C., M. M. Kundinger, D. R. Miller, and J. T. Keeton. 2005. Alternatives to antibiotics: chemical and physical antimicrobial interventions and foodborne pathogen response. Poult. Sci. 84:667–675.

Ronning, I. E., and H. A. Frank. 1987. Growth inhibition of putrefactive anaerobe 3679 caused by stringent-type response induced by protonophoric activity of sorbic acid. Appl. Environ. Microbiol. 53:1020–1027.

Saeed, A. I., V. Sharov, J. White, J. Li, W. Liang, N. Bhagabati, J. Braisted, M. Klapa, T. Currier, M. Thiagarajan, A. Sturn, M. Snuffin, A. Rezantsev, D. Popov, A. Ryltsov, E. Kostukovich, I. Borisovsky, Z. Liu, A. Vinsavich, V. Trush, and J. Quackenbush. 2003. TM4: a free, open-source system for microarray data management and analysis. BioTechniques 34:374–378.

Sambrook, J., and D. W. Russell. 2001. Molecular cloning: a laboratory manual. 3rd ed. Cold Spring Harbor Laboratory Press, New York, NY.

Sampathkumar, B., G. G. Khachatourians, and D. R. Korber. 2003. High pH during trisodium phosphate treatment causes membrane damage and destruction of *Salmonella enterica* serovar Enteritidis. Appl. Environ. Microbiol. 69:122–129.

San Martin, M. F., G. V. Barbosa-Cánovas, and B. G. Swanson. 2002. Food processing by high hydrostatic pressure. Crit. Rev. Food Sci. Nutr. 42:627–645.

Schechter, L. M., S. M. Damrauer, and C. A. Lee. 1999. Two AraC/XylS family members can independently counteract the effect of repressing sequences upstream of the *hilA* promoter. Mol. Microbiol. 32:629–642.

Schena, M., D. Shalon, R. W. Davis, and P. O. Brown. 1995. Quantitative monitoring of gene expression patterns with a complementary DNA microarray. Science 270:467–470.

Schulze, A., and J. Downward. 2001. Navigating gene expression using microarrays—a technology review. Nat. Cell. Biol. 3:E190–E195.

Singh, M., S. M. Simpson, H. R. Mullins, and J. S. Dickson. 2006. Thermal tolerance of acid-adapted and non-adapted *Escherichia coli* O157:H7 and *Salmonella* in ground beef during storage. Foodborne Pathog. Dis. 3:439–446.

Sirsat, S. A., S. E. Dowd, V. I. Chalova, A. Muthaiyan, and S. C. Ricke. 2007. Effect of temperature stress on gene expression of *Salmonella* Typhimurium. Abstract # 298P. Page 528 in 107th General Meeting American Society of Microbiology.

Smith, I. L., K. Halpin, D. Warrilow, and G. A. Smith. 2001. Development of a fluorogenic RT-PCR assay (TaqMan) for the detection of Hendra virus. J. Virol. Methods 98:33–40.

Stekel, D. 2003. Microarray bioinformatics. Press Syndicate of University of Cambridge, UK.

Suárez, M., and H. Rüssmann. 1998. Molecular mechanisms of *Salmonella* invasion: the type III secretion system of the pathogenicity island 1. Int. Microbiol. 1:197–204.

Taitt, C. R., Y. S. Shubin, R. Angel, and F. S. Ligler. 2004. Detection of *Salmonella enterica* serovar Typhimurium by using a rapid, array-based immunosensor. Appl. Environ. Microbiol. 70:152–158.

Tavazoie, S., J. D. Hughes, M. J. Campbell, R. J. Cho, and G. M. Church. 1999. Systematic determination of genetic network architecture. Nat. Genet. 22:281–285.

Terai, S., M. Yamasaki, S. Igimi, and F. Amano. 2005. Expression and degradation of SEp22, a pathogenicity-related protein of *Salmonella* Dps, in *Salmonella enterica* serovar Enteritidis isolated from the poultry farms in Japan. Bioscience Microflora 24:113–118.

Trayhurn, P. 1996. Northern blotting. Proc. Nutr. Soc. 55:583–589.

USDA-FSIS. 1996. The final rule on pathogen reduction and hazard analysis and critical control point (HACCP) systems. At http://www.fsis.usda.gov/oa/background/finalrul.htm#PERFORMANCE%20STANDARDS. Accessed August 19, 2007.

Vaitilingom, M., F. Gendre, and P. Brignon. 1998. Direct detection of viable bacteria, molds, and yeasts by reverse transcriptase PCR in contaminated milk samples after heat treatment. Appl. Environ. Microbiol. 64:1157–1160.

Van Asten, F. J. A. M., H. G. C. J. M. Hendriks, J. F. J. G. Koninkx, B. A. Van der Zeijst, and W. Gaastra. 2000. Inactivation of the flagellin gene of *Salmonella enterica* serotype Enteritidis strongly reduces invasion into differentiated Caco-2 cells. FEMS Microbiol. Lett. 185:175–179.

Wawrik, B., J. H. Paul, and F. R. Tabita. 2002. Real-time PCR quantification of *rbcL* (ribulose-1,5-bisphosphate carboxylase/oxygenase) mRNA in diatoms and pelagophytes. Appl. Environ. Microbiol. 68:3771–3779.

Weld, R. J., C. Butts, and J. A. Heinemann. 2004. Models of phage growth and their applicability to phage therapy. J. Theor. Biol. 227:1–11.

Wong, K.-K., M. McClelland, L. C. Stillwell, E. C. Sisk, S. J. Thurston, and J. D. Saffer. 1998. Identification and sequence analysis of a 27-kilobase chromosomal fragment containing a *Salmonella* pathogenicity island located at 92 minutes on the chromosome map of *Salmonella enterica* serovar Typhimurium LT2. Infect. Immun. 66:3365–3371.

Wood, M. W., M. A. Jones, P. R. Watson, S. Hedges, T. S. Wallis, and E. E. Galyov. 1998. Identification of a pathogenicity island required for *Salmonella* enteropathogenicity. Mol. Microbiol. 29:883–891.

Woods, L. F. J., J. M. Wood, and P. A. Gibbs. 1981. The involvement of nitric oxide in the inhibition of the phosphoroclastic system in *Clostridium sporogenes* by sodium nitrite. J. Gen. Microbiol. 125:399–406.

Xiang, Z., Y. Yang, X. Ma, and W. Ding. 2003. Microarray expression profiling: analysis and applications. Curr. Opin. Drug Discov. Devel. 6:384–395.

Zipper, H., H. Brunner, J. Bernhagen, and F. Vitzthum. 2004. Investigations on DNA intercalation and surface binding by SYBR Green I, its structure determination and methodological implications. Nucleic Acids Res. 32:e103.

■ ■ ■
Antibiotics and Antimicrobials in Food Safety:
Perspectives and Strategies

■ ■ ■

∎ 16 ∎

Quantitative Profiling of the Intestinal Microbiota of Drug-Free Broiler Chickens

Gregory R. Siragusa

Introduction

The withdrawal or outright ban on feeding subtherapeutic antibiotics to livestock for health and performance is an issue confronting the poultry production industries.* Following the initial discovery of the growth-enhancement effect by Jukes and Williams (Niewold 2007) the use of subtherapeutic growth-promoting antibiotics of antibiotic growth promotants (AGPs) has become standard practice in producing animal protein for human consumption. Since that time, however, there is increasing pressure to discontinue this practice (Levy and Marshall 2004); in fact such a ban took full effect within the European Union as of 2006. This action has spawned the need for both fundamental and applied research into the dynamics of intestinal microbial ecology as it relates to nutrition and health of food-animal species. Microbiologists are now at a stage of understanding more of the mechanisms at play in the growth performance effect due in part to the advent of molecular microbiology for profiling the makeup of complex microbial communities such as those found within the intestinal tract.

The effects of feeding AGPs to chickens have been studied by many workers. Recently a published proceedings included several aspects of this research into a single source (Donoghue 2006). Since presentation of this paper, two subsequent

---

* Disclaimer: Mention of trade names, opinions, or commercial products in this presentation does not imply recommendation or endorsement by the U.S. Department of Agriculture.

papers published are worth noting. First, an objective economic analysis of (Graham et al. 2007) antibiotics usage in poultry production from a large integrated poultry company concluded a net loss to producers from the use of subtherapeutic antibiotic growth promotants. Secondly, Niewold (2007) offered an interesting hypothesis on the role of AGPs in the avian in that they act mechanistically on the host immunity cells thereby reducing excretion of inflammatory mediators by the host resulting in a lessening of the pro-inflamatory response of the intestinal wall and a concomitant growth enhancement. This alternative hypothesis, while in its early stages of rigorous scientific testing, questions the relative role of the gut microbiota in the AGP performance-enhancement effect.

The growth-enhancement effect in the broiler chicken includes reduced mortalities, increased livability, and reduced clostridial diseases manifested as necrotic enteritis and reduced avian pathogenic *Escherichia coli*. Visek (1978) summarized the growth promotant effect as due to four general outcomes associated with AGP feeding: inhibition of subclinical infections, reduction of growth-depressing microbial metabolites, reduction of microbial use of nutrients, and enhanced uptake and use of nutrients through the thinner intestinal wall associated with antibiotic-fed animals. Mechanistically, the plethora of intestinal functions work in synchrony for nutritional physiology and efficient feed conversion by providing host defenses critical to survival (Donoghue 2006).

AGPs induce gut microbial populations that are taxonomically more homogenous, lower in both pathogenic clostridia (e.g., *Clostridium perfringens*) and mucolytic bacteria, and, in some cases, display higher levels of lactic acid bacteria (Collier et al. 2003). The so-called Inner Tube of Life (http://www.sciencemag.org/sciext/gut) and its resident microbiota has recently been found to modulate several seemingly disconnected physiological functions of the host such as the regulation of fat storage and gut antimicrobial lectin expression (Backhed et al. 2004; Cash et al. 2006). The host's major histocompatibility complex genetics has also been demonstrated to influence or correlate with differences in gut flora composition (Toivanen et al. 2001). There is much of a practical nature as well as of biologic importance to be learned by studying the gut microbiota. Obviously critical to the health and welfare of the animal, gut microbiota strongly influences production outcome of poultry flocks as well as the food safety of resulting products by dictating pathogen loads entering the processing plant. In addition, the gut is an unusually rich source of still-to-be-explored microbial biodiversity.

Several studies describing the gut flora in qualitative terms (Gong et al. 2002; Gong et al. 2007; Lan et al. 2005; Lu et al. 2003; Pedroso et al. 2006; Scupham 2007a, 2007b; van der Wielen et al. 2002) have led to the quantitative understanding of avian gut microbial populations in more specific terms by guiding the choice of primers for semi-quantitative PCR and fluorescent in situ hybridization (Amit-Romach et al. 2004; Dumonceaux, Hill, Briggs et al. 2006; Dumonceaux, Hill, Hemmingsen et al. 2006; Wise and Siragusa 2007; Zhu et al. 2002; Zhu and Joerger 2003). This chapter will present a brief review of intestinal microbiota community profiling and the subsequent research experiments that utilized quantitative real-time PCR techniques to study the dynamics of gut microbial populations in the avian. It will focus on a study of the quantitative relationships between major microbial groups in the avian gut derived from broilers reared with and without AGPs and vegetable and animal sources of protein.

Poultry Gut Microbial Ecology and Nontherapeutic Antibiotic Usage: Qualitative and Quantitative Population Aspects

Using cultural microbiology, Barnes (1979) reported a description of the broiler chicken gut microbial profile. A general progression from enterobacteriaceae in early life to a climax community at 42 days dominated by lactobacilli in the small intestine (duodenum, jejunum, ileum) to a higher proportion of clostridia and bacteroides in the cecum. Later, molecular techniques of community analyses using percent G+C profiling indicated the fed diet to be the major determining factor in the diversity of the broiler gut intestinal profile (Apajalahti et al. 1998, 2001).

Several subsequent workers utilized the highly conserved 16S rRNA-DNA gene as the basis of qualitative studies of the gut microbial community (Gong et al. 2002, 2007; Knarreborg et al. 2002; Lan et al. 2005; Lu et al. 2003; Pedroso et al. 2006; Scupham 2007a, 2007b; van der Wielen et al. 2002). Lu et al. (2003), in a study of broiler gut microbial constituents at days 3–7, 14–28, and 49 of life fed vegetarian antibiotic-free corn soy diets, reported 16S rRNA-DNA ileal sequences representing *Lactobacillus* (70% of clone sequences), *Clostrideaceae* (11%), *Strep - tococcus* (6.5%), and *Enterococcus* (6.5%) as opposed to cecal sequences that represented *Fusobacterium* (14%), *Lactobacillus* (8%), and *Bacteroides* (5%). The cecal qualitative profile reflected that of the ileal tract through two weeks of age then progresses to more individual-specific profiles as the bird matured to seven weeks of age.

As a consequence of AGP withdrawal, the most commonly reported poultry disease is necrotic enteritis (NE). Utilizing a chick model of NE, Collier et al. (2003), reported that the macrolide antibiotic tylosin reduced levels of intestinal *Clostridium perfringens*, NE lesion scores, and mucolytic bacteria.

Using DGGE (denaturing gradient gel electrophoresis) whole microbial community analysis based on 16S rRNA-DNA, van der Wielen et al. (2002) reported that in a population of commercially reared broiler chickens, each individual possessed a unique microbial profile at all sections of the gut, independent of age. However, with age, the complexity of the profile increased. The authors concluded that unknown factors specific to the host played an important role in the community profile itself.

Using temporal temperature gradient gel electrophoresis (TTGE) to study the broiler chicken cecum, Zhu and Joerger (2003) reported that 89% of 1,656 cloned sequences belonged to only four phylogenetic groups: *Clostridium leptum, Sporomusa* sp., *Clostridium coccoides,* and *Enterobacteriaceae.* Other taxa comprised less than 2% of the total and included *Bacteroides, Lactobacillus, Escherichia,* and *Pseudomonas* spp. Subsequently, Zhu and Joerger (2003), utilized FISH (fluorescent *in situ* hybridization) to enumerate microbiota associated with the broiler chicken cecae mucus and cecal content, in two-day- to six-week-old chickens using a set of six probes detecting: *Clostridium leptum* subgroup, *Clostridium coccoides-Eubacterium rectale, Bacteroides* group, *Bifidobacterium, Lactobacillus/ Streptococcus/Enterococcus,* and *Enterobacteriaceae.* They reported that at two days of age approximately 56%, 34%, and 3% of *Eubacteria* (detected by hybridization with the universal bacterial probe Bact 338) were hybridized to the *Enterobacteriaceae, Lactobacillus/Streptococcus/Enterococcus,* and *Bifidobacterium* probes as opposed to birds at six weeks which were reactive to all six probes and dominated by the two clostridial groups (54% total). Using a novel technique of feeding bromodeoxyuridine as a label for actively dividing populations of bacteria in the turkey gut, Scupham (2007a, 2007b) was able to demonstrate the differences in microbiota profiles imparted by feeding status and reported the Firmicute *Papillibacter* dominated the preadolescent fasting turkey gut.

Gut microbial profiling is now achievable to a higher degree of resolution of taxonomic groups hitherto not culturable by selective agar techniques. Based on descriptive studies of both low-resolution quantitative culture-based profiles and high-resolution qualitative molecular descriptions, we are now able to achieve a hybrid profile encompassing elements that are both quantitative and of a higher descriptive resolution (Wise and Siragusa 2007). Whereas previously workers have relied on strictly cultural methods to estimate functional or agar medium-selectable microbial groups, more recently quantitative noncultural DNA-based approaches have provided hitherto unknown views of the quantitative diversity present in the avian gut under different nutritional regimens over a time course (see Figure 16.1). These approaches have been used by performing FISH as well as using quantitative real-time PCR (Q-rtPCR).

Dumonceaux, Hill, Briggs et al. (2006) and Dumonceaux, Hill, and Hemmingsen et al. (2006) studied the quantitative microbial profile of broiler chickens fed dietary virginiamycin by quantitative real-time PCR (Q-rtPCR) analysis of *cpn*60

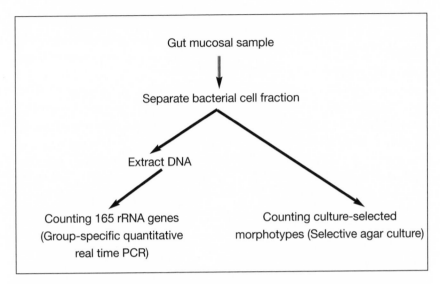

Figure 16.1. A generalized scheme for profiling the gut microbiota of sections of the avian gastrointestinal tract by cultural or colony counting methods and by molecular or gene signature sequence counting with quantitative real time PCR.

(chaperonin 60 a.n.a. heat shock protein 60 or Hsp60) gene copy numbers in the distal and cecal regions of the 49-day-old broiler gut. Clone libraries of the small intestine were found to be dominated by *Lactobacillales* (90%) while the cecal or distal regions were comprised of a more diverse population of *Clostridiales* (68%), *Lactobacillales* (25%), and *Bacteroides* (6%) of clones. Q-rtPCR indicated that differences in populations between virginiamycin fed and nonfed broilers were most significant in the proximal or ileal regions as opposed to the distal or ileocecal anatomy.

Quantitative Profile of Two Distinctive Broiler Gut Ecosystems

Wise and Siragusa (2007) utilized probes to DNA encoding 16S rRNA with specificities derived from poultry intestinal microbial profiles previously published in a quantitative microbial study of commercially reared broiler chickens fed conventional and vegetable-based antibiotic-free diets during the entire growout periods. Microbial groups quantified included Domain *Bacteria* (targeting V3 region), *Clostridium* 16S rRNA cluster IV (*Clostridium leptum* subgroup including *Faecalibacterium* (*Fusobacterium*) *prausnutzii*, *Clostridium* 16S rRNA clusters XIVa and XIVb (*Clostridium coccoides—Eubacterium rectale* subgroup), *Bacteroides* group (including *Prevotella* and *Porphyromonas*), *Bifidobacterium* genus, *Enterobacteriaceae* family *Lactobacillus* group (including *Leuconostoc, Pediococcus, Aerococcus,* and *Weissella* but not *Enterococcus* or *Streptococcus), Clostridium* 16S rRNA Cluster I (*Clostridium perfringens* subgroup) *Enterococcus* genus, *Veillonella* genus, *Atopobium* genus, *Campylobacter* genus and *Cl. Perfringens.* Briefly, chickens in the AGP-free or drug-free houses were fed a standard corn-soy diet with a strictly vegetarian protein source with all chicks receiving coccidial vaccine. The conventionally reared comparison populations were fed a standard corn-soy diet containing an animal protein source, no coccidial vaccine and at placement bacitracin methylene disalicylate (50g/t), nicarbazin (113 g/t), nitrophenylarsonic acid (45 g/t) and at day 17, monensin (100 g/t) and virginiamycin (10g/t). From each population, 10 birds were sampled at one, two, and three weeks of age. A graphical persentation of this quantitative progression is published indicating major bacterial group population changes over time in both conventionally reared and drug-free managed broiler flocks (Wise and Siragusa 2007). The ileal community was dominated by the *Enterobacteriaceae* and lactobacilli, with those groups more populous in AGP-free vegetable-based diet fed chickens than in chickens conventionally reared (receiving AGPs) at the 1- and 2-week time periods. Overall progressive changes in microbiota of the ceca in both AGP-free and conventional ceca were similar to each other, with the *Enterobacteriaceae* sequences dominating at day seven, transitioning to obligate anaerobic bacterial 16S-RRNA-DNA signature sequences by week two. It is noteworthy that in this study that in the population of chickens located geographically proximate, and under the auspice of the same integrator, all the week-two and -three replicate cecal samples from the AGP-free house were positive for *Campylobacter* spp. averaging >10^8 gene copies of *Campylobacter* spp. 16S rDNA per gram wet weight of cecal material. Five of 10 conventionally reared chicken cecal samples were positive for *Campylobacter* spp. 16S rDNAs at day seven, but no *Campylobacter* sequences were detected in any subsequent cecal sample after this time point in the conventionally reared flock samples. Within the AGP-free house, all replicate cecal samples at days 14 and 21 were positive for *Campylobacter* signature genes at levels ranging from approx 10^6–10^9 16S rRNA gene copies per gram cecal contents (moist weight). A higher rate of *Campylobacter* carriage in drug-free (organic) flocks (Luangtongkum et al. 2006) versus conventionally reared chickens was recently reported. The reported experiment was a small sample number and narrow scope of flock representation without cultural confirmation of *Campylobacter* presence and therefore should not be generalized to the AGP-free feeding practices. However, this finding clearly reinforces the need for continued experimentation in this field as more broilers are reared under AGP-free conditions.

Summary and Conclusions

Complex communities of microorganisms not practical, or even possible, to cultivate can now be measured reproducibly using Q-rtPCR. As our knowledge increases with regard to the biochemical importance of specific taxa within the gut ecosystem we will approach the time when manipulation and control of a gut population will offer predictable performance benefits for animal agriculture as well as AGP-free rearing and optimal animal welfare and health.

Perhaps the most interesting, if not complex, area of focus in this discipline is the interaction of gut microbiota with their host cells of the gastrointestinal tract. It is this vital interface that forms the basis of the host's gut immune response to its authocthonous microbiota and is likely to be a major driving factor in healthy growout. As written earlier in this chapter Niewold (2007) proposed a new function for antibiotics in the growth promotion effect in the chicken. In summary, it was proposed that antibiotics, administered as AGPs, achieve a consistently reproducible growth enhancement by inhibiting excretion of inflammatory factors from host intestinal immune cells thereby reducing the inflammatory response in the gut. The author questioned the strategy of managing gut microbiota from the point of it being a less reproducible growth enhancer than that achieved by direct feeding of subtherapeutic AGPs. It has been known for sometime that indeed the gut microbiota content or profile has profound effects on both the immune status and disease status of chickens (Donoghue 2006). Both host genetics, for example, indicated by major histocompatibility types, and gut microbiota together effect, if not modulate, seemingly unrelated host physiological functions such as obesity and gut microbiota composition. The new tools of gut microbial analysis, some of which were discussed herein, offer the methods to measure and compose a profile-in-time of the active host-microbiota interaction under conditions of health and disease. Application of this knowledge provides a basis with which to influence the host's immune status and ability to exclude or suppress pathogens by the rationale selection, design, and application of probiotics, direct-fed microbials, natural substances or molecules for optimal production performance and disease resistance.

Finally, the multiple defensive and microbial inhibitory systems of the avian gut function include cell-mediated and humoral immunity, anatomic integrity, host-antimicrobial peptides, and resident microbiota which collectively represent a set of antimicrobial defensive devices. These systems constitute what can be considered an "inhibitome" of the intestinal system. Our ability to readily quantify a major component of this gut inhibitome—namely, the resident microbiota— should lead to a more comprehensive understanding of the underlying biology of the avian gut function and its role in animal performance, food safety, and food animal growth.

References

Amit-Romach, E., D. Sklan, and Z. Uni. 2004. Microbiota ecology of the chicken intestine using 16S ribosomal DNA primers. Poult. Sci. 83:1093–1098.

Apajalahti, J. H., A. Kettunen, M. R. Bedford, and W. E. Holben. 2001. Percent G+C profiling accurately reveals diet-related differences in the gastrointestinal microbial community of broiler chickens. Appl. Environ. Microbiol. 67:5656–5667.

Apajalahti, J. H., L. K. Sarkilahti, B. R. Maki, J. P. Heikkinen, P. H. Nurminen, and W. E. Holben. 1998. Effective recovery of bacterial DNA and percent-guanine-plus-cytosine-based analysis of community structure in the gastrointestinal tract of broiler chickens. Appl. Environ. Microbiol. 64:4084–4088.

Backhed, F., H. Ding, T. Wang, L. V. Hooper, G. Y. Koh, A. Nagy, C. F. Semenkovich, and J. I. Gordon. 2004. The gut microbiota as an environmental factor that regulates fat storage. Proc. Natl. Acad. Sci. USA 101:15718–15723.

Barnes, E. M. 1979. The intestinal microbiota of poultry and game birds during life and after storage. Address of the president of the Society for Applied Bacteriology delivered at a meeting of the society on 10 January 1979. J. Appl. Bacteriol. 46:407–419.

Cash, H. L., C. V. Whitham, C. L. Behrendt, and L. V. Hooper. 2006. Symbiotic bacteria direct expression of an intestinal bactericidal lectin. Science 313:1126–1130.

Collier, C. T., J. D. van der Klis, B. Deplancke, D. B. Anderson, and H. R. Gaskins. 2003. Effects of tylosin on bacterial mucolysis, *Clostridium perfringens* colonization, and intestinal barrier function in a chick model of necrotic enteritis. Antimicrob. Agents Chemother 47:3311–3317.

Donoghue, A. M. 2006. Mechanisms of pathogen control in the avian gastrointestinal tract. Pages 138–155 in Avian gut function in health and disease: Poultry Science Symposium Series No 28. E. G. C. Perry, ed. CABI Publishers, Oxon, UK.

Dumonceaux, T. J., J. E. Hill, S. A. Briggs, K. K. Amoako, S. M. Hemmingsen, and A. G. Van Kessel. 2006. Enumeration of specific bacterial populations in complex intestinal communities using quantitative PCR based on the chaperonin-60 target. J. Microbiol. Methods 64:46–62.

Dumonceaux, T. J., J. E. Hill, S. M. Hemmingsen, and A. G. Van Kessel. 2006. Characterization of intestinal microbiota and response to dietary virginiamycin supplementation in the broiler chicken. Appl. Environ. Microbiol. 72:2815–2823.

Gong, J., R. J. Forster, H. Yu, J. R. Chambers, P. M. Sabour, R. Wheatcroft, and S. Chen. 2002. Diversity and phylogenetic analysis of bacteria in the mucosa of chicken ceca and comparison with bacteria in the cecal lumen. FEMS Microbiol. Lett. 208:1–7.

Gong, J., W. Si, R. J. Forster, R. Huang, H. Yu, Y. Yin, C. Yang, and Y. Han. 2007. 16S rRNA gene-based analysis of mucosa-associated bacterial community and phylogeny in the chicken gastrointestinal tracts: from crops to ceca. FEMS Microbiol. Ecol. 59:147–157.

Graham, J. P., J. J. Boland, and E. Silbergeld. 2007. Growth promoting antibiotics in food animal production: an economic analysis. Public Health Rep. 122:79–87.

Knarreborg, A., M. A. Simon, R. M. Engberg, B. B. Jensen, and G. W. Tannock. 2002.

Effects of dietary fat source and subtherapeutic levels of antibiotic on the bacterial community in the ileum of broiler chickens at various ages. Appl. Environ. Microbiol. 68:5918–5924.

Lan, Y., M. W. A. Verstegen, S. Tamminga, and B. A. Williams. 2005. The role of the commensal gut microbial community in broiler chickens. World's Poult. Sci. J. 61:95–104.

Levy, S. B., and B. Marshall. 2004. Antibacterial resistance worldwide: causes, challenges and responses. Nat. Med. 10:S122–129.

Lu, J., U. Idris, B. Harmon, C. Hofacre, J. J. Maurer, and M. D. Lee. 2003. Diversity and succession of the intestinal bacterial community of the maturing broiler chicken. Appl. Environ. Microbiol. 69:6816–6824.

Luangtongkum, T., T. Y. Morishita, A. J. Ison, S. Huang, P. F. McDermott, and Q. Zhang. 2006. Effect of conventional and organic production practices on the prevalence and antimicrobial resistance of *Campylobacter* spp. in poultry. Appl. Environ. Microbiol. 72:3600–3607.

Niewold, T. A. 2007. The nonantibiotic anti-inflammatory effect of antimicrobial growth promoters, the real mode of action? A hypothesis. Poult. Sci. 86:605–609.

Pedroso, A. A., J. F. Menten, M. R. Lambais, A. M. Racanicci, F. A. Longo, and J. O. Sorbara. 2006. Intestinal bacterial community and growth performance of chickens fed diets containing antibiotics. Poult. Sci. 85:747–752.

Scupham, A. J. 2007a. Examination of the microbial ecology of the avian intestine *in vivo* using bromodeoxyuridine. Environ. Microbiol. 9:1801–1809.

Scupham, A. J. 2007b. Succession in the intestinal microbiota of preadolescent turkeys. FEMS Microbiol. Ecol. 60:136–147.

Toivanen, P., J. Vaahtovuo, and E. Eerola. 2001. Influence of major histocompatibility complex on bacterial composition of fecal flora. Infect. Immun. 69:2372–2377.

van der Wielen, P. W., D. A. Keuzenkamp, L. J. Lipman, F. van Knapen, and S. Biesterveld. 2002. Spatial and temporal variation of the intestinal bacterial community in commercially raised broiler chickens during growth. Microb. Ecol. 44:286–293.

Visek, W. J. 1978. The mode of growth promotion by antibiotics. J. Anim. Sci. 46:1447–1469.

Wise, M. G., and G. R. Siragusa. 2007. Quantitative analysis of the intestinal bacterial community in one- to three-week-old commercially reared broiler chickens fed conventional or antibiotic-free vegetable-based diets. J. Appl. Microbiol. 102:1138–1149.

Zhu, X. Y., and R. D. Joerger. 2003. Composition of microbiota in content and mucus from cecae of broiler chickens as measured by fluorescent in situ hybridization with group-specific, 16S rRNA-targeted oligonucleotide probes. Poult. Sci. 82:1242–1249.

Zhu, X. Y., T. Zhong, Y. Pandya, and R. D. Joerger. 2002. 16S rRNA-based analysis of microbiota from the cecum of broiler chickens. Appl. Environ. Microbiol. 68:124–137.

17

Prevalence and Antimicrobial Resistance of Foodborne Pathogens in Conventional and Organic Livestock Operations

Orhan Sahin and Qijing Zhang

Foodborne diseases cause significant morbidity and mortality each year, and hundreds of millions of people worldwide suffer from these illnesses (Allos et al. 2004; Meng and Doyle 2002). According to the estimate by the Centers for Disease Control, foodborne pathogens account for approximately 76 million diseases, 325,000 hospitalizations, and 5,000 deaths annually in the United States alone (Mead et al. 1999). In England and Wales in 2000, an estimated 1.3 million cases, 20,759 hospitalizations, and 480 deaths were attributed to indigenous foodborne diseases (Adak et al. 2002). Among bacterial pathogens, *Campylobacter* and *Salmonella* are the leading cause of human foodborne illnesses in developed nations, and other important bacterial causes include *Clostridium perfringens, Shigella* spp., *Yersinia* spp., *Listeria monocytogenes, Staphylococcus aureus,* and Shiga toxin-producing *Escherichia coli* (STEC). Foodborne microbes are primarily intestinal organisms and are commonly present in livestock and poultry as well as in wild animals and birds. Since antibiotics are commonly used for modern animal production, foodborne microbes are often exposed to antibiotic selection pressure (Aarestrup 2005). A direct consequence of the exposure is the selection of antibiotic resistant pathogens and commensal organisms that can be potentially transmitted to humans via the food chain (Aarestrup 2005; McDermott et al. 2002; Teuber 2001).

In recent years, consumers' demand for organic foods has greatly increased, which has led to rapid expansion of organic farming practices worldwide (Dimitri and Greene 2002; USDA 2007; Van Overbeke et al. 2006). In the United States, the USDA National Organic Program defines and regulates the standard that organic production systems must meet (USDA 2002). These requirements for livestock whose meat, milk, or egg labeled as organic include the following: (a) organically raised herds and flocks may not be given any growth-promoting hormones or antibiotics and therapeutic use of antibiotics are very limited and restricted to only sick animals, but animals treated with an antimicrobial agent may not ever be labeled as organic; (b) animals of organic production must receive 100% organically produced feed; (c) animals in an organic livestock operation must have access to the outdoors and they must be maintained under conditions that are suitable for sufficient exercise and stress reduction based on the species; (d) producers must manage manure in a way not to contaminate the environment; (e) pesticides, fertilizers made with synthetic ingredients, or sewage sludge are not allowed for use in organic operations; (f) animals must be reared under organic conditions from the last third of gestation or from the second day of hatch for poultry; and (g) food sold as organic may not be subject to ionizing radiation for treatment.

In general, the scale of organic livestock production is much smaller than conventional production, but the number of organic production, especially dairy cows and laying hens, is rapidly increasing. In the United States in 2005, approximately 1% of dairy cows and 0.7% of the laying hens were produced under certified organic production, while approximately 0.12% broilers, 0.11% beef cows, 0.02% swine, and 0.07% sheep were certified organic animals (USDA 2007). California, Pennsylvania, Nebraska, Iowa, Wisconsin, North Carolina, Michigan, and Missouri were the leading states in organic livestock production in the United States (USDA 2007).

Foodborne pathogens, such as *Salmonella, E. coli,* and *Campylobacter* are present in both organic and conventional livestock productions. One of consumers' perceptions is that organic foods are microbiologically less loaded with pathogens and carry far less antibiotic resistant bacteria than foods from conventional production (Bailey and Cosby 2005; Hovi et al. 2003; Organic Trade Association 2007; Shepherd et al. 2005; Sundrum 2001). In response to the growing interest of consumers in organic products, we review the current literature to compare the prevalence and antibiotic resistance of pathogenic bacteria between organic production and conventional production, with a focus on bacterial organisms that are the most common causes of foodborne infections in humans.

Bacterial Prevalence in Conventional and Organic Productions

In general, the published studies indicate no consistent differences in the carriage levels of commensals, animal pathogens, and zoonotic bacteria between conventional and organic production systems. However, there are substantial variations among the reported carriage rates depending on the species of animals and bacteria investigated. A detailed comparison of microbiological status of organic and conventional systems by livestock production type is presented below.

Dairy Cattle

Several studies reported no major differences in the prevalence of certain diseases (mastitis, metabolic disorders, and fertility problems) between organically and conventionally managed herds (Bennedsgaard et al. 2006; Hovi et al. 2003; Roesch et al. 2007; Tikofsky et al. 2003), while other investigations reported significantly less disease incidences on organic dairy farms (Hardeng and Edge 2001; Pol and Ruegg 2007b; Thomsen et al. 2006). A recent report obtained from the database of all Danish dairy herds concluded that the risk of overall mortality was significantly lower in organic herds than conventional herds (Thomsen et al. 2006). A similar study from Norway also reported better health performance in organic dairy herds than conventional herds, especially with respect to clinical mastitis, ketosis, and milk fever (Hardeng and Edge 2001). Pol and Ruegg (2007b) evaluated the general health status of dairy farms from Wisconsin and found that conventional farms had significantly higher prevalence of diseases (clinical mastitis, respiratory disease, metritis, and foot infections) than organic farms. However, in a continuous study by the same investigators, analysis of milk samples from healthy animals indicated that significantly higher proportion of organic dairy herds had subclinical mastitis as evidenced by higher isolation of udder pathogens from milk samples of organic cows (Pol and Ruegg 2007a). Reports from Denmark (Bennedsgaard et al. 2006) and Switzerland (Roesch et al. 2007) revealed no significant differences in the risk of udder infections between the two production types, although a Swedish study found slightly better udder health in organic dairy than conventional herds (Hamilton et al. 2006). With the exception of work reported by Pol and Ruegg (2007a) (see above), there appears to be no obvious differences in prevalence and spectrum of mastitis pathogens isolated from milk samples of clinically healthy cows between conventional and organic dairy herds (Bennedsgaard et al. 2006; Hamilton et al. 2006; Roesch et al. 2007; Tikofsky et al. 2003).

Findings from comparative studies indicated that there were no fundamental differences in the prevalence of foodborne pathogens between conventional and organic dairy cattle. For example, several reports have shown that both generic *E. coli* (Sato et al. 2005) and pathogenic STEC prevalences were similar between different production systems (Cho et al. 2006; Franz et al. 2007; Kuhnert et al. 2005). Similarly, no differences in the fecal carriage rates of *Salmonella* and *Campylobacter* were observed between organic and conventional dairy herds surveyed in the United States (Fossler et al. 2005; Sato, Bartlett et al. 2004).

Swine

Similar levels of bacterial carriage rate including commensals and pathogens were reported for conventional and organic swine herds. Generic *E. coli* levels were found to be equally high in fecal samples of pigs from both conventional and organic farms located in multiple regions in the United States (Bunner et al. 2007; Mathew et al. 2001). Similarly, a study on the presence of *E. coli* O139 (a pig pathogen) conducted in Europe found no difference in the herd level prevalence between organic and conventional swine farms (Docic and Bilkei 2003). Mathew et al. (2001) reported similar levels of *Salmonella* both from conventionally and organically raised swine herds in the United States; however, very low numbers of *E. coli* O157 were cultured from some conventional pig farms, but not from organic swine farms. A recent study in Germany (Nowak et al. 2006) detected significantly higher prevalence of foodborne pathogen *Yersinia enterocolitica* in conventional pig herds than in organic pigs at the slaughter age. Thakur and Gebreyes (2005) found that prevalence of *Campylobacter coli* on the nursery farms was significantly higher in conventional herds (77%) than that in organic farms (27%), although no significant differences at the finisher farms and the slaughterhouse were observed between the two production systems.

Poultry

Studies comparing the prevalence of foodborne pathogens between conventionally and organically raised poultry reported higher prevalence rates of *Salmonella* and *Campylobacter* in organic poultry production than conventional production at the preharvest stage, but poultry meat at retail appears to be equally contaminated by the pathogens regardless of production types. A recent survey of 9 organic and 11 conventional broiler farms in Belgium found that although *Salmonella* prevalence between organic and conventional flocks were similar at slaughter, *Campylobacter* prevalence was significantly higher in the intestinal tract of organic chickens than in conventional birds (Van Oberbeke et al. 2006). The same study also measured lower antibody titers against infectious bronchitis and Newcastle disease in organic flocks, and suggested that the respiratory health status was better on organic farms than on conventional farms (Van Oberbeke et al. 2006). Luangtongkum et al. (2006) investigated the prevalence of *Campylobacter* on organic and conventional broiler or turkey farms tested in the United States, and the study revealed that the prevalence of *Campylobacter* on conventional and organic turkey farms were similar, but organic broiler farms had significantly higher prevalence of *Campylobacter* at slaughter than conventional broiler farms. In Denmark, microbiological testing of cloacal samples from 79 conventional and 22 organic broiler flocks showed that *Campylobacter* was more prevalent in organic flocks (100%) than in conventional

(37%) flocks at the time of slaughter (Heuer et al. 2001). Similarly, the flock-level *Campylobacter* prevalence was reported to be significantly higher in French organic broilers than conventional broilers as determined by testing of cecal samples collected at slaughterhouses (Avrain et al. 2003).

Reports from the United States and the United Kingdom indicate that both organic and conventional poultry meat in retail stores are often contaminated with *Campylobacter* at similar and usually high (75–100%) rates (Cui et al. 2005; Price et al. 2005; Soonthornchaikul et al. 2006). For *Salmonella,* organic chicken meat appears to be more often contaminated with this pathogen than conventional chicken meat. Cui et al. (2005) surveyed the prevalence of *Salmonella* in chicken meat and reported isolation rates of 61% and 44%, respectively, for organically and conventionally produced chicken meats at retail stores in Maryland. Bailey and Cosby (2005) reported 25–31% prevalence rates of *Salmonella* in free-range and all-natural chicken carcasses purchased from multiple producers and retail grocery stores in Georgia. The investigators compared this finding with the published data from the U.S. commercial broilers industry, which reported approximately 10% *Salmonella* prevalence and concluded that *Salmonella* was more prevalent in organic chicken meats (Bailey and Cosby 2005). Besides these reports on foodborne pathogens, a single study from Spain examined the level of carcass contamination by *Enterococcus* spp. (indicator organisms for fecal contamination of foods) in prepackaged poultry samples taken from supermarkets and butcher shops and showed that organic chicken meat had significantly higher numbers of these organisms than conventional chicken meat or conventional turkey meat (Miranda et al. 2007).

Antimicrobial Resistance in Conventional and Organic Operations

In general, conventional livestock products have significantly greater proportions of bacteria with increased resistance to antibiotics as compared with organic livestock and products. However, for some antibiotic-pathogen-host combinations, it appears that factors other than production systems also influence antimicrobial resistance as high rates of bacteria resistant to certain antibiotics were found in both organic and conventional livestock (Tikofsky et al. 2003; Sato et al. 2005; Bunner et al. 2007; Luangtongkum et al. 2006). A detailed comparison of antimicrobial resistance between organic and conventional systems by livestock production type is presented below.

Dairy Cattle

In general, the antimicrobial resistance rates do not differ significantly between production types for most antibiotics, but this varies substantially depending on the drug-pathogen combinations and among different studies (Bennedsgaard et al. 2006; Cho et al. 2007; Halbert et al. 2006; Ray et al. 2006; Roesch et al. 2006; Sato, Bartlett et al. 2004). It appears that mastitis pathogens from conventional dairy are less susceptible to pirlimycin, ampicillin, penicillin, and tetracycline than those from organic herds (Pol and Ruegg 2007a; Tikofsky et al. 2003). However, Pol and Ruegg (2007a) noted that although cephapirin was a widely used antimicrobial on conventional farms in their study, isolates from both conventional and organic herds were equally susceptible to this drug. Conversely, Tikofsky et al. (2003) found a high level of erythromycin resistance among *Staphylococcus aureus* isolated from both organic and conventional dairy herds, although this class of antibiotic was not used on the organic farms.

Testing of antimicrobial susceptibility of foodborne pathogen from dairy cows indicates that there are no apparent differences in resistance to many antibiotics between organic and conventional production systems. Although Sato et al. (2005) reported that generic *E. coli* from organic dairy farms had significantly less resistance to seven antimicrobials, the susceptibility to 10 other tested drugs was similar between organic and conventional dairy fecal isolates. Among STEC isolates, significantly higher resistance rates were found for spectinomycin and sulphadimethoxine in conventional farms, but resistance to 13 other antimicrobials did not differ significantly between organic and conventional dairy farms in Minnesota (Cho et al. 2007). Analysis of a large number of fecal *Salmonella* isolates from dairy cows in the Midwest and northeast United States indicated a significant increase in resistance rates to streptomycin and sulfamethoxazole on conventional farms as compared with organic farms, but the susceptibility to other antibiotics was not different (Ray et al. 2006). Similarly, *Campylobacter* isolates from organic and conventional dairy herds in Wisconsin were essentially the same in terms of their susceptibility to several antimicrobials including ciprofloxacin, gentamicin, erythromycin, and tetracycline (Sato, Bartlett et al. 2004).

Swine

Conventional swine production systems tend to harbor higher proportions of bacteria with antibiotic resistance than organic swine operations. However, the resistance rates and patterns vary by specific drug-bacterium combinations or the age of the animals from which the isolates were obtained. Mathew et al. (2001) reported that generic *E. coli* isolates from conventional pigs had significantly higher resistance rates to ampicillin, gentamicin, oxytetracycline, and sulfamethazine as compared with the isolates from organic pigs, but most *Salmonella* isolates were susceptible to all of the tested drugs regardless of the sources of isolation. Bunner et al. (2007) compared antimicrobial resistance in generic fecal *E. coli* isolated from large numbers of swine farms in the Midwest states and found significantly less resistance prevalence for ampicillin, sulfamethoxazole, tetracycline, and chloramphenicol in finisher pigs reared under organic conditions than in pigs from conventional herds. However, there were no significant differences in the susceptibility to many other antibiotics tested including cephalosporins, aminoglycosides, and fluoroquinolones between the two production systems (Bunner et al. 2007). Another study from Eastern Europe found pathogenic *E. coli* O139 isolates from conventional pig production were significantly more resistant to ampicillin, doxycycline, enrofloxacin, gentamicin, oxytetracycline, or sulfamethacin than the isolates from organic herds (Docic and Bilkei 2003). Analysis of a large number of *Campylobacter* isolates collected from multiple production stages on pig farms in North Carolina demonstrated that significantly higher rates of resistance to tetracycline and erythromycin were detected in conventional systems than in organic systems, although the susceptibility to chloramphenicol, ciprofloxacin, gentamicin, and nalidixic acid did not differ between the two production types (Thakur and Gebreyes 2005).

Poultry

In general, foodborne bacterial organisms (e.g., *Campylobacter, Salmonella,* and *Enterococcus*) from conventional poultry productions have increased rates of antimicrobial resistance to many antibiotics as compared with those from organic productions. Luangtongkum et al. (2006) showed that conventional poultry (especially turkey) farms had significantly higher prevalence of antibiotic-resistant *Campylobacter* than organic poultry farms. The difference in resistance was especially greater with fluoroquinolone, a key antibiotic used for treating human *Campylobacter* infections. It was interesting to notice from this study that none of the isolates from conventional chicken farms was resistant to erythromycin, but 9% of isolates from organic broilers were resistant to this drug (Luangtongkum et al. 2006). However, organic turkey farms had significantly less erythromycin resistant *Campylobacter* than conventional turkey farms (Luangtongkum et al. 2006). In contrast to these results, another study from France found no significant differences between conventional and free-range broilers in the percentage of resistant *Campylobacter* isolates to many antimicrobials including ampicillin, nalidixic acid, enrofloxacin, erythromycin, and gentamicin (Avrain et al. 2003). But the levels of tetracycline resistance of *C. coli* isolates between the two production types was significantly different (90% in conventional versus 51% in free-range broilers) in the same study (Avrain et al. 2003).

For retail poultry meat, the proportion of fluoroquinolone-resistant *Campylobacter* isolates was significantly less from organic chickens than from conventional broilers (Cui et al 2005; Price et al. 2005; Soonthornchaikul et al. 2006). However, prevalence of tetracycline resistant *Campylobacter* is usually high (about 75%) in meat from both production systems (Cui et al. 2005). Interestingly, with respect to erythromycin-resistant *Campylobacter*, organic chicken carcasses from retail stores surveyed in Maryland harbored significantly more resistance than conventional chickens (Cui et al. 2005). Another survey assessing the prevalence of antimicrobial-resistant *Campylobacter* from prepackaged chickens at London supermarkets found overall high levels of resistance (> 80%) for erythromycin and nalidixic acid in both organic and conventional products (Soonthornchaikul et al. 2006). Cui et al. (2005) reported that although *S. enterica* serovars Typhimurium and Kentucky from conventional retail chickens from Maryland stores were more resistant to many antimicrobials than those from organic chickens, several other *Salmonella* serovars such as Ohio, Agona, Braenderup, and Thompson were susceptible to all drugs tested regardless of the isolation source. When *Enterococcus* spp. (an indicator organism for dissemination of antimicrobial resistance) from retail chicken meat in Spain was tested for susceptibility, the prevalence of resistance to several antibiotics including ampicillin, chloramphenicol, doxycycline, ciprofloxacin, erythromycin, and vancomycin was significantly lower in the isolates from organic chickens than those from conventional chickens and turkeys (Miranda et al. 2007).

Conclusions

Despite the limited information and conflicting reports, it appears that foodborne pathogens are prevalent in both organic and conventional livestock productions. This observation is of no surprise considering the fact that transmission of foodborne bacteria and pathogens on farms can occur via multiple routes, and both production types are vulnerable to invasion by bacterial pathogens. There is a general trend that conventional productions harbor more antibiotic-resistant bacteria than organic productions, and this trend is more obvious in poultry and swine than in the bovine. It is also apparent from the published studies that organically grown livestock products still harbor bacteria that are resistant to antimicrobial agents that are not used in organic operations. This suggests that antibiotic usage (a common practice in conventional productions) is not the only factor in the development and spread of antimicrobial resistance. Several possible explanations exist for the presence of drug-resistant bacteria in organic productions. First, it is possible that the environment (such as soil or water) harbor a low level of antibiotic-resistant bacteria, which may continuously transfer resistant bacteria or resistance genes to livestock in the vicinity (Aminov and Mackie 2007). Second, acquisition of antimicrobial-resistance may not confer a fitness cost on the resistant organism, allowing it to transmit and persist in the environments where antibiotic selection pressure is absent (Zhang et al. 2006). Third, an antimicrobial-resistance gene is often located on mobile genetic elements such as integrons and plasmids, which may also carry genes encoding resistance to other drugs or genes involved in adaptation to certain niches (Aminov and Mackie 2007). In this case, co-selection of an irrelevant antibiotic-resistance gene may occur via nonantibiotic selection pressure. Finally, antibiotic-resistant bacteria may be transmitted across production systems via horizontal means such as flies, insects, birds, and other vectors. Thus, the open production system of organic operations is not immune to infection by pathogenic organisms with or without antibiotic resistance. But the lack of antibiotic selection pressure on organic farms indeed contributes to the reduced prevalence of antibiotic-resistant bacteria including both commensals and pathogens.

References

Aarestrup, F. M. 2005. Veterinary drug usage and antimicrobial resistance in bacteria of animal origin. Basic Clin. Pharma. Toxicol. 96:271–281.

Adak, G. K., S. M. Long, and S. J. O'Brien. 2002. Trends in indigenous foodborne disease and deaths, England and Wales: 1992 to 2000. Gut 51:832–841.

Allos, B. M., M. R. Moore, P. M. Griffin, and R. V. Tauxe. 2004. Surveillance for sporadic foodborne disease in the 21st century: the FoodNet perspective. Clin. Infect. Dis. 38 Suppl 3:S115–S120.

Aminov, R. I., and R. I. Mackie. 2007. Evolution and ecology of antibiotic resistance genes. FEMS Microbiol. Lett. 271:147–161.

Avrain, L., F. Humbert, R. L'Hospitalier, P. Sanders, C. Vernozy-Rozand, and I. Kempf. 2003. Antimicrobial resistance in *Campylobacter* from broilers: association with production type and antimicrobial use. Vet. Microbiol. 96:267–276.

Bailey, J. S., and D. E. Cosby. 2005. *Salmonella* prevalence in free-range and certified organic chickens. J. Food Prot. 68:2451–2453.

Bennedsgaard, T. W., S. M. Thamsborg, F. M. Aarestrup, C. Enevoldsen, M. Vaarst, and A. B. Christoffersen. 2006. Resistance to penicillin of *Staphylococcus aureus* isolates from cows with high somatic cell counts in organic and conventional dairy herds in Denmark. Acta Vet. Scand. 48:24–29.

Bunner, C. A., B. Norby, P. C. Bartlett, R. J. Erskine, F. P. Downes, and J. B. Kaneene. 2007. Prevalence and pattern of antimicrobial susceptibility in *Escherichia coli* isolated from pigs reared under antimicrobial-free and conventional production methods. J. Am. Vet. Med. Assoc. 231:275–283.

Cho, S. B., J. B. Bender, F. Diez-Gonzalez, C. P. Fossler, C. W. Hedberg, J. B. Kaneene, P. L. Ruegg, L. D. Warnick, and S. J. Wells. 2006. Prevalence and characterization of *Escherichia coli* O157 isolates from Minnesota dairy farms and county fairs. J. Food Prot. 69:252–259.

Cho, S. B., C. P. Fossler, F. Diez-Gonzalez, S. J. Wells, C. W. Hedberg, J. B. Kaneene, P. L. Ruegg, L. D. Warnick, and J. B. Bender. 2007. Antimicrobial susceptibility of Shiga toxin-producing *Escherichia coli* isolated from organic dairy farms, conventional dairy farms, and county fairs in Minnesota. Foodborne Pathogen Dis. 4:178–186.

Cui, S., B. Ge, J. Zheng, and J. Meng. 2005. Prevalence and antimicrobial resistance of *Campylobacter* spp. and *Salmonella* serovars in organic chickens from Maryland retail stores. Appl. Environ. Microbiol. 71:4108–4111.

Dimitri, C., and C. Greene. 2002. Recent growth patterns in the U.S. organic foods market. Pages 16–99 in Economic Research Service/USDA. Agriculture Information Bulletin-777.

Docic, M., and G. Bilkei. 2003. Differences in antibiotic resistance in *Escherichia coli* isolated from East-European swine herds with or without prophylactic use of antibiotics. J. Vet. Med. B 50:27–30.

Fossler, C. P., S. J. Wells, J. B. Kaneene, P. L. Ruegg, L. D. Warnick, L. E. Eberly, S. M. Godden, L. W. Halbert, A. M. Campbell, C. A. Bolin, and A. M. Zwald. 2005. Cattle and environmental sample-level factors associated with the presence of *Salmonella* in a multi-state study of conventional and organic dairy farms. Prev. Vet. Med. 67:39–53.

Franz, E., M. A. Klerks, O. J. De Vos, A. J. Termorshuizen, and A. H. C. van Bruggen. 2007. Prevalence of Shiga toxin-producing *Escherichia coli stx(1), stx(2), eaeA,* and *rfbE* genes and survival of *E. coli* O157:H7 in manure from organic and low-input conventional dairy farms. Appl. Environ. Microbiol. 73:2180–2190.

Halbert, L. W., J. B. Kaneene, P. L. Ruegg, L. D. Warnick, S. J. Wells, L. S. Mansfield, C. P. Fossler, A. M. Campbell, and A. M. Geiger-Zwald. 2006. Evaluation of antimicrobial susceptibility patterns in *Campylobacter* spp. isolated from dairy cattle and farms managed organically and conventionally in the midwestern and northeastern United States. J. Am. Vet. Med. Assoc. 228:1074–1081.

Hamilton, C., U. Emanuelson, K. Forslund, I. Hansson, and T. Ekman. 2006. Mastitis and related management factors in certified organic dairy herds in Sweden. Acta Vet. Scand. 48:11–17.

Hardeng, F., and V. L. Edge. 2001. Mastitis, ketosis, and milk fever in 31 organic and 93 conventional Norwegian dairy herds. J. Dairy Sci. 84:2673–2679.

Heuer, O. E., K. Pedersen, J. S. Andersen, and M. Madsen. 2001. Prevalence and antimicrobial susceptibility of thermophilic *Campylobacter* in organic and conventional broiler flocks. Lett. Appl. Microbiol. 33:269–274.

Hovi, M., A. Sundrum, and S. M. Thamsborg. 2003. Animal health and welfare in organic livestock production in Europe: current state and future challenges. Livest. Product. Sci. 80:41–53.

Kuhnert, P., C. R. Dubosson, M. Roesch, E. Homfeld, M. G. Doherr, and J. W. Blum. 2005. Prevalence and risk-factor analysis of Shiga toxigenic *Escherichia coli* in faecal samples of organically and conventionally farmed dairy cattle. Vet. Microbiol. 109:37–45.

Luangtongkum, T., T. Y. Morishita, A. J. Ison, S. Huang, P. F. McDermott, and Q. Zhang. 2006. Effect of conventional and organic production practices on the prevalence and antimicrobial resistance of *Campylobacter* spp. in poultry. Appl. Environ. Microbiol. 72:3600–3607.

Mathew, A. G., M. A. Beckmann, and A. M. Saxton. 2001. A comparison of antibiotic resistance in bacteria isolated from swine herds in which antibiotics were used or excluded. J. Swine Health Product. 9:125–129.

McDermott, P. F., S. Zhao, D. D. Wagner, S. Simjee, R. D. Walker, and D. G. White. 2002. The food safety perspective of antibiotic resistance. Anim. Biotech. 13:71–84.

Mead, P. S., L. Slutsker, V. Dietz, L. F. McCaig, J. S. Bresee, C. Shapiro, P. M. Griffin, and R. V. Tauxe. 1999. Food-related illness and death in the United States. Emerg. Infect. Dis. 5:607–625.

Meng, J., and M. P. Doyle. 2002. Introduction. Microbiological food safety. Microb. Infect. 4:395–397.

Miranda, J. M., M. Guarddon, A. Mondragon, B. I. Vazquez, C. A. Fente, A. Cepeda, and C. M. Franco. 2007. Antimicrobial resistance in *Enterococcus* spp. strains isolated from organic chicken, conventional chicken, and turkey meat: a comparative survey. J. Food Prot. 70:1021–1024.

Nowak, B., T. Von Mueffling, K. Caspari, and J. Hartung. 2006. Validation of a method for the detection of virulent *Yersinia enterocolitica* and their distribution in slaughter pigs from conventional and alternative housing systems. Vet. Microbiol. 117:219–228.

Organic Trade Association (OTA). Organic facts. Available at http://www.ota.com/organic.html. Accessed October 10, 2007.

Pol, M., and P. L. Ruegg. 2007a. Relationship between antimicrobial drug usage and antimicrobial susceptibility of gram-positive mastitis pathogens. J. Dairy Sci. 90:262–273.

Pol, M., and P. L. Ruegg. 2007b. Treatment practices and quantification of antimicrobial drug usage in conventional and organic dairy farms in Wisconsin. J. Dairy Sci. 90:249–261.

Price, L. B., E. Johnson, R. Vailes, and E. Silbergeld. 2005. Fluoroquinolone-resistant *Campylobacter* isolates from conventional and antibiotic-free chicken products. Environ. Health Perspect. 113:557–560.

Ray, K. A., L. D. Warnick, R. M. Mitchell, J. B. Kaneene, P. L. Ruegg, S. J. Wells, C. P. Fossler, L. W. Halbert, and K. May. 2006. Antimicrobial susceptibility of *Salmonella* from organic and conventional dairy farms. J. Dairy Sci. 89:2038–2050.

Roesch, M., M. G. Doherr, W. Scharen, M. Schallibaum, and J. W. Blum. 2007. Subclinical mastitis in dairy cows in Swiss organic and conventional production systems. J. Dairy Res. 74:86–92.

Roesch, M., V. Perreten, M. G. Doherr, W. Schaeren, M. Schallibaum, and J. W. Blum. 2006. Comparison of antibiotic resistance of udder pathogens in dairy cows kept on organic and on conventional farms. J. Dairy Sci. 89:989–997.

Sato, K., P. C. Bartlett, J. B. Kaneene, and F. P. Downes. 2004. Comparison of prevalence and antimicrobial susceptibilities of *Campylobacter* spp. isolates from organic and conventional dairy herds in Wisconsin. Appl. Environ. Microbiol. 70:1442–1447.

Sato, K., P. C. Bartlett, and M. A. Saeed. 2005. Antimicrobial susceptibility of *Escherichia coli* isolates from dairy farms using organic versus conventional production methods. J. Am. Vet. Med. Assoc. 226:589–594.

Sato, K., T. W. Bennedsgaard, P. C. Bartlett, R. J. Erskine, and J. B. Kaneene. 2004. Comparison of antimicrobial susceptibility of *Staphylococcus aureus* isolated from bulk tank milk in organic and conventional dairy herds in the midwestern United States and Denmark. J. Food Prot. 67:1104–1110.

Shepherd, R., M. Magnusson, and P. O. Sjoden. 2005. Determinants of consumer behavior related to organic foods. Ambio 34:352–359.

Soonthornchaikul, N., H. Garelick, H. Jones, J. Jacobs, D. Ball, and M. Choudhury. 2006. Resistance to three antimicrobial agents of *Campylobacter* isolated from organically- and intensively-reared chickens purchased from retail outlets. Int. J. Antimicrob. Agents 27:125–130.

Sundrum, A. 2001. Organic livestock farming: a critical review. Livest. Produc. Sci. 67:207–215.

Teuber, M. 2001. Veterinary use and antibiotic resistance. Curr. Opin. Microbiol. 4:493–499.

Thakur, S., and W. A. Gebreyes. 2005. Prevalence and antimicrobial resistance of *Campylobacter* in antimicrobial-free and conventional pig production systems. J. Food Prot. 68:2402–2410.

Thomsen, P. T., A. M. Kjeldsen, J. T. Sorensen, H. Houe, and A. K. Ersboll. 2006. Herd-level risk factors for the mortality of cows in Danish dairy herds. Vet. Rec. 158:622–626.

Tikofsky, L. L., J. W. Barlow, C. Santisteban, and Y. H. Schukken. 2003. A comparison of antimicrobial susceptibility patterns for *Staphylococcus aureus* in organic and conventional dairy herds. Microb. Drug Resist. 9:S39–S45.

USDA. National Organic Program. 2002. Available at http://www.ams.usda.gov/nop/NOP/standards.html. Accessed October 9, 2007.

USDA. Organic Production. Economic Research Service. 2007. Available at http://www.ers.usda.gov/Data/Organic/. 2007. Accessed October 9, 2007.

Van Overbeke, I., L. Duchateau, L. De Zutter, G. Albers, and R. Ducatelle. 2006. A comparison survey of organic and conventional broiler chickens for infectious agents affecting health and food safety. Avian Dis. 50:196–200.

Zhang, Q., O. Sahin, P. F. McDermott, and S. Payot. 2006. Fitness of antimicrobial-resistant *Campylobacter* and *Salmonella*. Microb. Infect. 8:1972–1978.

∎ 18 ∎

Fluoroquinolone-Resistant *Campylobacter jejuni* in Raw Poultry Products

Ramakrishna Nannapaneni, Omar A. Oyarzabal,
Steven C. Ricke, and Michael G. Johnson

Introduction

Campylobacter is a leading cause of foodborne illnesses, with 21.7 cases every 100,000 persons in the United States (Anonymous 1999; Altekruse et. al. 2006) and 30.2 cases every 100,000 persons in Canada (Galanis 2007). Although outbreaks of campylobacteriosis are rare and usually linked to the consumption of raw milk (Evans et al. 1996; Korlath et al. 1985) or contaminated water (Jones and Roworth 1996; Koenraad, Ayling et al. 1995; Koenraad, Jacobs-Reitsma et al. 1995; Mentzing 1981), there are between two and eight million estimated cases of campylobacteriosis, most as sporadic cases with no association to single-source outbreaks, and 200 to 800 estimated deaths in the United States annually (Moore et al. 2006). Infections with foodborne *Campylobacter* can lead to Guillain-Barré syndrome, one of the most common causes of flaccid paralysis in the United States in the last 50 years (Tam et al. 2006; Mishu et al. 1993).

Surveys of commercial broiler farms have shown that broilers frequently carry large numbers of *Campylobacter* in their intestinal contents, and during processing, highly contaminated carcasses can cross-contaminate carcasses of *Campylobacter*-free birds, thereby increasing the incidence of contaminated retail products (Miwa et al. 2003; Potturi-Venkata et al. 2007). In a recent U.S. study, the largest (24%) population attributable fraction (PAF) was associated to the consumption of poultry prepared in restaurants (Friedman et al. 2004). Therefore, poultry meat contaminated with *C. jejuni* is considered a significant risk factor for enteritis (Anonymous

1999, 2007; Newell and Wagenaar 2000), and consumption of undercooked poultry meat is considered a significant risk factor for human campylobacteriosis (Kramer et al. 2000).

A two-year study conducted by the Minnesota Department of Health found that 88% of raw poultry sampled from local supermarkets tested positive for *Campylobacter* (Smith et al. 1999). Recent studies suggest that 65 to 85% retail raw chicken carcasses are *Campylobacter* positive with approximately 1 to 4.82 $\log_{10}$ CFU/carcass of *Campylobacter* load per carcass rinse (Nannapaneni et al. 2005a, 2006a; Oyarzabal et al. 2005, 2007; Potturi-Venkata et al. 2007).

With the introduction of the Hazard Analysis and Critical Control Points systems in the United States, poultry processors have been required to meet performance standards for *Salmonella* in their products (Anonymous 1996). In the last few years, the Food Safety and Inspection Services of the U.S. Department of Agriculture (FSIS-USDA) has been evaluating the need for the implementation of similar performance standards for *Campylobacter* spp. The FSIS-USDA has not set a processing standard for *Campylobacter* (incidence or numbers) on raw or further processed poultry products yet. However, the FSIS-USDA is performing a new annual nationwide young chicken microbiological baseline data-collection program that may provide the necessary information to establish such a performance (FSIS-USDA Notice 31–07).

Fluoroquinolone-Resistant *Campylobacter* Strains

Until the early 1990s, most *C. jejuni* and *C. coli* were susceptible to fluoroquinolones (Fliegelman et al. 1985). However, the incidence of fluoroquinolone resistance (*FR*), and macrolide resistance, has been increasing in recent years and in some countries fluoroquinolones may, unfortunately, be of limited use for the empiric treatment of campylobacteriosis (Engberg et al. 2001; Reina et al. 1994; Sánchez et al. 1994). In other countries, the incidence of antimicrobial resistance has remained low over the years (Unicomb et al. 2003). Yet, the risk of acquiring *FR-Campylobacter* strains varies from country to country, even among countries with high incidence of *FR* strains. For instance, the risk for a tourist visiting Thailand is much higher than the same tourist visiting Spain, although both countries have an incidence of around 80% *FR-Campylobacter* strains (Hakanen et al. 2003; Hoge 1998; Ruiz 1998). Erythromycin, a macrolide antimicrobial, remains the drug of choice for the treatment of confirmed or presumptive campylobacteriosis (Allos 2001; Engberg et al. 2001; Kist 2002), and resistance to this antimicrobial may pose a more serious threat to public health.

For some, one of the reasons for the increase in *FR* among *Campylobacter* strains relates to the use of fluoroquinolones in food production animals, an event that has always been discussed in countries with high incidence of *FR-Campylobacter* strains (Unicomb et al. 2003). It has also been suggested that fluoroquinolone-resistant *C. jejuni* causes a more prolonged diarrheal disease than susceptible strains (Gupta et al. 2004; Smith et al. 1999). Nevertheless, a recent investigation summarizing the correlation between *FR-Campylobacter* infections and the severity of the diarrheal diseases produced by these *C. jejuni* strains has concluded that there is no data to support a prolonged, more severe disease hypothesis by these strains (Wassenaar et al. 2007).

Antimicrobial resistance has increased substantially in *Campylobacter* over the past two decades (FDA-CVM 2000). Molecular subtyping has revealed an association between *C. jejuni* strains isolated from chicken products and *C. jejuni* strains isolated from domestically acquired human cases of campylobacteriosis (Smith et al. 1999). Fluoroquinolone-resistant *C. jejuni* strains have been found to be ecologically competitive, rapidly replacing fluoroquinolone-susceptible strains in fluoroquinolone-treated chickens (Zhang et al. 2003), and persisting regardless of antimicrobial usage. In a five-year study conducted from 2002 to 2006, *Campylobacter* was detected on 57 to 96% of carcasses sampled with total *Campylobacter* load ranging from 0.90–4.82 $\log_{10}$ CFU/carcass. Ciprofloxacin-resistant *Campylobacter* CFU ($\log_{10}$ 0.90 or greater CFU/carcass) were found on 20 to 60% of sampled carcasses and total ciprofloxacin-resistant *Campylobacter* load ranging from 0.90–3.95 $\log_{10}$ CFU/carcass. While some reductions were seen for carcasses with higher loads of total *Campylobacter* or total ciprofloxacin-resistant *Campylobacter* during the five-year period from 2002 to 2006, random colony picks on CA and CCA confirmed the continued presence of high degree of ciprofloxacin-resistant *C. jejuni* (ciprofloxacin MIC's ranging from ≥16 to ≤32 µg/ml) in these retail raw chicken carcass rinses (Nannapaneni et al. 2005a, 2006a). A new method to enumerate fluoroquinolone-resistant *Campylobacter* has been developed by using a lethal dose of ciprofloxacin in *Campylobacter* selective agar plates to kill all ciprofloxacin-sensitive *Campylobacter* and selectively isolate naturally occurring ciprofloxacin-resistant *Campylobacter* (Nannapaneni et al. 2005a).

Virulence Mechanisms of *C. jejuni*

Several virulence factors have been characterized in *C. jejuni*. The most studied factor is the bipolar flagella, which comprises a system with an intricate regulatory network (Hendrixson and DiRita 2003; Wosten et al. 2004; Carrillo et al. 2004) and whose channels are apparently used to export adhesive proteins that are involved in cell invasion (Konkel et al. 2004). The presence of flagella is also important for the colonization of intestinal cells in chickens (Wassenaar et al. 1993). The adherence of *C. jejuni* to epithelial cells is mediated by several multiple factors, such as

CadF, PEB1, JlpA, Fibronectin, a 43 kDa major outer membrane protein and lipopolysaccharides (Fry et al. 2000; Jin et al. 2003; Konkel et al. 1997, 1999; Krause-Gruszczynska et al. 2007; Lara-Tejero et al. 2002; Lee et al. 2003; Moser et al. 1997; Pei et al. 1998). Three *cdt* genes of the cytolethal distending toxin (CDT) group are also important pathogenicity factors (Pickett et al. 1996; Pickett and Whitehouse 1999; Lara-Tejero and Galan 2001). In addition, it is possible that non-specific binding to lipids can mediate adherence (Szymanski and Armstrong 1996).

An invasion-associated marker (*imv*), a chromosomal marker associated with adherence and invasion in 85% of invasive and 20% of noninvasive *C. jejuni* isolates, has also been described (Carvalho et al. 2001), although the function of *imv* is not well understood. More recently, a lipoprotein termed CapA, for *Campylobacter* adhesion protein A, has also been reported to reduce the capacity of *C. jejuni* to attach and invade Caco-2 cells. *Campylobacter jejuni* strains with a *capA* insertion mutant failed to colonize and persist in chickens (Ashgar et al. 2007). An isogenic mutation in the gene virB11 encoding a putative component of a type-IV secretion system resulted in reduction in adherence and invasion (Bacon et al. 2000). The colonization ability of *C. jejuni* isolates was also compromised when some genes encoding methyl-accepting chemotaxis proteins, such as *docB* and *docC*, were mutated (Hendrixson and DiRita 2004).

Some *C. jejuni* cells also exhibit microtubule-, microfilament-, and caveolin-dependent mechanisms (Hu and Kopecko 1999; Konkel and Jones 1989; Oelschlaeger 1993) that play a critical role in pathogenesis (Biswas et al. 2003; Byrne et al. 2007). Although the pathogenic mechanisms of *C. jejuni* are still not completely understood, molecular tools based on microarrays are starting to elucidate the presence of virulence factors at the DNA level that could be used to potentially screen isolates for pathogenicity. *In vitro* models based on animals or cell lines, such as cell monolayers (Caco-2, Hep-2, HeLa or L-cells) or detached individual cells (β-lymphocyte derived hybridoma cells), will still play an important role in confirming the results found from microarray data (Fauchere et al. 1986; Friis et al. 2006; Wassenaar and Blaser 1999; Lee et al. 2003; Deun 2007).

To study attachment, recognition, and invasive mechanisms of *C. jejuni*, selected monoclonal antibodies have been developed that could be used to block selected binding sites on target Caco-2 and INT-cells, and to visualize *C. jejuni* invasion (Friis et al. 2005; Nannapaneni et al. 2006b; Qian et al. 2007). Various inhibitors of cells' functions, such as cytochalasin B (inhibitor of actin polymerization), vincristine, colchicine, nocodazole (inhibitor of microtubule polymerization), and filipin (caveolin inhibitor), have been used to examine internalization of *C. jejuni* into human intestinal cells (Byrne et al. 2007; Konkel and Jones 1989). *C. jejuni* isolates possessing CadF and binding fibronectin have been detected by fibronectin-coated coverslips with and without the addition of an anti-fibronectin antibody, and the increase in tyrosine phosphorylation by *C. jejuni* isolates have been detected by an immunoprecipitation assay with anti-paxillin conjugated to Protein A beads.

Hu et al. (2006) utilized the inhibitors fillipin III, pertussis toxin, and cholera toxin to demonstrate that *C. jejuni* 81–176 interacts with G proteins in caveolae of the host cell membrane. Fillipin III disrupts the formation of caveolae while cholera toxin and pertussis toxin are G protein inhibitors. As in the microfilament-dependent pathway for invasion, the microtubule-dependent pathway relies on the phosphorylation of host cell proteins. Various inhibitors can be utilized to block phosphorylation such as staurosporine, a broad spectrum kinase inhibitor, or genistein, an inhibitor of tyrosine kinase phosphorylation. Specific inhibitors can also be incorporated such as Wortmannin and LY294022 to inhibit PI 3-kinase, PD98059 to indirectly inhibit ERK MAP kinase, and SB203580 to inhibit P38 MAP kinase. The MAP kinase pathway is thought to trigger factors involved in cytoskeletal rearrangement of the host cell resulting in bacterial invasion (Friis et al. 2005; Hu et al. 2006).

Although *C. jejuni* with different degrees of resistance against ciprofloxacin can be frequently isolated from rinses from retail raw chicken carcasses, their virulence differences and the effects of environmental stresses on invasiveness of *C. jejuni* have not been uncharacterized. A diverse set of *C. jejuni* isolates collected from rinses of retail poultry products were tested for the presence of the *iam* gene, and their virulence against Caco-2 and INT-407 cells evaluated (Nannapaneni et al. 2005b). These results showed that *C. jejuni* expressed high levels of invasiveness, and most of the isolates infected the cells within 2 hours after the challenge. Only 12% of the isolates were found to be minimally invasive. Invasiveness in Caco-2 increased with infection time and infection dose for selected ciprofloxacin-sensitive and ciprofloxacin-resistant *C. jejuni* tested.

Effects of Temperature on *C. jejuni*

C. jejuni cells are commonly exposed to temperature changes and low temperatures (~4°C) during food processing and storage. The colonization of the intestinal tract of chickens by *C. jejuni* occurs at a temperature of 42°C, but the invasion of human cells occurs at 37°C. A two-component regulatory system termed RacR-RacS (reduced ability to colonize) appears to be involved in a temperature-dependent signaling pathway. *C. jejuni* with a mutation of the response regulator gene *racR* had a reduced ability to colonize the chicken intestinal tract and resulted in temperature-dependent changes in its protein profile and growth characteristics (Brás et al. 1999).

C. *jejuni* is able to sense, adapt, and respond to temperature fluctuations, and heat shock proteins have been demonstrated to help this organism colonize intestinal surfaces and survive at higher temperatures (Stintzi 2003). Different gene expression patterns were observed in C. *jejuni* in response to sudden shifts in temperature, but their relation to differences in virulence gene expression are presently unknown. There may be either an induction or repression of different genes in response to a temperature shift from 37 to 42°C, but little is known about C. *jejuni* responses to low-temperature shocks (Stintzi 2003), although actively growing cells transferred to saline remained competent to invasion of Caco-2 cells when exposed to 4°C for up to four weeks (Nannapaneni et al., unpublished data).

During poultry processing and storage, C. *jejuni* is exposed to cold stress and the adaptation to low-temperature shocks may play a critical role in its survival and virulence mechanisms. We still do not understand if any of the known pathogenicity factors is activated by temperature shocks or other environmental changes occurring during poultry processing. It is also unclear if adherence and invasiveness among FR-C. *jejuni* strains changes after exposure to temperature shifts. These studies would reveal whether or not a connection exists between pathogenicity and environmental conditions present during poultry processing in C. *jejuni*.

Conclusions and Future Perspectives

There are serious data gaps in the risk-assessment models for C. *jejuni* (FDA-CVM 2000). For quantitative risk-assessment purposes, a better understanding of the counts and virulence properties of poultry isolates of C. *jejuni* is needed. There appears to be considerable diversity and virulence determinants of floroquinolone-resistant and -sensitive C. *jejuni* occurring on raw poultry products. This issue becomes particularly critical since antibiotic–resistant strains of C. *jejuni* can enter the poultry product chain at any point.

Rapid and accurate detection of virulent strains of C. *jejuni* isolated from food products is essential in any food-safety monitoring program. The question remains whether the highly invasive phenotypes isolated from raw poultry products have the same virulence toward humans as those C. *jejuni* strains isolated from human infections. The binding, invasion, and translocation of C. *jejuni* in Caco-2 and INT-407 cell models may increase or decrease when C. *jejuni* cells are exposed to temperature shocks, and examining these conditions will provide a better understanding of the potential risks associated with different sources of C. *jejuni* from poultry during preharvesting and processing (temperature shock).

References

Allos, B. M. 2001. *Campylobacter jejuni* infections: update on emerging issues and trends. Clin. Infect. Dis. 32:1201–1206.

Altekruse, S. F., N. J. Stern, P. I. Fields, and D. L. Swerdlow. 2006. *Campylobacter jejuni*— An emerging foodborne pathogen. Emerg. Infect. Dis. 5:28–35.

Anonymous. 1999. Incidence of foodborne illnesses: Preliminary data from the Foodborne Diseases Active Surveillance Network (FoodNet)—United States, 1998. MMWR 48 (09):189–194.

Anonymous. 2007. Dirty birds, even premium chickens harbor dangerous bacteria. Consumer Reports, January 2007: 20–23.

Ashgar, S. S. A., N. J. Oldfield, K. G. Wooldridge, M. A. Jones, G. J. Irving, D. P. J. Turner, and D. A. A. Ala'Aldeen. 2007. CapA, an autotransporter protein of *Campylobacter jejuni*, mediates association with human epithelial cells and colonization of the chicken gut. J. Bacteriol. 189:1856–1865.

Bacon, D. J., R. A. Alm, D. H. Burr, L. Hu, D. J. Kopecko, C. P. Ewing, T. J. Trust, and P. Guerry. 2000. Involvement of a plasmid in virulence of *Campylobacter jejuni* 81–176. Infect. Immun. 68:4384–4390.

Biswas, D., K. Itoh, and C. Sasakawa. 2003. Role of microfilaments and microtubules in the invasion of INT-407 cells by *Campylobacter jejuni*. Microbiol. Immunol. 47:469–473.

Brás, A. M., S. Chatterjee, B. W. Wren, D. G. Newell, and J. M. Ketley. 1999. A novel *Campylobacter jejuni* two-component regulatory system important for temperature-dependent growth and colonization. J. Bacteriol. 181:3298–3302.

Byrne, C. M., M. Clyne, and B. Bourke. 2007. *Campylobacter jejuni* adhere to and invade chicken intestinal epithelial cells *in vitro*. Microbiology 153:561–569.

Carrillo, C. D., E. Taboada, J. H. Nash, P. Lanthier, J. Kelly, P. C. Lau, R. Verhulp, O. Mykytczuk, J. Sy, W. A. Findlay, K. Amoako, S. Gomis, P. Willson, J. W. Austin, A. Potter, L. Babiuk, B. Allan, and C. M. Szymanski. 2004. Genome-wide expression analyses of *Campylobacter jejuni* NCTC11168 reveals coordinate regulation of motility and virulence by *flhA*. J. Biol. Chem. 279:20327–20338.

Carvalho, A. C. T., G. M. Ruiz-Palacios, P. Ramos-Cervantes, L. Cervantes, X. Jiang, and L. K. Pickering. 2001. Molecular characterization of invasive and noninvasive *Campylobacter jejuni* and *Campylobacter coli* isolates. J. Clin. Microbiol. 39:1353–1359.

Deun, K. V., F. Haesebrouck, M. Heyndrickx, H. Favoreel, J. Dewulf, L. Ceelen, L. Dumez, W. Messens, S. Leleu, F. V. Immerseel, R. Ducatelle, and F. Pasmans. 2007. Virulence properties of *Campylobacter jejuni* isolates of poultry and human origin. Med. Microbiol. 56:1284–1289.

Engberg, J., F. M. Aarestrup, D. E. Taylor, P. Gerner-Smidt, and I. Nachamkin. 2001. Quinolone and macrolide resistance in *Campylobacter jejuni* and *C. coli*: resistance mechanisms and trends in human isolates. Emerg. Infect. Dis. 7:24–34.

Evans, M., R. Roberts, C. D. Ribeiro, D. Gardner, and D. Kembrey. 1996. A milk-borne *Campylobacter* outbreak following an educational farm visit. Epidemiol. Infect. 117:457–462.

Fauchere, J. L., A. Rosenau, M. Veron, E. N. Moyen, S. Richard, and A. Pfister. 1986. Association with HeLa cells of *Campylobacter jejuni* and *C. coli* isolated from human feces. Infect. Immun. 54:283–287.

FDA-CVM. 2000. Human health impact of fluoroquinolone resistant *Campylobacter* attributed to the consumption of chicken. Online at http://www.fda.gov/cvm/Documents/Risk_asses.pdf. Accessed July 7, 2009.

Fliegelman, R. M., R. M. Petrak, L. J. Goodman, J. Segreti, G. M. Trenholme, and R. L. Kaplan. 1985. Comparative in vitro activities of twelve antimicrobial agents against *Campylobacter* species. Antimicrob. Agents Chemother. 27:429–430.

Friedman, C. R., R. M. Hoekstra, M. Samuel, R. Marcus, J. Bender, B. Shiferaw, S. Reddy, S. D. Ahuja, D. L. Helfrick, F. Hardnett, M. Carter, D. Anderson, and R. V. Tauxe. 2004. Risk factors for sporadic *Campylobacter* infection in the United States: a case-control study in FoodNet sites. Clin. Infect. Dis. 38:S285–S296.

Friis, L. M., C. Pin, B. M. Pearson, and J. M. Wells. 2005. *In vitro* cell culture methods for investigating *Campylobacter* invasion mechanisms. J. Microbiol. Methods 61:145–160.

Fry, B. N., S. Feng, Y. Y. Chen, D. G. Newell, P. J. Coloe, and V. Korolik. 2000. The *galE* gene of *Campylobacter jejuni* is involved in lipopolysaccharide synthesis and virulence. Infect. Immun. 68:2594–2601.

Galanis, E. 2007. *Campylobacter* and bacterial gastroenteritis. CMAJ 177:570–571.

Gupta, A., J. M. Nelson, T. J. Barrett, R. V. Tauxe, S. P. Rossiter, C. R. Friedman, K. W. Joyce, K. E. Smith, T. F. Jones, M. A. Hawkins, B. Shiferaw, J. L. Beebe, D. J. Vugia, T. Rabatsky-Her, J. A. Benson, T. P. Root, and F. J. Angulo for the NARMS Working Group. 2004. Antimicrobial resistance among *Campylobacter* strains, United States, 1997–2001. Emerg. Infect. Dis. 10:1102–1109.

Hakanen, A., H. Jousimies-Somer, A. Siitonen, P. Huovinen, and P. Kotilainen. 2003. Fluoroquinolone resistance in *Campylobacter jejuni* isolates in travelers returning to Finland: association of ciprofloxacin resistance to travel destination. Emerg. Infect. Dis. 9:267–270.

Hendrixson, D. R., and V. J. DiRita. 2003. Transcription of sigma54-dependent but not sigma28-dependent flagellar genes in *Campylobacter jejuni* is associated with formation of the flagellar secretory apparatus. Mol. Microbiol. 50:687–702.

Hendrixson, D. R., and V. J. DiRita. 2004. Identification of *Campylobacter jejuni* genes involved in commensal colonization of the chick gastrointestinal tract. Mol. Microbiol. 52:471–484.

Hoge, C. W., J. M. Gambel, A. Srijan, C. Pitarangsi, and P. Echeverria. 1998. Trends in antibiotic resistance among diarrheal pathogens isolated in Thailand over 15 years. Clin. Infect. Dis. 26:341–345.

Hu, L., and D. J. Kopecko. 1999. *Campylobacter jejuni* 81–176 associates with microtubules and dynein during invasion of human intestinal cells. Infect. Immun. 67:4171–4182.

Hu, L., J. P. McDaniel, and D. J. Kopecko. 2006. Signal transduction events involved in human epithelial cell invasion by *Campylobacter jejuni* 81–176. Microbial Pathogenesis 40:91–100.

Jin, S., Y. C. Song, A. Emili, P. M. Sherman, and V. L. Chan. 2003. JlpA of *Campylobacter jejuni* interacts with surface-exposed heat shock protein 90α and triggers signaling pathways leading to the activation of NFk B and p38 MAP kinase in epithelial cells. Cell Microbiol. 5:165–174.

Kist, M. 2002. Impact and management of *Campylobacter* in human medicine—European perspective. Int. J. Infect. Dis. 6:44–48.

Koenraad, P. M., R. Ayling, W. C. Hazeleger, F. M. Rombouts, and D. G. Newell. 1995. The speciation and subtyping of campylobacter isolates from sewage plants and waste water from a connected poultry abattoir using molecular techniques. Epidemiol. Infect. 115:485–494.

Koenraad, P. M., W. F. Jacobs-Reitsma, T. Van der Laan, R. R. Beumer, and F. M. Rombouts. 1995. Antibiotic susceptibility of campylobacter isolates from sewage and poultry abattoir drain water. Epidemiol. Infect. 115:475–483.

Konkel, M. E., S. G. Garvis, S. L. Tipton, D. E. Anderson Jr., and W. Cieplak Jr. 1997. Identification and molecular cloning of a gene encoding a fibronectin-binding protein (CadF) from *Campylobacter jejuni*. Mol. Microbiol. 24:953–963.

Konkel, M. E., S. A. Gray, B. J. Kim, S. G. Garvis, and J. Yoon. 1999. Identification of the enteropathogens *Campylobacter jejuni* and *Campylobacter coli* based on the *cadF* virulence gene and its product. J. Clin. Microbiol. 37:510–517.

Konkel, M. E., and L. A. Jones. 1989. Adhesion to and invasion of HEp-2 cells by *Campylobacter* spp. Infect. Immun. 57:2984–2990.

Konkel, M. E., J. D. Klena, V. Rivera-Amill, M. R. Monteville, D. Biswas, B. Raphael, and J. Mickelson. 2004. Secretion of virulence proteins from *Campylobacter jejuni* is dependent on a functional flagellar export apparatus. J. Bacteriol. 186:3296–3303.

Korlath, J. A., M. T. Osterholm, L. A. Judy, J. C. Forfang, and R. A. Robinson. 1985. A point-source outbreak of campylobacteriosis associated with consumption of raw milk. J. Infect. Dis. 152:592–596.

Kramer, J. M., J. A. Frost, F. J. Bolton, and D. R. A. Wareing. 2000. *Campylobacter* contamination of raw meat and poultry at retail sale: identification of multiple types and comparison with isolates from human infection. J. Food Prot. 63:1654–1659.

Krause-Gruszczynska, M., L. B. Van Alphen, O. A. Oyarzabal, T. Alter, I. Hänel, A. Schliephake, W. König, J. P. M. Van Putten, M. E. Konkel, and S. Backert. 2007. Expression patterns and role of CadF protein in *Campylobacter jejuni* and *Campylobacter coli*. FEMS Microbiol. Lett. 274:9–16.

Lara-Tejero, M., and J. E. Galán. 2001. CdtA, CdtB, and CdtC form a tripartite complex that is required for cytolethal distending toxin activity. Infect. Immun. 69:4358–4365.

Lara-Tejero, M., and J. E. Galán. 2002. Cytolethal distending toxin: limited damage as a strategy to modulate cellular functions. Trends Microbiol. 10:147–152.

Lee, R. B., D. C. Hassane, D. L. Cottle, and C. L. Pickett. 2003. Interactions of *Campylobacter jejuni* cytolethal distending toxin subunits CdtA and CdtC with HeLa cells. Infect. Immun. 71:4883–4890.

Jones, I. G., and M. Roworth. 1996. An outbreak of *Escherichia coli* O157 and Campylobacteriosis associated with contamination of a drinking water supply. Public Health 110:277–282.

Mentzing, L. O. 1981. Waterborne outbreaks of campylobacter enteritis in central Sweden. Lancet. 2(8242):352–354.

Mishu, B., A. A. Ilyas, C. L. Koski, F. Vriesendorp, S. D. Cook, F. A. Mithen, and

M. J. Blaser. 1993. Serological evidence of previous *Campylobacter jejuni* infection in patients with the Guillain-Barré syndrome. Ann. Intern. Med. 118:947–953.

Miwa, N., Y. Takegahara, K. Terai, H. Kato, and Y. Takeuchi. 2003. *Campylobacter jejuni* contamination on broiler carcasses of *C. jejuni*-negative flocks during processing in a Japanese slaughterhouse. Int. J. Food Microbiol. 84:105–109.

Moore, J. E., M. D. Barton, I. S. Blair, D. Corcoran, J. S. Dooley, S. Fanning, I. Kempf, A. J. Lastovica, C. J. Lowery, M. Matsuda, D. A. McDowell, A. McMahon, B. C. Millar, J. R. Rao, P. J. Rooney, B. S. Seal, W. J. Snelling, and O. Tolba. 2006. The epidemiology of antibiotic resistance in *Campylobacter*. Microbes Infect. 8:1955–1966.

Moser, I., W. Schroeder, and J. Salnikow. 1997. *Campylobacter jejuni* major outer membrane protein and a 59-kDa protein are involved in binding to fibronectin and INT 407 cell membranes. FEMS Microbiol. Lett. 157:233–238.

Nannapaneni, R., R. Story, K. C. Wiggins, and M. G. Johnson. 2005a. Concurrent quantitation of total *Campylobacter* load and total ciprofloxacin-resistant *Campylobacter* loads in rinses from retail raw chicken carcasses from 2001 to 2003 by direct plating method at 42°C. Appl. Environ. Microbiol. 71:4510–4515.

Nannapaneni, R., R. Story, K. C. Wiggins, and M. G. Johnson. 2005b. Highly virulent *Campylobacter jejuni* in retail raw chicken carcass rinses. Abst. No. P5–35 in 2005 IAFP Ann Mtg, Aug. 14–17, Baltimore, MD.

Nannapaneni, R., R. Story, K. C. Wiggins, and M. G. Johnson. 2006a. Total *Campylobacter* and total ciprofloxacin-resistant *Campylobacter* loads in rinses from retail raw chicken carcasses in 2005, 5p, in Food Safety Consortium Annual Report, Oct. 1–3, 2006. University of Arkansas, Fayetteville.

Nannapaneni, R., R. Story, K. C. Wiggins, and M. G. Johnson. 2006b. New monoclonal antibody probes against *Campylobacter* and *Campylobacter jejuni*, 6p, in Food Safety Consortium Annual Report, Oct. 1–3, 2006. University of Arkansas, Fayetteville.

Newell, D. G., and J. A. Wagenaar. 2000. Poultry infection and their control at the farm level. Pages 497–509 in *Campylobacter*. I. Nachamkin and M. J. Blaser, eds. American Society for Microbiology, Washington, DC.

Oelschlaeger, T. A., P. Guerry, and D. J. Kopecko. 1993. Unusual microtubule-dependent endocytosis mechanisms triggered by *Campylobacter jejuni* and *Citrobacter freundii*. Proc. Natl. Acad. Sci. 90:6884–6888.

Oyarzabal, O. A. 2005. Reduction of *Campylobacter* spp. by commercial antimicrobials applied during the processing of broiler chickens: a review from the United States perspective. J. Food Prot. 68:1752–1760.

Oyarzabal, O. A., R. Rad, and S. Backert. 2007. Conjugative transfer of a chromosomally-encoded antibiotic resistance from *Helicobacter pylori* into *Campylobacter jejuni*. J. Clin. Microbiol. 45:402–408.

Pei, Z., C. Burucoa, B. Grignon, S. Baqar, X. Z. Huang, D. J. Kopecko, A. J. Bourgeois, J. L. Fauchere, and M. J. Blaser. 1998. Mutation in the *peb1A* locus of *Campylobacter jejuni* reduces interactions with epithelial cells and intestinal colonization of mice. Infect. Immun. 66:938–946.

Pickett, C. L., E. C. Pesci, D. L. Cottle, G. Russell, A. N. Erdem, and H. Zeytin. 1996. Prevalence of cytolethal distending toxin roduction in *Campylobacter jejuni* and relatedness of *Campylobacter* spp. *cdtB* genes. Infect. Immun. 64:2070–2078.

Pickett, C. L., and C. A. Whitehouse. 1999. The cytolethal distending oxin family. Trends Microbiol. 7:92–297.

Potturi-Venkata, L-P., S. Backert, S. L. Vieira, and O. A. Oyarzabal. 2007. Evaluation of logistic processing to reduce cross-contamination of commercial broiler carcasses with *Campylobacter* spp. J. Food Prot. 70:2549–2554.

Qian, H., E. Pang, Q. Du, J. Chang, J. Dong, S. L. Toh, F. K. Ng, A. L. Tan, and J. Kwang. 2008. Production of a monoclonal antibody specific for the major outer membrane protein of *Campylobacter jejuni* and characterization of the epitope. Appl. Environ. Microbiol. 74:833–839.

Reina, J., M. J. Ros, and A. Serra. 1994. Susceptibilities to 10 antimicrobial agents of 1,220 *Campylobacter* strains isolated from 1987 to 1993 from feces of pediatric patients. Antimicrob. Agents Chemother. 38:2917–2920.

Ruiz, J., P. Goñi, F. Marco, F. Gallardo, B. Mirelis, T. Jimenez De Anta, and J. Vila. 1998. Increased resistance to quinolones in *Campylobacter jejuni:* a genetic analysis of *gyrA* gene mutations in quinolone-resistant clinical isolates. Microbiol. Immunol. 42:223–226.

Sánchez, R., V. Fernández-Baca, M. D. Díaz, P. Muñoz, M. Rodríguez-Créixems, and E. Bouza. 1994. Evolution of susceptibilities of *Campylobacter* spp. to quinolones and macrolides. Antimicrob. Agents Chemother. 38:1879–1882.

Smith, K. E., J. M. Besser, C. W. Hedberg, F. T. Leano, J. B. Bender, J. H. Wicklund, B. P. Johnson, K. A. Moore, and M. T. Osterholm. 1999. Quinolone-resistant *Campylobacter jejuni* infections in Minnesota 1992–1998. NEJM 340:1525–1532.

Stintzi, A. 2003. Gene expression profile of *Campylobacter jejuni* in response to growth temperature variation. J. Bacteriol. 185:2009–2016.

Szymanski, C. M., and G. D. Armstrong. 1996. Interactions between *Campylobacter jejuni* and lipids. Infect. Immun. 64:3467–3474.

Tam, C. C., S. J. O'Brien, and L. C. Rodrigues. 2006. Influenza, campylobacter and mycoplasma infections, and hospital admissions for Guillain-Barré syndrome, England. Emerg. Infect. Dis. 12:1880–1887.

USDA-FSIS. 1996. Pathogen reduction; hazard analysis and critical control point (HACCP) systems; final rule. Federal Register. Online at http://www.fsis.usda.gov/OA/fr/haccp_rule.htm.

USDA-FSIS. 2007. May 22, 2007 Notice 31–07. Nationwide Young Chicken Microbiological Baseline Data Collection Program Update. U.S. Department of Agriculture. Washington, DC.

Unicomb, L., J. Ferguson, T. V. Riley, and P. Collignon. 2003. Fluoroquinolone resistance in *Campylobacter* absent from isolates, Australia. Emerg. Infect. Dis. 9:1482–1483.

Wassenaar, T. M., and M. J. Blaser. 1999. Pathophysiology of *Campylobacter jejuni* infections of humans. Microbes Infect. 1:1023–1033.

Wassenaar, T. M., M. Kist, and A. de Jong. 2007. Review. Re-analysis of the risks attributed to ciprofloxacin-resistant *Campylobacter jejuni* infection. Int. J. Antimicrobial Agents 30:195–201.

Wassenaar, T. M., B. A. M. van der Zeijst, R. Ayling, and D. G. Newell. 1993. Colonization of chicks by motility mutants of *Campylobacter jejuni* demonstrates the importance of flagellin A expression. J. Gen. Microbiol. 139:1171–1175.

Wosten, M. M., J. A.Wagenaar, and J. P. van Putten. 2004. The FlgS/FlgR two-component signal transduction system regulates the fla regulon in *Campylobacter jejuni*. J. Biol. Chem. 279:16214–16222.

Zhang, Q., J. Lin, and S. Pereira. 2003. Fluoroquinolone-resistant *Campylobacter* in animal reservoirs: dynamics of development, resistance mechanisms and ecological fitness. Anim. Health Res. Rev. 4:63–71.

▌ 19 ▌

Plant Extracts, Natural Antimicrobials, and Irradiation to Improve Microbial Safety and Quality of Meat Products

Satchi Eswaranandam, Navam S. Hettiarachchy, Theivendran Sivarooban, Taha M. Rababah, Ken Over, and Michael G. Johnson

Introduction

Foodborne pathogens are a great concern to consumers and cause significant economic losses for the food industry. For example, the economic loss associated with *Salmonella* (nontyphoidal serotypes only) and *E. coli* O157:H7 was estimated to be $2.8 billion by the Economic Resource Service (ERS-USDA 2005). Each year in the U.S. foodborne illnesses affect an estimated 14 million persons and causes 1,800 deaths (CDC 2003). Five bacterial pathogens—*Campylobacter* (all serotypes), *Salmonella* (nontyphoidal serotypes only), *E. coli* O157:H7, *E. coli* non-O157:H7 (STEC), and *Listeria monocytogenes*—cause the majority of these illnesses (Naidu et al. 2003). Outbreaks of foodborne illness attributed to *Listeria monocytogenes*, *E. coli* O157:H7, and *Salmonella* Typhimurium are of great concern to the food industry and the general public (Eswaranandam et al. 2004; Goff et al. 1996). Recent recalls of contaminated meat products included ground beef produced by a Washington firm containing possible *E. coli* O157:H7 (March 2, 2007; 16,743 pounds), sausage products produced by a Colorado Firm containing possible *Listeria* (January 5, 2007; 15,514 pounds), beef products produced by a Michigan firm containing possible *E. coli* O157:H7 (May 11, 2007; 129,000 pounds), and beef trim produced by a Minnesota firm containing possible *E. coli* O157:H7 (May 10, 2007; 117,500 pounds) (FSIS-USDA 2007).

Recent recalls of meat and poultry products included cooked ham and turkey products produced by an Ohio Firm containing possible *Listeria* monocytogenes (November 24, 2006; 46,941 pounds); chicken salad products produced by a Pennsylvania firm containing possible *Listeria* (November 10, 2005; 5,523 pounds); ready-to-eat meat and poultry products produced by a Massachusetts firm containing possible *Listeria* (October 22, 2005; 11,200 pounds); and chicken frankfurters by a New York firm containing possible *Listeria monocytogenes* (September 20, 2005; 23,040 pounds) (FSIS-USDA 2007).

In 2005 *Salmonella, E. coli* O157:H7, and *L. monocytogenes* infections caused 45,322, 2,621, and 896 cases of illness, respectively, in the United States (CDC 2007). *Salmonella* infection causes diarrhea, fever, and abdominal cramps that occur 12 to 72 hours after infection. Depending on the immunity of the patient *Salmonella* infection can result in mild to severe illness. *L. monocytogenes* and *E. coli* O157:H7 can cause severe, often devastating maladies (Doyle 2000). Consumption of food contaminated with *L. monocytogenes* and *E. coli* O157:H7 can lead to fatal diseases of listeriosis and hemorrhagic colitis, respectively (FDA 2001). Recent reports on the outbreak of foodborne pathogens in raw and cooked meat products have created a major concern for consumers, the food industry, and regulatory agencies. The economic impact of foodborne illness and the short shelf life of refrigerated meat products demand the development of effective controls for microbial contamination.

Antimicrobial Edible Film

Edible films and coatings can be produced with casein, collagen, corn zein, gelatin, soy protein, wheat gluten, calcium alginate, and methyl cellulose and a plasticizer (glycerol) (Hettiarachchy and Eswaranandam 2007, 2005; Eswaranandam et al. 2004; Were 1998; Were et al. 1999; Hoffman et al. 1998). These edible films can be used as carriers for antimicrobials. Incorporation of antimicrobial compounds into edible films may provide additional safety and shelf life for meat and poultry products. Finding potent and efficacious food-grade antimicrobials for use in edible films/coatings to inhibit foodborne pathogens is a continuing opportunity. Such compounds include plant extracts, organic acids, bacteriocins, and peptides.

The current positive health images of grape seed, tea, and other herbal extracts (used in a variety of food applications for their nutraceuticals and health beneficial effects) created positive consumer perception and acceptability of these natural antimicrobials in food products. However, dilution of these antimicrobials in the food matrix will reduce their antimicrobial activity. Incorporation of these antimicrobials in edible coatings allows them to remain in contact with food surfaces. Antimicrobial activities of grape seed extract (GSE) and green tea extract

(GTE) have been demonstrated in growth medium, films, and meat coatings. Sivarooban et al. (2006) reported that GSE (1%) and nisin (10,000 IU/mL) in phosphate-buffered saline (PBS) medium caused a 9-log cycle reduction of the *L. monocytogenes* population after 3 hours incubation at 37°C. He also reported that in the meat system, the *L. monocytogenes* population (7.1 CFU/g) was decreased by more than 2 log cycle after 28 days at 4°C and 10°C, in the samples coated with soy protein film containing nisin (10,000 IU) combined with either GSE (1%) or GTE (1%). There is no appreciable reduction in log numbers observed in coating containing nisin (10,000 IU) or GSE (1%) or GTE (1%) alone. Sivarooban et al. (2007) also reported that a combination of nisin (6,400 IU/ml) and GSE (1%) gave the greatest inhibitory activity in both tryptic soy broth containing 0.6% yeast extract and on turkey frankfurters with reductions of *L. monocytogenes* populations to nondetectable levels after 15 hours and 21 days, respectively. They showed that the combination of nisin with GSE, GTE, or pure phenolic compounds (epicatechin, catechin, caffeic acid) altered cell membrane and condensed the cytoplasm of *L. monocytogenes* as observed by electron microscope. Phenolics in GSE and GTE could bind to bacterial extra cellular proteins and enzymes and outer cell membrane and cause leakage of cellular compounds and a change in the cell's osmotic pressure. Phenolics also bind to membrane-bound ATPase and inhibit its enzymatic activity thereby affecting energy metabolism and resulting in cell death.

Organic acids and their salts are promising antimicrobial agents because of their general consumer acceptance in food products and low cost (Miller et al. 1996). Organic acids are either naturally present in fruits and vegetables or synthesized by microorganisms as a result of fermentation. Partial replacement of glycerol in the film with malic and lactic acids (2.6%) increased its antimicrobial activity in film model system. Soy films containing malic acid yielded 2.8, 6.0, and 2.2 log CFU/ml population reductions of *L. monocytogenes, Salmonella,* and *E. coli* O157:H7, respectively, compared to controls without film of 8.3, 9.0, and 8.9 log CFU/ml (Hettiarachchy and Eswaranandam 2007; Eswaranandam et al. 2004). Adams et al. (2005) studied the inhibitory activity of soy protein edible films containing grape seed extract (1%), nisin (10,000 IU/g), and malic acid (1%) and their combinations against *L. monocytogenes* in PBS and found the highest population reductions of log 3.7 CFU/ml was achieved in films containing combinations of GSE, nisin, and malic acid compared to control film (6.6 log CFU/ml) devoid of antimicrobial agents.

The edible films produced weigh approximately 1–2mg/sq inch (2.54 cm) and are relatively inexpensive. An advantage of edible films is their applicability as carriers for antimicrobial agents to provide protection from recontamination by pathogenic bacteria on the surface of ready-to-eat poultry products. Edible films can also serve as an anchor for the controlled release of antimicrobials for an extended period of time to maintain product quality and extend shelf life.

Activated Lactoferrin as a Naturally Safe Antimicrobial

Lactoferrin is a safe and natural antimicrobial approved by the FDA and the USDA to use on fresh beef to protect consumers from pathogenic bacteria (FDA 2003; Fallon and Welty 2003). Lactoferrin is derived from milk and currently marketed as a dietary supplement (immune stimulant) in the United States and is used in a variety of food products including infant formulas and sports beverages (Britigan 1997).

Lactoferrin (80 kDa) is an iron-binding glycoprotein, capable of binding two molecules of Fe^{3+} per protein molecule. Lactoferrin has a bacteriostatic effect against over 30 different types of Gram-positive and Gram-negative bacteria (*Escherichia coli, Salmonella* Typhimurium, *Shigella dysenteriae, L. monocytogenes, Campylobacter jejuni, Streptococcus mutans, Bacillus stearothermophilus, and Bacillus subtilis*) (Berkhout et al. 2003).

Lactoferrin prevents proliferation, attachment of bacteria to the surfaces of meat by preventing formation of essential attachment structures, and neutralizes their toxic substances from meat surfaces (Naidu and Nimmagudda 2003). Unlike lactoferrin, other antimicrobial substances currently used by food processors to destroy microorganisms may leave behind toxic substances (Trenev 1998). Lactoferrin is also an effective inhibitor of the functioning of microbial colonization factors such as fimbria on the surface of *E. coli* cells. These structures serve as attachment appendages in meat tissues. Prevention of pathogen attachment and growth is critical to ensure food safety at the retailer and consumer level. Any bacterial control method should ideally prevent bacterial adherence to the food matrix and multiplication (Naidu et al. 2003).

Bacteriostatic activity of lactoferrin is attributed to its binding to bacteria and iron-sequestering capabilities. As a consequence of binding, the bacteria may lose cell membrane integrity resulting in death. Iron, which has the ability to alternate between two valence states, Fe^{+2} and Fe^{+3}, participates in many pathways of bioenergy synthesis including the electron transport system that forms ATP by phosphorylation. Lactoferrin binds iron effectively. Two homologous lobes in lactoferrin consisting of two sub-lobes, which form a cleft where the ferric ion (Fe^{3+}) is tightly bound in synergistic cooperation with bicarbonate (HCO_3^- anion). This molecule deprives bacteria of iron, resulting in the inhibition of their multiplication and leading to growth arrest (Naidu 2002). However, commercially available lactoferrin is saturated with iron (143 ppm) and possesses no antimicrobial activity. Eswaranandam et al. (2006) developed a simple process to activate lactoferrin by removing iron with citric/malic/lactic acids and/or EDTA and dialyzing against de-ionized water. Activated lactoferrin solution (1%) had antimicrobial activity against *L. monocytogenes* and *E. coli* O157:H7.

L. monocytogenes was most sensitive to lactoferrin activated by citric/malic/ lactic acids (0.5, 1.0, and 2.0%) in combination with EDTA (10mM). Activated lactoferrin inhibited the growth of *L. monocytogenes* to nondetectable levels (minimum detection limit is 10 CFU/ml) whereas the controls without lactoferrin had populations of 8.2 log CFU/ml. Citric/lactic acid (0.5–2.0%) alone activated the lactoferrin against *L. monocytogenes* yielding 1.0–2.9 and 0.1–2.9 log CFU/ml reductions in population, respectively. Lactoferrin activated by EDTA at concentrations of 1, 10, and 100 mM resulted in reductions of 5.2, 6.1, and 6.5 log CFU/ml, respectively, compared with the control. *E. coli* O157:H7 was less sensitive to lactoferrin compared with *L. monocytogenes.* Lactoferrin activated by EDTA (10mM) and citric (1.0%)/malic (1.0%)/lactic (0.5%) acids inhibited the growth of *E. coli* O157:H7 yielding reductions in populations of 5.6, 4.9, and 5.1 log CFU/ml, respectively. Organic acids alone did not activate lactoferrin against *E. coli* O157:H7. EDTA alone at the concentrations of 1, 10, and 100 mM activated lactoferrin and resulted in *E. coli* O157:H7 population reductions of 1.2, 1.4, and 3.3 log CFU/ml, respectively.

Lactoferrin treated with citric/malic/lactic acids and/or EDTA did not show appreciable inhibitory activity against *S.* Typhimurium. Less than one log reduction was observed in all samples tested. Research is in progress to improve the antimicrobial activity of lactoferrin in meat systems by using it in combinations with other natural antimicrobials such as nisin and plant extracts.

Fallon and Welty (2003) reported that a combination of activated lactoferrin and lactic acid (2%) rinse gave better protection against *E. coli* O157:H7 contamination on beef. Incorporation of lactoferrin into film-forming solutions that are then applied to meat surfaces provides a continuous barrier to inhibit these pathogens on the surface of meat products. Lactoferrin and malic/lactic acid which both exert a broad spectrum of activity against Gram-positive and Gram-negative bacteria can be used as antimicrobial agents in meat coatings to provide a continuous barrier against microbial pathogens and to prolong shelf life.

Branen and Davidson (2000) reported pepsin-hydrolyzed lactoferrin to be effective against *L. monocytogenes,* enterohaemorrhagic *E. coli,* and *S.* Enteritidis in 1.0% glucose medium. The addition of EDTA enhanced the activity of hydrolyzed lactoferrin. Since lactoferrin has a bacteriostatic effect against these major pathogens, it can be used to prevent contamination on food.

Lactoferrin incorporated film coatings are hypothesized to obstruct the microbial colonization and multiplication on meat surfaces by its antimicrobial blocking activity and by acting as a physical barrier. Protein films help to prevent diffusion of lactoferrin into the meat and localize it to the meat surface where it will serve as a barrier to the bacteria and thus prevent colonization with pathogens and spoilage organisms. This provides a continuous barrier to these pathogens until the product is consumed. Thus, it is important to further investigate the antimicrobial activity of lactoferrin-incorporated film coatings applied to the surface of poultry meat products. Oh et al. (2003) reported lactoferrin (7/14/28 mg) and lysozyme (6/12/24 mg) applied on the surface of the casein and zein film significantly inhibited the growth of *E. coli* and formed clear inhibitory zones.

Irradiation and Plant Extracts to Improve Food Safety and Quality

In 2005, 8.59 billion broilers with a production value of $21 billion were produced in the United States. Arkansas has been the number-one poultry-producing state for the past 25 years and produced 1.2 billion broilers in 2005 (USDA 2007). There are two main factors in the deterioration of raw and cooked poultry: lipid oxidative deterioration and microbial contamination. Irradiation and thermal processing are the two most effective techniques to control and destroy pathogenic and spoilage microorganisms while preserving product quality. Electron beam irradiation may be the best method available to penetrate (4.5 cm) deep muscle to destroy poultry pathogens. Injury to foodborne microorganisms can be induced by irradiation (Jay 2000). This fact can result in the existence of metabolically injured pathogenic organisms that can recover during product storage. Any *Listeria monocytogenes* cell (psychotropic, capable of growing at refrigerated temperatures) not killed during irradiation can subsequently grow if conditions permit.

Ionizing irradiation is an effective tool in a food company's HACCP program for killing microbial pathogens associated with meat products. The FDA has approved irradiation of raw chicken and turkey at doses between 1.5 and 3.0 kGys to eliminate pathogens and reduce microbial contamination (DHHS-FDA 1990). Numerous studies have shown that irradiation effectively reduces the population of microorganisms. Of the common vegetative cell pathogens, *Salmonella* spp. is the most resistant to irradiation. For irradiation treatments, the D or decimal reduction value is the dose in kGy required to kill 90% of the cells. For *L. monocytogenes,* D values range from 0.42 to 0.55 kGy compared to 0.62–0.80 kGy for *Salmonella* spp., 0.25–0.45 kGy for *E. coli* O157:H7, and 0.18–0.24 kGy for *Camp - ylobacter* (DHHS/FDA 1990). When *L. monocytogenes* is exposed to a sublethal irradiation dose (1–10 kGy), it will experience a sublethal metabolic injury, and may subsequently recover if the growth conditions are favorable (Tarte et al. 1996). The sensitivity of five strains of *Listeria* to electron beam irradiation in ground pork as well as the extent of sublethal radiation injury exhibited by each was investigated by Tarte et al. (1996). Three strains were found to be susceptible to radiation-induced sublethal injury, with the populations of injured organisms increasing with irradiation dose. However, the two pathogenic strains of *Listeria monocytogenes* were not injured significantly at the tested dose levels used in treating ground pork containing a high fat percentage.

Irradiation can induce free radical formation and lipid oxidation that will affect the physicochemical, sensory, overall quality, and consumer acceptability. Plant extracts containing antioxidants can inhibit lipid oxidation by scavenging free radicals produced by irradiation thereby maintaining overall quality. Research from our laboratory has shown that the natural fenugreek (spice) is an effective antioxidant in ground beef patties, beef cubes, and in cooked and uncooked poultry (Hettiarachchy et al. 1996; Al-ameri 2001; Armitage et al. 2002).

Due to toxicological concerns over the use of synthetic antioxidants, natural antioxidants and plant extracts having antioxidant properties are preferred by the consumer and the industry (Bandyopadhyay et al. 2007). Rababah et al. (2004a) evaluated total phenolics content and antioxidant activity of fenugreek, green tea, black tea, grape seed, ginger, rosemary, gotu kola, and ginkgo extracts. Total phenolics in these extracts ranged from 24.8 to 92.5 mg of chlorogenic acid equivalent/g of dry material. The individual phenolic contituents in grape seed and green tea extracts ranged from 15.3 to 1158.5 and 18.3 to 1087.0 mg/100 g extract, respectively. Caffeic acid (830.1 mg/100 g extract) and epicatechin (1087.0 mg/100 g extract) were the main phenolics found in green tea, while in grape seed extracts epicatechin (1158.5 mg/100 g extract), catechin (887.4 mg/100 g extract), and gentistic acid (472.8 mg/100 g extract) were the predominant phenolics. The antioxidant activities of green tea and grape seed extracts and their combinations (at 200, 500, and 1,000 ppm concentrations) ranged from 41 to 88.8, 74.2 to 89.2, and 41 to 91.7% inhibition of linoleic acid oxidation (Heinonen et al. 1998), respectively. The grape seed and green tea extracts at 6,000 ppm were more effective in minimizing lipid oxidation in chicken meat. Rababah, Hettiarachchy, Horax et al. (2004) also reported that compared to controls the thiobarbituric acid reactive substances (TBARS) values (after nine days of storage; raw irradiated and nonirradiated 7.00 and 7.56 mg malondialdehyde/100 g chicken, respectively; cooked irradiated and nonirradiated 32.46 and 42.69 mg malondialdehyde/100 g chicken, respectively) of antioxidant-infused irradiated and nonirradiated chicken breast meat were generally lower and decreased with increasing antioxidant concentration (after nine days of storage; raw irradiated and nonirradiated 3.31 and 3.17 mg malondialdehyde/100 g chicken, respectively; cooked irradiated and nonirradiated 5.52 and 6.49 g malondialdehyde/100 g chicken, respectively). Lower TBARS values indicate higher antioxidant activity. The addition of grape seed and green tea extracts/combinations also minimized the carbonyl contents in meat lipids and proteins in both nonirradiated and irradiated chicken breast samples.

Rababah et al. (2005) found that irradiation did not affect the sensory flavor attributes except for the creation of a brothy flavor. However, irradiation increased the texture attributes of hardness, cohesiveness, and hardness and cohesiveness of mass based on descriptive sensory analysis conducted with six trained panelists. Instrumental data showed that irradiation increased maximum shear force, shear work, hardness, and chewiness of cooked meats. Addition of green tea extract, but not of grape seed extract, improved the color of raw and cooked meats.

Rababah et al. (2005) concluded that infusion of chicken breasts with green tea extracts at 3,000 ppm is effectively prevented and minimized major sensory changes of the meat during irradiation. Rababah et al. (2006) also found that irradiation increased the major volatiles (hexanal and pentanal) of the controls (as-is and water infused) and plant extracts infused meat. Cooking the samples significantly increased the amounts of volatiles (acetylaldehyde, propanol, propanal, butanal, butanol, pentanol, heptanal, 1-octen-3-ol, and nonanal). Addition of plant extracts decreased the amount of hexanal and pentanal concentrations. Although irradiation tends to increase lipid oxidation, infusion of chicken meat with plant extracts reduces lipid oxidation caused by irradiation.

Conclusion

Plant extracts such as green tea and grape seed extracts having high phenolic content can be used to inhibit microbial growth and retard lipid oxidation in a variety of food products. Our research has demonstrated that the use of an edible film coating containing malic acid, nisin, and green tea and grape seed extracts is a promising means to reduce the population load of *L. monocytogenes* on ready-to-eat meat products. Citric/malic/lactic acid and EDTA can be used to activate commercial lactoferrin. The activated lactoferrin demonstrated antimicrobial activity in BHI medium against *L. monocytogenes* and *E. coli* O157:H7. Infusion of chicken meat with green tea and grape seed extracts is an effective method to minimize lipid oxidation and development of volatiles by irradiation. Infusion of chicken breasts with GT extracts at 3,000 ppm was effective in preventing and minimizing major sensory changes during the irradiation of meat.

Acknowledgments

The financial support for this research study by the Food Safety Consortium—a special grant from the USDA—is greatly appreciated.

References

Adams, B., T. Sivarooban, N. S. Hettiarachchy, and M. G. Johnson. 2005. Inhibitory activity against *Listeria monocytogenes* by soy-protein edible film containing grape seed extract, nisin, and malic acid. Discovery 6:3–9.

Al-ameri, F. M. 2001. Physical properties of edible films with and without antioxidants and the effectiveness of selected film coatings on the shelf life stability of beef cubes. MSc thesis, Fayetteville, University of Arkansas. 85p. Available from Mullins Library, Fayetteville, AR.

Armitage, D. B., N. S. Hettiarachchy, and M. A. Monsoor. 2002. Natural antioxidants as a component of an egg albumen film in the reduction of lipid oxidation in cooked and uncooked poultry. J. Food Sci. 67(2):631–634.

Bandyopadhyay M., R. Chakraborty, and U. Raychaudhuri. 2007. Incorporation of herbs into sandesh, an Indian sweet dairy product, as a source of natural antioxidants, Inter. J. Dairy Tech. 60(3):228–233.

Berkhout, B., R. Floris, I. Recio, and S. Visser. 2003. Antibacterial effects of the milk protein lactoferrin. Natural antimicrobial for food safety. AGRO Food industry hi-tech 32–33. Available from http://www.teknoscienze.com/agro/pdf/may_june03/antibacterial_effects.PDF. Accessed June 3, 2007.

Branen, J. K., and P. M. Davidson. 2000. Activity of hydrolyzed lactoferrin against foodborne pathogenic bacteria in growth media: the effect of EDTA. Lett. Appl. Microb. 30(3):233–237.

Britigan, S. 1997. Pages 211–232 in Lactoferrin: interactions and biological functions. T. W. Hutchens and B. Lönnerdal, eds. Humana Press, Totowa, NJ.

CDC. 2007. Centers for Disease Control and Prevention. Summary of Notifiable Diseases, United States. Morbidity and Mortality Weekly Reports 54(53):18, 19. Available from http://www.cdc.gov/ncidod/dbmd/diseaseinfo/listeriosis_g.htm.

CDC. 2003.Centers for Disease Control and Prevention update: listeriosis disease information. At http://www.cdc.gov/mmwr/PDF/wk/mm5453.pdf.

DHHS-FDA. 1990. Department of Health and Human Services. U.S. Food and Drug Administration, 21 CFR. Part 179. 55(85):18538–18544.

Doyle, M. P. 2000. Reducing food borne disease. Food Tech. 54(11):130.

ERS-USDA. 2005. Economic Resource Service. USDA food-borne illness cost calculator Available from http://www.ers.usda.gov/Data/FoodBorneIllness/. Accessed May 18, 2007.

Eswaranandam, S., N. S. Hettiarachchy, and M. G. Johnson. 2004. Antimicrobial activity of citric, lactic, malic, or tartaric acids and nisin-incorporated soy protein film against *Listeria monocytogenes, Escherichia coli* O157:H7, and *Salmonella gaminara*. J. Food Sci. 69(3):79–84.

Eswaranandam, S., N. S. Hettiarachchy, and M. G. Johnson. 2006. Activation of lactoferrin by citric/lactic/malic acid alone and in combination with EDTA and the effect of activated lactoferrin on the antimicrobial activity against *Listeria monocytogenes* and *Escherichia coli* O157:H7. Institute of Food Technologists Annual Meeting and Food Expo, June 24–28, Orlando, FL. Book of Abstracts # 054A-19.

Fallon, J., and K. Welty. 2003. USDA grants additional approval for activated lactoferrin, News National beef. August 27, 2003. Available from http://www.nationalbeef.com/newsDetail.asp?ID=42. Accessed June 3, 2007.

FDA U.S. Food and Drug Administration. 2003. Lactoferrin considered safe to fight *E. coli*, FDA News, August 22, 2003. Available from http://www.fda.gov/bbs/topics/NEWS/2003/NEW00935.html. Accessed June 3. 2007.

FDA U.S. Food and Drug Administration. 2001. Food borne pathogenic microorganisms and natural toxins hand book. Available from http://vm.cfsan.fda.gov/. Accessed June 3, 2007.

FSIS-USDA. 2007. Food Safety and Inspection Service. United States Department of Agriculture, Washington, DC.. Available from http://www.fsis.usda.gov/Fsis_Recalls/index.asp. Accessed June 3, 2007.

Goff, J. H., A. K. Bhunia, and M. G. Johnson. 1996. Complete inhibition of low levels of *Listeria monocytogenes* on refrigerated chicken meat with pediocin AcH bound to heat killed *Pediococcus acidilactici* cells. J. Food Prot. 59:1187–1191.

Heinonen, I. M., P. J. Lehtonen, and A. Hopia. I. 1998 Antioxidant activity of berry and fruit wines and liquors. J. Agric. Food Chem. 46:25–31.

Hettiarachchy, N. S., and S. Eswaranandam. 2005. Edible films and coatings from soybean and other protein sources. Page 519 in Bailey's industrial oil and fat products, Volume 6, Industrial and nonedible products from oils and fats. Fereidoon Shahidi, ed. Wiley Publishing, Indianapolis, IN.

Hettiarachchy, N. S., and S. Eswaranandam. 2007. Organic acids incorporated edible antimicrobial films, Patent No. US 7,160,580, B2 issued on January 9, 2007.

Hettiarachchy, N. S., K. C. Glenn, R. Gnanasambandam, and M. G. Johnson. 1996. Natural antioxidant extract from fenugreek (*Trigonella foenumgraecum*) for ground beef patties. J. Food Sci. 61(3):516–519.

Hoffman, K. L., P. L. Dawson, J. C. Acton, I. Y. Han, and A. A. Ogale. 1998. Film formation effects on nisin activity in corn zein and polyethylene films. Research and Development Associates for military food packaging systems. 49/50:238–244.

Jay, J. M. 2000. Modern food Microbiology. 6th ed. Aspen Publishers, Gaithersburg, MD.

Miller, A. J., J. E. Call, and B. L. Bowles. 1996. Sporostatic, sporocidal and heat sensitizing action of malic acid against spores of proteolytic *Clostridium botulinum* J. Food Prot. 59(2):115–120.

Naidu, A. S. 2002. Activated lactoferrin—A new approach to meat safety. Food Tech. 56(3):40–46.

Naidu, A. S., and R. Nimmagudda. 2003. Activated lactoferrin, Part 1: a novel antimicrobial formulation. AGRO Food industry hi-tech 47–50. Available from http://www. teknoscienze.com/agro/pdf/march_april03/activated_lactoferrin.PDF. Accessed June 1, 2007.

Naidu, A. S., J. Tulpinski, K. Gustilo, R. Nimmagudda, and J. B. Morgan. 2003. Activated lactoferrin, Part 2: natural antimicrobial for food safety. AGRO Food industry hi-tech 27–31. Available from http://www.teknoscienze.com/agro/pdf/may_june03/activated_lactoferrin.PDF. Accessed June 3, 2007.

Oh, J. H., B. Wang, A. Dessai, and H. Aglan. 2003. The efficacy of natural anti-microbial agent lactoferrin in protein-based edible films against *E. coli*. Abstract, IFT annual meeting 2003, IFT Chicago 45D-8.

Rababah, T., N. S. Hettiarachchy, S. Eswaranandam, J. F. Meullenet, and B. Davis. 2005. Sensory evaluation of irradiated and non-irradiated poultry breast meat infused with plant extracts. J. Food Sci. 70(3):S228–235.

Rababah, T., N. S. Hettiarachchy, and R. Horax. 2004. Total phenolics and antioxidant activities of fenugreek, green tea, black tea, grape seed, ginger, rosemary, gotu kola, and ginkgo extracts, vitamine E and TBHQ. J. Agric. Food Chem. 52:5183–5186.

Rababah, T., N. S. Hettiarachchy, R. Horax, M. J. Cho, B. Davis, and J. Dickson. 2006. Thiobarbituric acid reactive substances and volatile compounds in chicken breast meat infused with plant extracts and subjected to electron beam irradiation. Poult. Sci. 85:1107–1113.

Rababah, T., N. S. Hettiarachchy, R. Horax, S. Eswaranandam, A. Mauromoustakos, J. Dickson, and S. Niebuhr. 2004. Effect of electron beam irradiation and storage at 5°C on thiobarbituric acid reactive substances and carbonyl contents in chicken breast meat infused with antioxidants and selected plant extracts. J. Agric. Food Chem. 52:8236–8241.

Sivarooban, T., N. S. Hettiarachchy, and M. G. Johnson. 2006. Inhibition of *Listeria monocytogenes* by nisin combined with grape seed extract or green tea extract in soy protein film coated on turkey frankfurters. J. Food Sci. 71:2–39.

Sivarooban, T., N. S. Hettiarachchy, and M. G. Johnson. 2007. Inhibition of *Listeria monocytogenes* using nisin with grape seed extract on turkey frankfurters stored at 4 and 10°C. J. Food Prot. 70(4):1017–1020.

Tarte, R. R., E. A. Murano, and D. G. Olson. 1996. Survival and injury of *Listeria monocytogenes, Listeria inocua* and *Listeria ivanovii* in ground pork following electron beam irradiation. J. Food Prot. 59(6):596–600.

Trenev, N. 1998. Page 272 in Probiotics: nature's internal healers. Avery Publishing Group, Garden City, NY.

USDA. 2007. United States Dept. of Agriculture, Arkansas, Agricultural Statistic Service. Poultry: value of production by year and US broiler production by state, US. At http://www.nass.usda.gov/Charts_and_Maps/Poultry/index.asp. Accessed July 7, 2009.

Were, L. 1998. Soy-based protein films with cysteine modification and added gluten. MS thesis, Department of Food Science, University of Arkansas, Fayetteville.

Were, L., N. S. Hettiarachchy, and M. Coleman. 1999. Properties of cysteine-added soy protein-wheat gluten films. J. Food Sci. 64(3):514–518.

Emerging Issues
in Food Safety

∎ 20 ∎

Assessing Consumer Concerns and Perceptions of Food-Safety Risks and Practices: Methodologies and Outcomes

Corliss A. O'Bryan, Philip G. Crandall,
and Christine M. Bruhn

Introduction

Consumers living in the United States have access to the most abundant and one of the safest food supplies in the world, and until 2007, over 80% of those surveyed indicated that they were confident in the safety of the food supply. The 2006 recall of spinach due to the presence of *Esherichia coli* O157:H7 and the discovery of an adulterant in pet food appear to have lessened consumer confidence (Food Marketing Institute 2007). Some consumers express a great deal of concern about the safety of their food (Piggot and Marsh 2004). Heightened concern may be related to extensive, often sensational media coverage of foodborne outbreaks. In 2002, *USA Today* (Schlosser 2002) reported on a recall of ground beef with the headline "Hamburger with those fries? Buyers beware." The author stated that the recall of 140 tons of ground beef was "enough to make more than a million tainted quarter-pound hamburgers." A 2004 article in *Prevention* magazine reporting on an outbreak of hepatitis A associated with apples was headlined "Poisoned Apples" and the lead sentence was "Tainted fruits and vegetables are a hidden health threat to you and your family" (Kamps 2004). Weise (2006a) reported on the 2006 outbreak of *E. coli* O157:H7 associated with spinach in an article titled "It came from beneath the earth." Her lead lines were "The first rule of public health is one most of us learn in kindergarten: Don't eat poop. But that's what the people were eating who were struck down with *E. coli* in the late summer outbreak tied to bagged spinach." Weise (2006b) later recapped food news of 2006 with the title "What a scary stretch it was" and proclaimed the top story was that "the foods that are supposed to keep us in top form . . . might instead kill us." These sensationally negative news stories when associated with one food negatively affect sales of that food but can also negatively impact the whole food supply and the entire food industry (Sennauer 1992). It is important for researchers in the area of food safety and novel processing methods to have a good grasp of what consumers are most concerned with in these areas. Several different methods have been developed to gain an understanding of consumer attitudes and concerns.

Survey Methods

Postal Surveys

Various methods have been used to survey the opinions of consumers about food safety. A mail survey is widely used. Typically researchers purchase names and addresses with specific demographic characteristics, such as age of head of household, urban or rural residency, or ethnicity. The typical procedure is described in detail by Dillman (1978). Tucker et al. (2006) modified the usual practice by sending a preliminary letter to potential participants explaining the purpose of the study and asking for participants. Those who volunteered were sent a cover letter, the questionnaire, a return envelope, and $2 to encourage participation. Over the next several weeks this was followed with a postcard to remind them to fill out the questionnaire, a second questionnaire package, and another reminder postcard. These types of surveys often have low response rates, and Tucker et al. (2006) had a total response rate of 56%. Sapp and Bird (2003) used another variation of the mail survey. They contacted random persons via telephone to recruit them to take the survey. The questionnaires were then mailed to those who had agreed to the survey with $15 as an incentive to finish and return the questionnaire. They had a response rate of 73% of those who had agreed by telephone to complete the survey.

Customer Intercept Surveys

A more direct method of surveying consumer attitudes and knowledge is to use customer intercept surveys that involve face-to-face contact between the surveyor and the person being interviewed. For example, Batte et al. (2007) selected six traditional grocery stores, two in the inner city of a large metropolitan area, two in suburban areas, and two in small towns in a rural area. Random customers were approached as they entered the stores during the hours of 1 P.M. to 6 P.M. Monday through Thursday. About one-third of the customers approached agreed to fill out the survey. Maciorowski et al. (1999a, 1999b) also used the customer intercept method. They set up a table in the stores they surveyed where they placed the questionnaires, pencils, souvenir cups, and pamphlets with information about food safety and irradiation. Stores were surveyed on Friday afternoons and Saturday mornings. Two or three workers solicited willing volunteers to fill out the surveys. Workers offered to read the surveys to the customers, and questionnaires were also available in Spanish. The advantage of this type of approach is that the surveys can continue until a target number of questionnaires are filled out. A disadvantage is that the sample is not collected in a controlled systematic random fashion, and there is no monitoring to assure that people only complete the survey once.

Telephone Surveys

To avoid face-to-face contact, questions can also be asked of consumers by way of telephone surveys, such as that conducted by Roseman and Kurzynske (2006). They contacted households by telephone to find willing participants who were at least 18 years of age and considered themselves to be the main food preparer in the home. Random digit dialing was used to obtain the sample, which was to assure that every household in the state including those with unlisted phone numbers was equally likely to be contacted. A Computer Assisted Telephone Interviewing system dials the numbers at different times of the day and different days of the week until contact is made. Each number was called 15 times or until an answer was received. Only 47.5% of those contacted agreed to take part in the survey.

Focus Groups

Focus groups have been traditionally used in marketing to examine consumers' reaction to a new product or a product concept (Best 1991). They can also be used, however, to determine consumer attitudes to food-safety measures and hazards. Focus groups were demonstrated by Stewart et al. (1994) to be a reliable means of understanding consumer behaviors. Numbers of participants can vary, with experts in the field recommending between 8 and 12 persons per group (Kreuger 2000). A trained moderator in charge of the group follows a planned list of questions and works to keep the discussion on topic. These groups can provide in-depth information on a topic of choice but also allows new topics and new ideas to be brought up by the participants. The moderator must also be able to encourage everyone in the group to participate and not let the discussion be dominated by one or two persons. The comments of the focus group are typically recorded using audio or video equipment to assure comments and reactions can be reviewed and studied. Analysis of results also presents a challenge since results are qualitative in nature. Focus groups are also too small to be considered representative of a population as a whole (Wan et al. 2007). Ideas explored through a focus group may be quantified by a follow-up interview or mail survey.

Risk Perception

Discovering the attitudes consumers have and the specific concerns they have about food safety and food processors in general give us a good indication that the average consumer makes decisions using a perception of risk that may be far different from what professional risk assessors believe to be the "true" risk or the one with the largest quantitative likelihood of causing harm. The majority of people rely on their intuition when making judgments about risk, often questioning or ignoring expert risk assessments. Scientists, especially risk analysts, have said that consumers are acting irrationally, stemming from what the experts believe to be a lack of knowledge and understanding. In actuality Hansen et al. (2003) have found that "lay risk assessments . . . are in fact complex, situationally [*sic*] sensitive expressions of a person's value system." Many Americans believe that they are more at risk today than they were in the past and they also believe that the risk will only increase in the future (Spencer and Crossen 2003). Personal intuition of risk perception can lead to underestimation or overestimation of risk (Slovic 1987). Many psychological factors influence consumers' intuition of perceived risks, including ethics, trust or distrust of scientists and regulatory agencies, and whether the people feel they have some control of their exposure to the risk (Frewer 2000). Slovic and his coworkers (Slovic 1992) have verified that when people see a risk as being involuntary, as being catastrophic or unmanageable, these attributes greatly increase their perception of the risk. Frewer et al. (1998a) also found that people are less likely to be tolerant of any risk from a new technology if they do not see a strong benefit to offset the potential exposure to a new hazard. In order for the public to be accepting of a new technology they also must see that there are clear benefits for consumers or the environment, not just for the manufacturer alone (Frewer et al. 1998b). Mireaux et al. (2007) found that consumers worry that new processing technologies might have "consequences that are unknown" and risks that cannot be determined without extensive testing. Consumers will underestimate some risks that have a high probability of negative health consequences (smoking and obesity) while greatly overestimating other risks (pesticides and food additives) (Miles and Frewer 2001).

Optimistic bias refers to the pervasive phenomenon of perceiving oneself as being less likely than an *average other* to experience negative events (Weinstein 1980). Optimistic bias has been demonstrated for many years in many contexts. In the case of food poisoning, optimistic bias leads people to believe that they are less at risk for illness than other people, and therefore they do not need to pay attention to food-safety information because they think that information is intended for other, more vulnerable people (Miles and Scaife 2003). Parry et al. (2004) studied the difference in risk perception related to foodborne illness between persons who had experienced a verifiable food-poisoning event and those who had not. They found that those who had personal experience with foodborne illness rated their personal risk to a second illness higher than persons who had never had a food-borne illness, although optimistic bias did not disappear entirely.

Consumer Food-Handling Practices

Researchers have determined that consumers perceive their personal risk of food-borne illness from home-cooked food as being low; consumers also believe that they are personally knowledgeable of food-safety risks, that they are in control of their exposure to foodborne illness, and that there is an optimistic bias in effect (it won't happen to me) (Food Marketing Institute 2007; Frewer et al. 1994; Miles et al. 1999). While consumers may believe that they have extensive personal knowl-edge of safe food-handling practice, Meer and Misner (2000) found in a survey of 268 people that most respondents (67%) have very little actual correct food-safety information. Maciorowski et al. (1999b) conducted surveys in Texas in which they asked participants where they mainly received their food-safety information. They found that adult Caucasian consumers in these cities mainly received information from television (60%) or magazines (52%). In contrast, children received most of their information from family (71%) or school (57%). Schafer et al. (1993) devel-oped a Health Belief Model that indicates individuals who perceive that they are very unlikely to be at risk for foodborne illness are the same consumers who are less apt to use good food-handling practices. Roseman and Kurzynske (2006) found this also to be true in their survey of consumers in Kentucky. They found that people who were confident in the safety of the nation's food supply were more likely to engage in risky food handling or unsafe consumer behaviors. Females were more likely than males to perceive that foodborne illnesses are quite common, and they also practiced better food-handling practices and were less likely to con-sume highest risk foods (raw eggs or underdone hamburgers). Byrd-Bredbenner et al. (2007) did a nationwide survey of college students for their self-reported food-handling behaviors as well as their perceived knowledge of safe food-handling practices. They found that the scores on "best practices" for food handling were poor, averaging below 50% in most cases.

Specific problem areas for food handling include subsets for cross-contami-nation prevention, safe time and temperatures for cooking and storing food, com-mon food sources of foodborne illness, and hand washing. Many of the studies into consumer food-handling practices are self-reporting such as the one by Byrd-Bredbender et al. (2007). A few studies have used observational techniques to observe food-handling errors as the person makes them. Anderson et al. (2004) randomly recruited volunteers via telephone to participate in one such study. The volunteers were told that the study was for marketing purposes in order to avoid bias for food safety. The participants were videotaped while preparing an entrée and a salad in their own kitchens. On average the participants failed to wash their hands on seven occasions when they should have done so; the most common scenario for failing to wash hands was when switching between raw meat, seafood, or egg and preparing a salad. Most participants also failed to adequately clean surfaces, especially after raw meat had contacted the counter. Very few of the participants used a thermometer to determine doneness of the entrée, and nearly 50% did not know the recommended final internal temperatures for various meats. Few of the subjects had ever checked the temperature of their refrigerators, and close to half of the refrigerators had temperatures above 4°C (some had temperatures above 7°C).

Kendall et al. (2004) also did an observational study to determine use of safe food handling by participants, although their study took place in community kitchens rather than in private homes. They found that most of the participants washed their hands "correctly" (according to recommended guidelines) only before beginning to prepare food; after working with raw hamburger only half washed their hands before slicing a raw tomato. Only one-third of the participants washed their hands after cutting up raw chicken before slicing an apple. Thermometers were available, although no instructions were given to use a thermometer; instead, participants were asked to tell the observer when chicken or ground beef was done to their preference. More than 80% did not use a thermometer to determine the doneness of the meat, although around 90% did cook the meat to the recommended temperature or higher.

Consumer Food-Safety Concerns

The Food Marketing Institute conducts annual polls of consumers in America, and one of the parameters they check on is food safety. In the 2006 survey (Food Marketing Institute 2007) shoppers were given a list of food-related risks and asked to rank them according to their concern. Bacterial contamination (germs) was identified by 49% of the respondents as being a serious threat to their health and safety. Pesticide/herbicide residues in food were classified as a serious risk by 37% of those polled, GMOs by 20%, irradiated foods by 18%, and additives/preservatives by 16%.

Maciorowski et al. (1999a) observed that demographics can also determine the type of response. They found that out of three Texas cities, those with large Caucasian populations believed that poultry presented the highest food-safety risk when asked to choose among the risk from eating poultry, beef, and pork. However, Hispanics tended to believe that pork had the most harmful bacteria. A majority of consumers who had graduated from college believed that they had become ill from foodborne bacteria but most of the consumers with just a high school degree did not.

Consumer Concerns about Microbiological Contamination

Public opinion of the importance of microbial contamination of foods is increasing. Historically the public has been more concerned about the potential hazards of pesticides or food additives in contrast to scientists and risk assessors who judged microbial pathogens to be the greatest health risk from food (Lechowich 1992). Nearly ten years ago Maciorowski et al. (1999a) also found that 42% of their Texas respondents felt that chemical residues were the most important food-safety problem, while 38% thought it was bacteria and only 16% thought it was food additives or preservatives. Recently, however, almost half of consumers surveyed now believe "germs" in food to be the most serious health risk (Food Marketing Institute 2007). Ralston et al. (2002) believe that this change in perception is due in part to the extensive media coverage of foodborne illness as well as to efforts of food-safety educators promoting the benefits of following food-safety recommendations. The inflammatory tone of some of the media pieces may also contribute to consumer concern about microbiological hazards. Peter Sandman spoke at a food-safety education conference in 1997 and said, "the public responds to outrage. When people are outraged they tend to think the hazard is serious" (Hingley 1997).

Miles and Frewer (2001) found that participants in an interview on food safety were aware that eggs and poultry were particularly at risk for *Salmonella*. They felt that they had good personal knowledge of how to handle, cook, and store these products, but they were concerned about eating eggs and poultry from a food-service establishment. However, several also felt that the press exaggerates the dangers associated with *Salmonella*-caused foodborne illness.

Anderson et al. (2004) noted that consumers tend to believe that their own food-handling processes do not put them at risk for foodborne illness. Most (80%) think foodborne illness is a failure in the food-processing system. While watching food-preparation behavior with cameras they found that nearly all the persons observed cross-contaminated food, primarily through unwashed hands. Many undercooked meat and poultry, and very few consumers ever used a meat thermometer. When Maciorowski et al. (1999b) asked Texas consumers about their preferred method of thawing poultry they found that the majority of certain groups still thawed their poultry outside of refrigeration (without a high school degree 55%, Hispanics 57%, and minors 79%).

Schroeter (2001) found that consumers were worried about the cleanliness of meat-processing plants. She also noted that consumers believed the ground beef they served in their own kitchens was safe because they were well versed in food handling for safety. However, many of the participants in the survey believed that contaminated meat always looks and smells bad rather than understanding that a food may contain a large number of microbial pathogens and still smell, taste, and look fine. This lack of understanding is compounded by the failure to use food-safety information for safer preparation of food in the home. For example, internal meat temperature is a food-safety factor that consumers can control, yet the vast majority of persons in this survey did not use a meat thermometer because of "inconvenience," "laziness," or the "hassle."

Consumer Perception of Novel Food-Processing Technologies and Food Safety

Consumers' attitude studies are useful to assess consumer concerns and responses to information, but these studies should not be considered to directly reflect consumers' responses in the marketplace. This is because consumer attitude research methods typically focus on one specific technology or ingredient while consumers respond to the matrix of stimulus from the product as a whole. The research approach encourages the volunteer to focus on the issue being examined, and this focus may in itself raise concerns where none existed before. For example, when consumers are asked to indicate their concern about processing technologies, some will state that they are very concerned about pasteurization and other technologies commonly used in food preservation (Cardello et al. 2007).

New food-processing methods and new packaging and coating materials for food are of great concern to consumers. As stated in the risk-perception portion of this chapter, consumers feel great concern about these new technologies because they view them to be an involuntary risk and out of their control. They are afraid that these novel methods will only benefit the industry while they can have unknown and potentially fatal effects on their health and safety. Let's now take a look at consumers' perception of some novel food-processing technologies.

Edible Films and Coatings

Wan et al. (2007) conducted focus group studies to determine consumer attitudes and concerns about edible films and coatings (EFCs). Of the 27 participants in the groups, most were not aware of EFCs. When EFCs were described the typical response was that they would need to know the type of product being coated to determine whether the risk would be acceptable. In general most felt that a coating would be more acceptable on a type of product that has a natural outer layer (peel) that could be removed before consumption, such as a banana or an orange. The safety of the coating was a concern, mainly in reference to the ingredients and the handling of the product. The products would need to be labeled as having been coated as well as having an ingredient statement about the coating. They would prefer that the coating ingredients be "natural" and not artificial. If the stated purpose of the coating was to extend shelf life the participants preferred the coating to have no taste or color and to be transparent. However, they had a favorable response to the use of EFCs as carriers of flavors for development of new products. Using the coatings to extend shelf life was viewed as being of greater benefit to the food manufacturers and the retailers than to the consumers. In addition they were worried that the coating might encourage the producers, shippers and/or retail employees to be less careful about cleanliness. When comparing coated and uncoated products, most respondents preferred to buy the cheaper uncoated product unless they saw a real advantage to themselves from the coatings.

High-Pressure Processing

Butz et al. (2003) interviewed 3,000 consumers in France, Germany, and the United Kingdom regarding their attitudes toward high-pressure processing (HPP). They set an acceptability threshold termed "Most Advanced Yet Acceptable" of 60%. Information was provided to the panelists by means of a statement card with information about HPP. The total of conditional and unconditional purchasers was determined to be 74% in Germany, 71% in France, and 55% in the United Kingdom. Most of the potential buyers were conditional buyers, meaning that they would buy products produced using HPP if they were not more expensive and if they felt that there was a personal health benefit. Interest in benefits varied by country with French consumers most interested in improved flavor and German consumers most interested in enhanced health properties.

Cardello (2003) found that somewhat less than 30% of participants in a taste test of chocolate pudding preserved by HPP expressed any level of concern about HPP. Expected desire for the product increased when a description of the process was added and when the pudding was seen before tasting.

Irradiation

Tauxe (2001) has estimated that if only half of the ground beef, poultry, pork, and processed meat produced in the United States each year was irradiated at least 900,000 cases of foodborne illness, 8,500 hospitalizations, and 350 deaths could be prevented. This would be a huge benefit to society. However, many consumers are unfamiliar with these benefits. When asked to indicate level of concern toward a range of issues, fewer than 30% indicate they are concerned about irradiation (Food Marketing Institute 2006).

When the term "irradiation" is the focus of a research project, some consumers respond negatively. Cardello (2003) used pre-tests and post-tests in conjunction with seeing then tasting chocolate puddings they were told were irradiated. A majority, 65%, of the respondents expressed some level of concern about irradiation prior to the test. Expected desire for chocolate pudding was assessed by using the name "chocolate pudding" only. "Liking" was then assessed for chocolate pudding processed with irradiation, processed with irradiation plus a description of the irradiation process, or processed with irradiation plus the description of the process plus a stated benefit. Expected "liking" decreased when irradiation was mentioned as the processing method, and did not recover even with the description of the process plus a benefit statement. After having rated their expectation for liking the product they were allowed to taste it and there was a direct correlation between how bad they expected the product to be and how low they rated the taste of the product.

Hoefer et al. (2006) used a telephone survey to assess the knowledge and attitudes of more than 3,000 consumers at the FoodNet sites. While the vast majority of respondents knew that cross-contamination could occur from raw meat to other foods and most of them knew that irradiation kills harmful bacteria that could cross-contaminate other foods in the kitchen, only 60% believed that irradiation of meat was safe. Only slightly more than 40% were aware that irradiated meat was available for purchase in supermarkets. A very small percentage thought that they would be exposed to radioactive material from eating irradiated meat.

When Maciorowski et al. (1999a) surveyed the acceptability of irradiation for Texas consumers from three cities, 42% of Caucasians surveyed said that they would eat irradiated poultry. Men (44%) were more likely than women (27%) to say they would eat irradiated poultry. In general, the more educated the respondent the more likely they were to say they would consume irradiated poultry. Many consumers will accept irradiation when they are provided with specific information about the advantages and about the safeguards in place to protect the environment as well as the workers in the irradiation industry (Bruhn 1995). When consumers hear that irradiation provides protection from foodborne illness, from 60% to 99% report that they would choose irradiated meat and poultry (Aiew et al. 2003; Johnson et al. 2004; Fox 2002; Vickers and Wang 1999).

Hoefer et al. (2006) noted that grocery store chains that carry irradiated food can assist in educating the public in advertising and with point-of-sale displays of information. Providing training about food irradiation to community health educators also increased the availability of information to the public about this food-processing method (Thompson and Knight 2006). Actual marketplace experience indicates that consumers valued the longer shelf life of irradiated produce offered in a Chicago area store (Pszczola 1992). Further, consumers have chosen the added safety provided by irradiated meat and poultry sold in select regions of the country (Eustice and Bruhn 2006).

The food industry and regulatory authorities need to develop, evaluate, and facilitate safe and effective technologies to improve the safety of the food supply.

Educational efforts should perhaps focus on people's prior attitudes and the credibility of the information source, and public beliefs about the uncertainties inherent in the risk-assessment process.

Pesticide Residues in Foods

Pesticide residues have been at or near the top of lists of consumer concerns for food safety for decades (Byrne et al. 1991; Govindasamy et al. 1997). This could possibly be due to the uncertainty associated with exactly how much pesticide each person is actually exposed to each day. Govindasamy and Italia (1998) wanted to determine whether consumer concern about pesticide residues had a sociodemographic basis. They found that females, especially those with children, and persons above the age of 35 were more concerned about risk from pesticides. Higher income and education levels led to less feeling of being more at risk.

Miles and Frewer (2001) found that consumers they interviewed associated pesticides with chemicals in general and had a negative opinion of chemicals. They were also concerned with the impact on the environment and impact on animals, especially pets. They were in favor of alternatives to pesticides and had a high opinion of organic food, but interestingly did not buy organic food because of their perception that organic foods were more expensive.

Tucker et al. (2006) conducted a mail survey in Ohio to find out Ohioans perceptions of food-safety risks and to seek to determine what factors might influence their opinions. Ohio is a large and populous state with large areas of farmland as well as many small towns and three major metropolitan areas. Respondents were asked to rank the risk from various items from 1 (no risk) to 7 (serious risk). They found that over 70% of the respondents to the survey felt that pesticides in food were of moderate to serious risk.

Conclusions

The information that consumers ask for is quite different than that requested by scientists. Scientists tend to ask questions such as "how high is the risk?" or "am I in immediate danger?" Callers to food-safety experts mostly asked "What can I do?" or "What can I eat without being in danger?" (Renn 2005). These consumers are also concerned about whom to trust, who is telling the truth. Tucker et al. (2006) found that most of their respondents most trusted physicians or other health-care professionals and university scientists to give them the best information about food safety and environmental issues. Interestingly, consumer advocacy groups were at the bottom of the list of trusted sources.

It appears that many consumers are not aware of the level of control they can actively exercise to minimize their risk of foodborne illness. Educators need to emphasize practical steps consumers can take to reduce their risks from food.

Educators also need to be aware that it is not always simple lack of information or lack of understanding by consumers. They need to also assess the aims and values of the consumers and understand that consumers do not necessarily evaluate risks and benefits separately. People may also have been misled to believe that certain risks are exceptionally high. For food additives and novel processing methods, acceptance will likely depend on whether the consumer sees a benefit to themselves that they perceive to outweigh any perceived loss of quality or danger from the additive or process. Benefit will need to be perceived as directed at the consumer or the environment rather than purely to benefit the food industry. Television and magazines have been reported to be the source of most people's food-safety information (Tucker et al. 2006; American Dietetic Association 2000). Risk communicators and food-safety specialists can work together to put together stories and publications that will attract consumers without creating panic and outrage. Many consumers remain unsure about food-safety risks and new technologies. Science-based information should be presented to them in a manner that they can easily understand and presented to them via mass media outlets.

References

Aiew, W., N. Rudolfo, and J. Nichols. 2003. The promise of food irradiation: will consumers accept it? Choices (fall 2003):31–34.

American Dietetic Association. 2000. Nutrition and you: trends 2000: what do Americans think, need, expect? J. Am. Diet. Assoc. 100(6):626–627.

Anderson, J. B., T. A. Shuster, K. E. Hansen, A. S. Levy, and A. Volk. 2004. A camera's view of consumer food-handling behaviors. J. Am. Diet. Assoc. 104(2):186–191.

Batte, M. T., N. H. Hooker, T. C. Haab, and J. Beaverson. 2007. Putting their money where their mouth is: consumer willingness to pay for multi-ingredient, processed organic food products. Food Policy 32145–159.

Best, D. 1991. Designing new products from a market perspective. Pages 1–28 in Food product development. E. Graf and I. S. Saguy, eds. AVI Book, New York, NY.

Bruhn, C. M. 1995. Consumer attitudes and market response to irradiated food. J. Food Prot. 58(2):175–181.

Butz, P., E. C. Needs, A. Baron, O. Bayer, B. Geisel, B. Gupta, U. Oltersdorf, and B. Tauscher. 2003. Consumer attitudes to high pressure food processing. Food Ag. Environ. 1(1):30–34.

Byrne, P., C. Gempesawll, and U. Toensmeyer. 1991. An evaluation of consumer pesticide residue concerns and risk perceptions. S. J. Agric. Econ. 23(2):167–174.

Cardello, A. V. 2003. Consumer concerns and expectations about novel food processing technologies: effect on product liking. Appetite 40:217–233.

Cardello, A. V., H. G. Schutz, L. L. Lesher. 2007. Consumer perception of foods processed by innovative and emerging technologies: a conjoint analysis study. Innov. Food Sci. Emerg. 8(1):73–83.

Dillman, D. 1978. Mail and telephone surveys. The Total Design Method. John Wiley, New York, NY.

Eustice, R., and C. M. Bruhn. 2006. Consumer acceptance and marketing of irradiated foods. In Food irradiation research and technology. Christopher Sommers and Xuetong Fan, eds. Blackwell Publications, New York, NY.

Food Marketing Institute. 2007. US grocery shopper trends 2007: food safety. Available at http://www.fmi.org/foodsafety/presentations/trends_food_safety_chapter.pdf. Accessed August 2007.

Fox, J. A. 2002. Influences on purchases of irradiated foods. Food Technol. 56(11):34–37.

Frewer, L. J. 2000. Risk perception and risk communication about food safety issues. Nutr. Bull. 25:31–33.

Frewer, L. J., C. Howard, and R. Shepherd. 1998a. Development of a scale to assess attitudes towards technology. J. Risk Res. 1: 221–237.

Frewer, L. J., C. Howard, and R. Shepherd. 1998b. Understanding risk perceptions associated with different food processing technologies used in cheese production—a case study using conjoint analysis. Food Qual. Pref. 8: 271–290.

Frewer, L. J., R. Shepherd, and K. Sparks. 1994. The interrelationship between perceived knowledge, control and risk associated with a range of food-related hazards targeted at the individual, other people and society. J. Food Safety 14:19–40.

Govindasamy, R., and J. Italia. 1998. Predicting consumer risk perceptions towards pesticide residue: a logistic analysis. Appl. Econ. Letts. 5:793–796.

Govindasamy, R., J. Italia, and C. Liptak. 1997. Quality of agricultural produce: consumer preferences and perceptions. New Jersey Agricultural Experiment Station, Rutgers University, P-02137-1-97, February.

Hansen, J., L. Holm, L. Frewer, P. Robinson, and P. Sandoe. 2003. Beyond the knowledge deficit: recent research into lay and expert attitudes to food risks. Appetite 41:111–121.

Hingley, A. 1997. Rallying the troops to fight food-borne illness. Available at http://www.fda.gov/fdac/features/1997/797_food.html. Accessed August 2007.

Hoefer, D., S. Malone, P. Frenzen, R. Marcus, E. Scallan, and S. Zansky. 2006. Knowledge, attitude, and practice of the use of irradiated meat among respondents to the FoodNet population survey in Connecticut and New York. J. Food Prot. 69(10):2441–2446.

Johnson, A., R. A. Estes, C. Jinru, and A. V. A. Resureccion. 2004. Consumer attitudes toward irradiated food: 2003 vs. 1993. Food Prot. Trends, 24(6):408–418.

Kamps, L. 2004. Poisoned apples. Prevention 56(9):142–151.

Kendall, P. A., A. Elsbernd, K. Sinclair, M. Schroeder, G. Chen, V. Bergmann, V. N. Hillers, and L. C. Medeiros. 2004. Observation versus self-report: validation of a consumer food behavior questionnaire. J. Food Prot. 67(11):2578–2586.

Kreuger, R. A., and M. A. Casey. 2000. Focus groups: a practical guide for applied research. Sage Publications, Thousand Oaks, CA.

Lechowich, R. V. 1992. Current concerns in food safety. ACS Symposium Series 484:232–242.

Maciorowski, K. G., S. C. Ricke, and S. G. Birkhold. 1999a. Consumer food safety awareness and acceptance of irradiated raw poultry in three Texas cities. Dairy Food Environ. Sanit. 19(1):18–28.

Maciorowski, K. G., S. C. Ricke, and S. G. Birkhold. 1999b. Consumer poultry meat handling and safety education in three Texas cities. Poult. Sci. 78:833–840.

Meer, R. R., and S. L. Misner. 2000. Food safety knowledge and behavior of expanded food and nutrition education program participants in Arizona. J. Food Prot. 63(12):1725–1731.

Miles, S., D. S. Braxton, and L. J. Frewer. 1999. Public perceptions about microbiological hazards in food. Br. Food J. 101(10):744–762.

Miles, S., and L. J. Frewer. 2001. Investigating specific concerns about different food hazards. Food Qual. Pref. 12:47–61.

Miles, S., and V. Scaife. 2003. Optimistic bias and food. Nutr. Res. Rev. 16(1):3–19.

Mireaux, M., D. N. Cox, A. Cotton, and G. Evans. 2007. An adaptation of repertory grid methodology to evaluate Australian consumers' perceptions of food products produced by novel technologies. Food Qual. Pref. 18:834–848.

Parry, S. M., S. Miles, A. Tridente, S. R. Palmer, South and East Wales Infectious Disease Group. 2004. Differences in perception of risk between people who have and have not experienced *Salmonella* food poisoning. Risk Anal. 24(1):289–299.

Piggott, N. E., and T. L. Marsh. 2004. Does food safety information impact US meat demand? Am. J. Agr. Econ. 86(1):154–174.

Pszczola, D. E. 1992. Irradiated produce reaches Midwest market. Food Technol. 46(5):89–92.

Ralston, K., C. P. Brent, Y. Starke, T. Riggins, and C. T. J. Lin. 2002. Consumer food safety behavior: a case study in hamburger cooking and ordering. Agricultural Economic Report No. 804. Available at: http://www.ers.usda.gov/Publications/aer804/. Accessed August 2007.

Renn, O. 2005. Risk perception and communication: lessons for the food and food packaging industry. Food Addit. Contam. 22(10):1061–1071.

Roseman, M., and J. Kurzynske. 2006. Food safety perceptions and behaviors of Kentucky consumers. J. Food Prot. 69(6):1412–1421.

Sapp, S. G., and S. R. Bird 2003. The effects of social trust on consumer perceptions of food safety. Soc. Behav. Personal 31(4):413–422.

Schafer, R. B., E. Schafer, G. L. Bultena, and E. O. Hoiberg. 1993. Food safety—an application of the health belief model. J. Nutr. Educ. 25(1):17–24.

Schlosser, E. 2002. Hamburger with those fries? Buyers beware. USA Today, July 10, 2002: 11a.

Schroeter, C. 2001. Consumer attitudes towards food safety risks associated with meat processing. MS thesis, Kansas State University.

Senauer, B. 1992. Consumer food safety concerns. Cereal Foods World 37:298–303.

Slovic, P. 1987. Perception of risk. Science 236:280–285.

Slovic, P. 1992. Perceptions of risk: reflections on the psychometric paradigm. Pages 117–152 in Social theories of risk. D. Golding and S. Krimsky, eds. Greenwood, Westport, CT.

Spencer, J., and C. Crossen. 2003. Fear factors: why do Americans feel that danger lurks everywhere? Wall Street Journal, April 24, 2003: A1.

Stewart, B., D. Olson, C. Goody, A. Tinsley, R. Amos, N. Betts, C. Georgiou, S. Hoerr, R. Ivaturi, and J. Voichick. 1994. Converting focus group data in food choices into a quantitative instrument. J. Nutr. Educ. 26:159–168.

Tauxe, R. V. 2001. Food safety and irradiation: protecting the public from foodborne infections. Emerg. Inf. Dis. 7(3):516–521.

Thompson, B. M., and S. L. Knight. 2006. The effect of multicomponent professional development training on the beliefs and behaviors of community health educators concerning food irradiation. Health Ed. Behav. 33(5):703–713.

Tucker, M., S. R. Whaley, and J. S. Sharp. 2006. Consumer perceptions of food-related risks. Int. J. Food Sci. Technol. 42:135–146.

Vickers, Z., and J. Wang. 1999. Acceptability of irradiated fresh ground beef patties; influence of information and product identification. University of Minnesota.

Wan, V.C.-H., C. M. Lee, and S-Y Lee. 2007. Understanding consumer attitudes on edible films and coatings: focus group findings. J. Sens. Stud. 22: 353–356.

Weinstein, N. D. 1989. Effects of personal experience on self-protective behavior. Psycholog. Bull. 105:31–50.

Weise, E. 2006a. It came from beneath the earth. USA Today, October 31, 2006: 8d.

Weise, E. 2006b. What a scary stretch it was. USA Today, December 26, 2006: 4d.

▮ 21 ▮

Food Safety of Natural and Organic Poultry

Philip G. Crandall, Corliss A. O'Bryan, Steven C. Ricke,
Frank T. Jones, Steven C. Seideman, Ron Rainey,
Elizabeth A. Bihn, Teresa Maurer, and Anne C. Fanatico

Introduction

The 1,600 members of the Organic Trade Association (OTA) that grow and process organic foods recently released preliminary findings from its survey of manufacturers (OTA 2007). Organic foods accounted for nearly 3% of total U.S. food sales and meat/fish/poultry sales showed the largest annual market growth percentage (Table 21.1). The growth rate for organic foods in 2006 was 22.1%, exceeding the prediction of 20.7% (Table 21.2). This rate of growth is more than seven times the typical rate of growth, 3%, of most other retail food categories. In a 2005 survey OTA members predicted 10–15% growth rate in the organic food sector from 2006 through 2010 and 5–10% additional growth from 2011 to 2025 (OTA 2005). These predictions seem quite reasonable given that the annual growth in organic food purchases has maintained a sustained annual increase of 16–21% per year from 1997 to 2004. This predicted growth will add about $2 billion in annual sales to the total projected organic food sales of $50 billion to $70 billion in 2025. Predictions are that by 2025, 68.3% of all U.S. food companies will offer organic foods as part of their product line (Givens and Bell 2005). Another key prediction by OTA members was that by 2025, the average consumer will purchase some organic foods on a regular basis. For this to occur, organic foods will need to be available at almost every retail food outlet. Members thought that young, affluent shoppers, especially Millennials (born between 1977 and 1994), Gen X's (born between 1965 and 1980), and the 76 million Boomers (born between 1946 to 1964) plus Asian, black, and Hispanic Americans will purchase organic foods in greater amounts than the overall population.

Table 21.1. 2005 Organic Food Sales, Market Share, and Growth by Category

Food Category	Sales ($ mil.)	Percent of Total Organic Sales (%)	Annual Market Growth (%)
Dairy	2,140	15	23.6
Bread & Grains	1,360	10	19.2
Beverages	1,940	14	13.2
Fruit & Vegetables	5,369	39	10.9
Snack Foods	667	5	18.3
Packaged/Prepared Foods	1,758	13	19.4
Sauces/Condiments	341	2	24.2
Meat/Fish/Poultry	256	2	55.4
	13,831	15.7	

Source: Adapted from OTA 2007.

Part of the reason for the recent surge in sales of organic foods is the strong and increasing regulatory support for organic production, processing and enforcement of organic standards by USDA agencies. As recently as June 2007, USDA released another Interim Final Rule as an addition to the National Approved List of Chemicals that can be used in organic foods (USDA-AMS 2007a, 2007c). This regulatory oversight and continual additions keep organic production viable. The 2008 Farm Bill when passed contained a five-fold increase in mandatory funding for organic programs over funds mandated in the previous legislation, and authorized additional funding for many new and existing programs.

Another USDA agency heavily involved in organic foods is the Economic Research Service, which estimated that the organic food market sector is the fastest-growing sector in the total U.S. food industry. It attributed this rapid growth in part to increased consumer concerns about the safety of conventional foods combined with the evolution of strong, new organic production and marketing systems. From 1997 to 2001, sales of organic foods more than doubled (Table 21.2) growing from $3.5 billion to over $7 billion. From 2003 to 2006 sales more than doubled again (USDA-ERS 2005).

Regulatory and Consumer's Perceived Restrictions for Organics

To appreciate the demand and opportunities in this rapidly growing area it is important to understand the legal definitions and consumers' perceptions that have had a profound effect on the market and future growth of organics. There are also multiple levels of complexity in the production of organic foods due in part to the large number and small size of the current organic producers.

Table 21.2. Total Food and Organic Sales Data.

Year	Sales ($ bil.)	Organic Food Growth (%)	Total Food Sales ($ bil.)	Organic Penetration
1997	$3,594	—	443,724	0.81
1998	4,286	19.2	454,140	0.94
1999	5,039	17.6	474,790	1.06
2000	6,100	21.0	498,380	1.22
2001	7,360	20.7	521,830	1.41
2002	8,635	17.3	530,612	1.63
2003	10,381	20.2	535,406	1.94
2004	11,902	14.6	544,141	2.19
2005	13,831	16.2	556,791	2.48
2006*	$16,700	20.7	598,136	2.79

Source: Adapted from OTA (2006) and USDA-ERS (2006b).
*Preliminary estimate

According to USDA's Economic Research Service (Greene 2006), total certified organic crop acreage increased 11% between 2001 and 2003. This increase was paced by large increases for fruits, vegetables, and hay crops used by certified organic dairies. Farmers in 49 states dedicated 2.2 million acres of cropland and pasture to organic production systems in 2003, with over 68% of this acreage being devoted to growing crops.

In addition to this background on the current demand situation, it is critical to understand the regulatory side of the organic foods picture. The USDA publishes strict guidelines for USDA certified organic production as part of the National Organic Program (NOP) Rule and the National Organic Standards. These regulations were established by the Organic Foods Production Act of 1990, which became effective on October 21, 2002 (USDA-AMS 2000). The USDA organic seal can be used to identify organic foods whose production systems have been inspected, approved by third-party certifying agencies, and found in compliance with the USDA's organic regulatory requirements.

USDA's organic regulations require that USDA certified 100% organic foods be grown according to a written Organic System Plan without antibiotics or growth hormones in livestock production and that the animals' feed be certified as organic. In crop production, crops must be grown without the use of synthetic fertilizers or synthetic pesticides. For both crops and livestock producers there is a 36-month period when an organic grower must operate under strict organic guidelines but cannot sell their product as USDA certified organic until the completion of this "wash-out" time period. The only exception is when a grower can prove that no synthetic inputs have been used in the previous 36 months of production. Before organic production and processing facilities can be considered USDA "Certified Organic" an accredited certifying organization, contracting with third-party auditors, must make on-site inspections. Typically the state departments of agriculture contracts with a third-party organization to provide organic inspectors for organic producers in their state. Third-party auditors conduct these annual inspections according to 7 CFR 205.403(a). Legally imposed restrictions carry a $10,000 fine for each misleading violation of USDA's organic standards (USDA-AMS 2006a).

Definition of Organic and Natural Foods

Organic Foods

USDA has mandated a three-tiered nomenclature for organic foods. Foods labeled as "100% Organic" contain no nonorganic ingredients. Foods labeled as "Organic" must contain more than 95% organic ingredients. Finally, foods labeled as having been "made with organic ingredients" will contain between 70 and 95% of the ingredients as organic. Additional background for these concepts are on line at the 7 CFR § 205, "The National Organic Program" (NOP) (USDA-AMS 2006b). The USDA's press release on the latest changes to the National List of Allowed and Prohibited Substances can be found at USDA-AMS (2007a). Additional information is also available from Pittman (2004).

Natural Foods including Natural Poultry

Unfortunately, there have been years of delays before governmental regulators imposed these legal definitions for organic foods. In this void, marketers exploited the terms "natural or all natural" on their labels to answer consumer demands in this area. The USDA is considering mandating changes in what has become a very broadly defined market. The current USDA definition for "natural" is quoted below and can be found in 7 CFR § 205.2.

> *Non-synthetic (natural).* A substance that is derived from mineral, plant, or animal matter and does not undergo a synthetic process as defined in section 6502(21) of the Act (7 U.S.C. 6502(21)). For the purposes of this part, non-synthetic is used as a synonym for natural as the term is used in the Act.

In addition, the USDA defines natural for the purposes of labeling as

> a product containing no artificial ingredient or added color and is only minimally processed (a process which does not fundamentally alter the raw product) may be labeled natural. The label must explain the use of

the term natural (such as—no added colorings or artificial ingredients; minimally processed). (USDA-FSIS 2006b)

The USDA has also proposed a "Naturally raised" marketing claim standard that would be voluntary. It would allow livestock producers to make claims associated with production practices in order to distinguish their products in the marketplace (USDA-AMS 2007b).

The Penetration of Organic into U.S. Mainstream Food Markets

Today there are more than 20,000 natural food stores, and 73% of conventional grocery stores carry some organic food products. Organic foods have made deep penetration into mainstream retail markets since 1997, nearly tripling their sales from conventional retail outlets. Mainstream food markets provide access to the majority of consumers, beyond those consumers buying strictly from natural food stores. In 2005, conventional retail outlets accounted for nearly 40% of all organic food sales. This growth has laid a foundation for even deeper market penetration by reducing the price differential between conventional and organic, continuing to increase the demand for quality organics and further expansion of organic foods into new distribution channels (OTA 2005).

Organic Fruits, Vegetables, and Meats

The leading category, with 42% of total organic foods sales, is labeled, "Produce, fresh fruit and vegetables." Sales of fruit and vegetables dwarf prepared foods at 13% of organic food sales and organic meat with current sales of just 1%. On March 1, 2006, the Food Marketing Institute (FMI) together with American Meat Institute (AMI) published a survey of 1,750 consumers' purchasing behaviors. They reported that price was the most significant factor driving meat purchases. For organic meats, 17.4% of these consumers (or about 1 in 5) purchasers bought organic meat in the past 90 days and 48% of these purchases of organic meats took place in a conventional supermarket. The reasons given most often by organic consumers for purchasing organic meats included superior taste, better nutritional value, long-term health benefits, enhanced product freshness, and curiosity about the differences between organic and nonorganic meats. These consumer perceptions persist despite an almost complete lack of scientific research to support these positions (FMI/AMI 2006).

The Ready-to-Eat Segment

In 2004, the ready-to-eat foods market segment of conventional foods was $150 billion in annual sales. This conventional foods RTE market segment is by itself three times the size of the total market for organics, and organic RTE foods comprise only a tiny fraction of the organic market. Biggest gainers in the conventional RTE market share were shelf-stable dinners with meat (up 46%), skillet and bowl dinners with meat (up 36%), salad kits (up 11%), and the "luncheon solutions" category that grew $387 million in supermarket sales (Sloan 2004). If the same growth in consumer demand transfers to the organic RTE market, there could be a significant increase in the demand for organic meats, especially poultry.

Organic Eggs and Poultry

According to Oberholtzer et al. of the USDA- ERS (2006), organic eggs and organic poultry are the fastest-growing portion of the U.S. organic market. The cost to consumers for organic poultry is very high compared with conventional poultry and will remain so for sometime because of a lack of organic feed grains and very limited organic processing capacity. Oberholtzer et al. (2006) reported that the meat portion of organic sales is the fastest-growing organic food segment and two-thirds of the demand in the meat sector is for poultry. When consumers first venture into the organic arena, there are certain foods they are more likely to try first, known as "gateway foods." Organic poultry was first named as a gateway food in 2004 (Demeritt 2004).

Organic Poultry and Meats

Popularity

Chicken is the most popular source of protein in the American diet where almost 40 kg per capita of chicken are consumed each year (USDA-ERS 2005), up from 31 kg in 1993. In contrast, beef consumption decreased to 29 kg and pork consumption had dropped to 23 kg in 2003 (NRA 2005). Focusing more specifically on organically grown poultry, Laux (2006) published an updated review of "Organic Poultry" in which she stated that in 2005, organic fresh poultry, meat, and seafood sales were the fastest-growing market segment at an annual increase of 67.4% in sales for 2005. Organic poultry holds the greatest share of the organic meat market (Figure 21.1).

Future Growth for Organic Poultry, Value Added, and RTE

It is anticipated that organic poultry processing will follow a similar "logical market development" taken by conventional poultry: first developing a strong production base, then integrating production/processing (which is taking place now), and finally moving strongly into producing higher margin, value-added products. Researchers in food safety must anticipate this future demand and develop novel food-safety protocols and products that are sufficiently robust to maintain efficacy under further processing conditions that likely pose increased risks. If not antici-

Organic meat sales by type

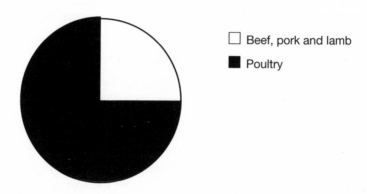

☐ Beef, pork and lamb
■ Poultry

Figure 21.1. Chart created using information from the Organic Trade Association 2006 Manufacturer Survey.

pated and addressed, food pathogens may well persist despite the organic industry's best efforts to minimize microbial pathogens on organic food products.

Much of the growth in conventional poultry consumption has been from "value-added" and further-processed poultry products. Value-added poultry products are those in which additional ingredients or processes have made the consumers' preparation more convenient or faster such as adding batter and breading before being set by partially frying in oil.

In a recent interview, a leader of the poultry industry stated his company will increase its profitability by focusing on adding value to products, "selling easy-to-fix chicken meals instead of basic (raw, dressed) chicken in a package." Value-added poultry products generate more dollars of revenue per dollar of poultry produced. Value-added margins run 2–2½ times that of raw, dressed chicken (Souza 2006).

The Regulations

The regulations for the production of organic livestock including poultry are divided into the following areas: origin of livestock (205.236), livestock feed (205.237), livestock health-care practices (205.238), and livestock living conditions (205.239) (USDA-AMS 2000; USDA-AMS 2006b).

According to the Organic Trade Association (2005) the philosophy of organic poultry production is to provide growing conditions that meet the health needs and natural behavior of the animal. Thus, organic livestock are given access to the outdoors, fresh air, sunshine, grass, and pasture and are fed 100% organic feed (OTA 2005). In addition, any shelter provided must be designed to allow the animal comfort and the opportunity to exercise.

Problems Unique to Organic Poultry Production and Processing

Both conventional and many organic poultry companies are vertically integrated. This means the parent company controls all aspects of production and processing from the breeder flock, egg laying, the hatchery, broiler feed supplies, and hygiene during grow-out, through transportation, slaughter, and further processing in facilities completely managed by the integrating parent company (Laux 2006; Petaluma Poultry 2007). Preharvest is the term given to poultry production steps from the fertilized egg to the point the broilers are transported to the processing plant. All of the processing steps from slaughter to the finished consumer product are called processing and further processing. Large, vertically integrated processors tend to have better control over their entire process than do small producers or processors who only accomplish a portion of the entire process. Because many organic poultry grow-out operations tend to be small or very small, there are increased inherent risks.

Good Agricultural Practices for Organic Poultry Production

Understanding and controlling the physical, chemical, and biological hazards involved in growing, harvesting, and processing meat and poultry are critical to reducing food-safety risks. Conventional and organic meat and poultry producers must be aware of and concerned about food safety. Foodborne illness outbreaks are bad for business and devastating for the consumer who becomes ill. Meat and poultry products have served as vehicles for the transmission of foodborne pathogens that have resulted in numerous foodborne illness outbreaks (Tauxe 2002).

Developing a food-safety plan depends on many variables. For those involved in meat and poultry slaughter and processing, the Food Safety and Inspection Service of the USDA has published a final rule entitled, "Pathogen Reduction: Hazard Analysis and Critical Control Point (HACCP) Systems." It was published on July 25, 1996, and specifies four areas of compliance to help insure safety: (1) Implement Standard Operating Procedures for Sanitation; (2) Test of certain products for generic E. coli; (3) Implementation of a HACCP system; and (4) Meeting USDA-FSIS Salmonella performance standards. This rule is highlighted to acknowledge all federal rules and state regulations must be complied with by processors at meat and poultry slaughter and processing operations. The detailed specifics of this rule can be found at USDA-FSIS (2008).

It is vital to raise awareness of persons considering entering organic production and encourage a careful review of current practices. Producing, transporting, slaughtering, and processing of poultry is a complex system with many opportunities for microbial pathogens to enter and persist. The more producers understand about microbial food safety, the more opportunities there will be to develop strategies to reduce the risks and improve the safety of the flocks as well as the safety of the food for consumers. This section provides an overview of areas to consider as well as suggestions for how to implement effective food-safety practices.

Farm activities related to preharvest poultry production are currently not mandated by regulations. However, there are several management strategies that are available to assist in developing a food-safety risk-reduction plan including Good Agricultural Practices, Good Production Practices, Good Management Practices, or commodity-specific Quality Assurance Programs. The FSIS has established many state partnerships to support programs and local expertise to help producers understand and implement food-safety protocols. The most important beginning step is to review current practices and identify areas where risks exist. In poultry production, food-safety practices should be incorporated into all aspects of production.

Animal living conditions, including house sanitation and pest control, are important to providing and maintaining animal health. Healthy animals are less likely to contract illnesses and infections that pose both animal and human health risk. Sanitation, including scheduled cleaning and sanitation of drinkers, feeders, houses, range areas, and transportation vehicles, is important for reducing pathogen load and preventing buildup. Pest (rodents, flies, birds, etc.) control is important whether it is considered part of the sanitation program or an independent program. Because these pests can carry pathogens into poultry houses, preventative pest controls are also needed.

Fanatico (2008) prepared guidelines for organic poultry grow-out and notes that for organic production cycles, allowing two weeks between flocks helps control host-specific pathogens. After rearing a flock, the poultry house needs to be cleaned of all organic matter, sprayed with a detergent using a high-pressure sprayer, then rinsed and allowed to dry. Approved disinfectants include chlorine, iodine, hydrogen peroxide, and organic acids. Water lines need to be flushed with citric acid or vinegar then sanitized with iodine or hydrogen peroxide.

Horizontal transmission of pathogens can result from contaminated feed and water. Hoover et al. (1997) documented that 14.8% of feed shipments to a turkey operation were contaminated with *Salmonella*. This contamination could move through the turkey flock. Ensuring incoming feed is not contaminated can be achieved by working with suppliers who test feed to enhance safety. Cleaning feeders and controlling rodents will reduce the risk of feed being contaminated once feed is distributed to the feeders. Rodents may be a reservoir for *Salmonella* spp., transferring the pathogen to feed troughs that in turn infect young poultry (Tauxe 2002).

Providing clean drinking water to poultry as well as cleaning and sanitizing drinking water reservoirs is also critical to reducing contamination risks. Many pathogens can be transmitted through water and, depending on sanitation practices, may persist or form biofilms in drinkers or water lines (CAST 2004). The opportunity for water contamination from this source can be minimized through management practices that incorporate standard operating procedures for cleaning, sanitizing, and replenishing drinking water to flocks.

Managing litter and manure is another way to reduce microbial risks on the farm. There is a positive association between litter dampness and *Salmonella* risk, so ventilating houses to reduce the relative humidity is beneficial (Mallinson et al. 2001). Improved airflow patterns also result in improved chicken carcass quality (Weaver and Meijerhof 1991). There are other ways to reduce dampness in poultry houses such as using nipple drinkers instead of trough-type drinkers (Mallinson et al. 2001). This may also lower the chance of water contamination by feces and other contaminants because the water reservoir is closed. In this way, one modification may reduce risks in both of these areas.

Reviewing the quality of chicks introduced into a grow-out operation is also important. Chicks can be contaminated with *Salmonella* while they are still eggs since contamination with *Salmonella* can result from direct ovarian transmission. Young chickens are also more susceptible to *Salmonella* colonization due to lack of mature gut flora, so with any exposure to *Salmonella* at this early stage, chicks are more likely to become contaminated (Williams 1984). Any time new animals are introduced into a flock, the health of the animal can impact the health of the group, thus requiring that day-old chicks from a hatchery be certified by the National Poultry Improvement Plan is ideal. Details can be found at http://www.aphis.usda.gov/animal_health/animal_dis_spec/poultry/.

Different production strategies may result in different challenges for controlling food-safety risks. Heuer et al. (2001) isolated *Campylobacter* spp. from 100% of organic broiler flocks, but only found *Campylobacter* in 36.7% of conventional broiler flocks and only 49.2% of extensive indoor broiler flocks examined were contaminated. Extensive indoor or barn-reared chickens should not be grown at a density of more than 15 birds to 10 square feet of living space and they should not be slaughtered before they are 56 days old. It is not immediately evident why organic production resulted in higher prevalence of *Campylobacter* spp., but it could be related to increased horizontal transmission from the environment, longer raising times for the broilers, increased contamination from the feed, or increased contamination in the production system, since many of these inputs vary widely in each case (Heuer et al. 2001).

There are certain challenges to reducing microbial risks that are difficult to control because they are not directly related to the operation. For instance, there are seasonal as well as regional variations in risks (CAST 2004; Hoover et al. 1997; Wedderkopp et al. 2000). Many pathogens are more prevalent in flocks in the summer months, while some pathogens tend to occur more frequently in certain locations. In addition, high temperatures can induce stress and depress the immune response in poultry, making them more susceptible to illness and more likely to shed pathogenic organisms such as *Salmonella* Enteritidis (Hoover et al. 1997). These variations are mentioned to highlight the need to be aware of not only the operation but also where and when production is occurring and what risks may exist as a result.

Consumer's Beliefs about Food Safety

Many consumers believe that organically grown poultry is safer in terms of microbial contamination. For example, some consumers believe that organic poultry has less *Salmonella* contamination, even though organic poultry processors and the USDA make no food-safety claims for organically grown chicken. Despite this consumer confidence, however, specific concerns have been expressed regarding organic poultry production. Such concerns are founded on potentially higher microbial risks from longer production cycles in smaller, organic production facilities with access to out of doors; prohibition of antimicrobial and antibiotic use during grow-out and slaughter that typically occurs in small facilities with less sophisticated sanitation systems.

The USDA's certified organic standards (USDA-AMS 2006b) do not permit the use of antibiotics such as virginiamicin or bacitracin as a means of controlling diseases or as a growth promoter during live poultry production (Bedford 2000). Conventional broiler production can use these antibiotics as needed and typically uses strong, synthetic chemical cleaning and disinfection techniques to clean and sanitize the broiler houses. Conventional cleaning of poultry houses between flocks may also include phenolic type disinfectants followed by fogging with formaldehyde to minimize *Salmonella* transfer to the incoming flock (Davies et al. 2001).

Contaminated feed from the feed mill is another route of spread for *Salmonella* (Davies et al. 1997; Maciorowski et al. 2004, 2005, 2007; Ricke 2005).

When litter beetles are present in broiler houses, flocks may become contaminated with *Salmonella* (Skov et al. 2004). Contamination can also occur during the transportation to the slaughter facility via *Salmonella* contaminated transport crates (Slader et al. 2002). For this rapidly growing organic segment of the market, improved hygiene must be required in the production facilities to eliminate horizontal transmission of *Salmonella* (Heyndrickx et al. 2002) and routine detection methods must be used in the field (Maciorowski et al. 2005, 2006).

Salmonella and Organic Poultry

In 2006, the USDA-FSIS reported the percentage of positives for *Salmonella* in conventionally grown broilers was inversely correlated to the size of the slaughter facility. Typically, large, vertically integrated poultry processors have better *Salmonella* control than do small and very small producers and processors. Many organic processors tend to be small operations. The prevalence rates for *Salmonella* in 2005 were 14.7, 18.6, and 32.9% for large, small, and very small establishments, respectively (USDA-FSIS 2006a; USDA-FSIS 2007). In 2006 the USDA-FSIS started categorizing plants and posting quarterly nationwide data for *Salmonella*, presented by product class, on their website. The best plants, Category I plants, do not have more than 23% (12 out of 51) carcasses that test positive for *Salmonella*. Some processors of organic chicken have expressed the fear that they will not be able to meet even the current FSIS *Salmonella* Performance Standards. Poultry industry spokespersons have pointed out the USDA's "self-fulfilling" prophesy because under the current standards FSIS will test more Category 3 plants, those with the poorest control and the most positive *Salmonella* carcasses, more often and will likely have more *Salmonella* control failures to report (USDA-FSIS 2006a).

Consumer Reports (Anonymous 2007) published the results of testing of 525 raw, whole broiler carcasses from a range of retail outlets. *Salmonella* was present in 15% of all brands. To quote their findings, organic poultry, "raised without antibiotics and costing $3 to $5 per pound were more likely to harbor *Salmonella* than conventionally produced broilers that cost more like $1 per pound." One large conventional poultry supplier only had a 5% positive *Salmonella*, while the USDA certified organic poultry producer had more than 5 times that level of *Salmonella*, 27%. Additionally, 84% of the *Salmonella* were resistant to one or more of the common antibiotics used to treat human infections (Anonymous 2007).

In an earlier study, Bailey and Cosby (2005) purchased three lots of free-range, organically grown broilers and tested them for *Salmonella*. All three lots contained carcasses that were positive for *Salmonella*. Sixty percent, 15 of 25 individual carcasses, were positive for *Salmonella*. In a separate study, Cui et al. (2005)

measured *Salmonella* in conventional and organically grown chicken by taking samples from Maryland retail stores. Their results showed that 61% of the organic and 44% of the conventional chicken carcasses were positive for *Salmonella*. On the organically grown poultry, two *Salmonella* species, Kentucky (59%) and Heidelberg (33%), were predominant.

However, given the relative small size and wide variation in organic rearing conditions, focusing research efforts on the processing plant for making improvements in the food safety of natural and organic poultry may pay the best dividends because this is the most economical point of control of *Salmonella*. To accomplish this, poultry processing plants require antimicrobials and sanitizers that are not only equally effective as synthetic compounds, but meet the "all natural and organic requirements." Although a wide variety of natural antimicrobials have been promoted, many are not currently commercially available, are not Generally Recognized as Safe (GRAS), or lack a production and established production system to make them economically viable. What is needed are sanitizers and antimicrobials that are co-products of large food-processing systems already in place which produce hundreds of tons of raw starting materials for a family of all-natural antimicrobial compounds that will have efficacy against the most common foodborne illness strains of *Salmonella*.

Fanatico (2007) has guidelines for organic poultry processing plants that are already complying with federal or state regulations. Typically these plants process organic poultry during the first shift after using approved organic detergents and sanitizers and pest-control methods. It is vital to prevent contamination and prevent commingling organic poultry with nonorganic products. Chlorine materials are commonly used for sanitizing facilities and equipment in both conventional and organic processing facilities. Sanitizers that are permitted in organic processing include hydrogen peroxide and organic acids, including lactic acid and acetic acid. In general, many certifiers permit highly chlorinated water to come in contact with food products in immersion chilling and for sanitizing surfaces, but the final rinse should be with a chlorine level less than the limit under the Safe Drinking Water Act (4 ppm).

Conclusions

The goal of this chapter was to raise awareness and encourage a review of current practices. Producing, transporting, slaughtering and processing organic poultry is a complex system with many opportunities for microbial pathogens to enter and persist. The more producers understand about microbial food safety, the more opportunities there will be to develop strategies to reduce the risks and improve the safety of the flocks as well as the safety of the food for consumers. This section provides an overview of areas to consider as well as suggestions for how to implement effective food-safety practices. A great deal of caution and effort must be exercised before entering into an organic or all natural poultry operation. Writing a sustainable business plan and extensive interviews with current producers is usually the first step. Additional reading materials and on-line references are available from Pittman (2004).

References

Anonymous. 2007. CR investigates dirty birds. Cons. Rep. 72(1):20–23.

Bailey, J. S., and D. E. Cosby. 2005. *Salmonella* prevalence in free-range and certified organic chickens. J. Food Prot. 68(11):2451–2453.

Bedford, M. 2000. Removal of antibiotic growth promoters from poultry diets: implications and strategies to minimize subsequent problems. World's Poult. Sci. J. 56(4):347–365.

Council for Agricultural Science and Technology (CAST). 2004. Intervention strategies for the microbiological safety of foods of animal origin. Issue Paper No. 24, January 2004.

Cui, S., B. Ge, J. Zheng, and J. Meng. 2005. Prevalence and antimicrobial resistance of *Campylobacter* spp and *Salmonella* serovars in organic chickens from Maryland retail stores. Appl. Environ. Microbiol. 71(7):4108–4111.

Davies, R., M. Breslin, J. E. L. Corry, W. Hudson, and V. M. Allen. 2001. Observations on the distribution and control of *Salmonella* species in two integrated broiler companies. Vet. Rec. 149:227–232.

Davies, R. H., R. A. Nicholas, I. M. McLaren, J. D. Corkish, D. G. Lanning, and C. Wray. 1997. Bacteriological and serological investigation of persistent *Salmonella enteritidis* infection in an integrated poultry organization. Vet. Microbiol. 58:277–293.

Demeritt, L. 2004. Organic pathways. *[N]Sight*. Hartman Group, Inc., Bellevue, WA.

Fanatico, A. C. 2007. Organic poultry processing. Unpublished draft. ATTRA, National Center for Appropriate Technology, Fayetteville, AR.

Fanatico, A. C. 2008. Organic poultry production: ATTRA in the United States. ATTRA publication. National Center for Appropriate Technology, Fayetteville, AR. http://www.attra.ncat.org/attra-pub/PDF/organicpoultry.pdf.

Food Marketing Institute (FMI) and American Meat Institute (AMI). 2006. The power of meat: an in-depth look at meat through the shopper's eyes. March 29, 2006. A Food Marketing Institute Publication.

Foodnavigator-USA. 2006. Organic meat market rockets, but supply lags behind. Available at: http://www.foodnavigator-usa.com/news/ng.asp?n=67914-organic-meat-organic-beef-organic-poultry. Accessed December 2006.

Givens, H., and L. Bell. 2005. The past, present and future of the organic industry: a retrospective of the first 20 years, a look at the current state of organic and forecasting the next 20 years. At http://www.ota.com/pics/documents/Forecasting2005.pdf. Accessed December 2006.

Greene, C. 2006. U.S. organic sector continues to expand. Amber Waves. Economic Research Service, United States Department of Agriculture. April 2006. http://www.ers.usda.gov/AmberWaves/April06/Findings/Organic.htm. Accessed December 2006.

Heuer, O. E., K. Pedersen, J. S. Andersen, and M. Madsen. 2001. Prevalence and antimicrobial susceptibility of thermophilic *Campylobacter* in organic and conventional broiler flocks. Lett. Appl. Microbiol. 33:269–274.

Heyndrickx, M., D. Vandekerchove, L. Herman, I. Rollier, K. Grijspeerdt, and L. Dezutter. 2002. Routes for *Salmonella* contamination of poultry meat: epidemiological study from hatchery to slaughterhouse. Epidemiol. Infect. 129:253–265.

Hoover, N. J., P. B. Kenney, J. D. Amick, and W. A. Hypes. 1997. Preharvest sources of *Salmonella* colonization in turkey production. Poult. Sci. 76:1232–1238.

Laux, M. 2006. Organic poultry. Available at http://www.agmrc.org/agmrc/commodity/livestock/poultry/poultry+organic.htm. Accessed June 2007.

Maciorowski, S. G., P. Herrera, F. T. Jones, S. D. Pillai, and S. C. Ricke. 2006. Cultural and immunological detection methods for *Salmonella* spp in animal feeds—a review. Vet. Res. Commun. 30:127–137.

Maciorowski, S. G., P. Herrera, F. T. Jones, S. D. Pillai, and S. C. Ricke. 2007. Effects on poultry and livestock of feed contamination with bacteria and fungi. Anim. Feed Sci. Technol. 133:109–136.

Maciorowski, S. G., F. T. Jones, S. D. Pillai, and S. C. Ricke. 2004. Incidence and control of food-borne *Salmonella* spp. in poultry feeds—A review. World's Poult. Sci. J. 60:446–457.

Maciorowski, S. G., F. T. Jones, S. D. Pillai, and S. C. Ricke. 2005. Polymerase chain reaction detection of foodborne *Salmonella* spp. in animal feeds. Crit. Rev. Microbiol. 31:45–53.

Mallinson, E. T., S. W. Joseph, C. L. E. deRezende, N. L. Tablante, and L. E. Carr. 2001. *Salmonella* control and quality assurance at the farm end of the food safety continuum. J. Am. Vet. Med. Assoc. 218(12):1919–1922.

National Restaurant Association (NRA). 2005. Poultry consumption gaining on red meat. At http://www.restaurant.org/research/news/story.cfm?ID=319. Accessed December 2006.

Oberholtzer, L., C. Greene, and E. Lopex. 2006. Organic poultry and eggs capture high price premiums and growing share of specialty markets. USDA-ERS Pub. No. LDP-M-150–01, December 2006.

Organic Trade Association (OTA). 2005. The past, present and future of the organic industry: a retrospective of the first 20 years, a look at the current state of organics and forecasting the next 20 years. At http://www.ota.com/index.html. Accessed December 2006.

Organic Trade Association (OTA). 2006. OTA's 2006 Manufacturer survey. At http://www.ota.com/pics/documents/short%20overview%20MMS.pdf. Accessed November 2006.

Organic Trade Association (OTA). 2007. Press release—U.S. organic sales show substantial growth. At http://www.organicnewsroom.com/2007/05/us_organic_sales_show_substant_1.html. Accessed December 2007.

Petaluma Poultry. 2007. Our Products. http://www.petalumapoultry.com/products/products.php. Accessed December 2007

Pittman, H. M. 2004. A legal guide to the national organic program, an agricultural law research article. http://www.nationalaglawcenter.org/assets/articles/pittman_organicprogram.pdf. Accessed December 2007.

Ricke S. C. 2005. Ensuring the safety of poultry feed. Pages 174–194 in Food safety control in the poultry industry. G. C. Mead, ed. Woodhead, Cambridge, UK.

Skov, M. N., A. G. Spencer, B. Hald, L. Petersen, B. Nauerby, B. Carstensen, and M. Madsen. 2004. The role of litter beetles as potential reservoir for *Salmonella enterica* and thermophilic *Campylobacter* spp. between broiler flocks. Avian Dis. 48(1):9–18.

Slader, J., G. Dominque, F. Jorgensen, K. McAlpine, R. J. Owen, F. J. Bolton, and T. J. Humphrey. 2002. Impact of transport crate reuse and of catching and processing on *Campylobacter* and *Salmonella* contamination of broiler chickens. Appl. Environ. Microbiol. 68(2):713–719.

Sloan, A. E. 2004. New rules for ready-to-eat. Food Technology 58(2):16.

Souza, K. 2006. The future of Tyson. The Morning News November 26, 2006 D1+.

Tauxe, R. V. Emerging foodborne pathogens. 2002. Int. J. Food Microbiol. 78:31–41.

United States Department of Agriculture-Agricultural Marketing Service (USDA-AMS). 2000. National Organic Program Final Rule. At www.ams.usda.gov/nop. Accessed December 2006.

United States Department of Agriculture-Agricultural Marketing Service (USDA-AMS). 2006a. The National Organic Program. At http://www.ams.usda.gov/nop/indexIE.htm. Accessed December 2006.

United States Department of Agriculture-Agricultural Marketing Service (USDA-AMS). 2006b. The National Organic Program. Organic food standards and labels: the facts. At www.ams.usda.gov/nop/Consumers/brochure.html. Accessed December 2006.

United States Department of Agriculture-Agricultural Marketing Service (USDA-AMS). 2007a. USDA publishes amendments to list of substances used in organic handling. At http://www.ams.usda.gov/news/133–07.htm. Accessed December 2006.

United States Department of Agriculture-Agricultural Marketing Service (USDA-AMS). 2007b. United States standards for livestock and meat marketing claims, naturally raised claim for livestock and the meat and meat products. At http://www.ams.usda.gov/lsg/stand/NR1107.txt. Accessed December 2007.

United States Department of Agriculture-Agricultural Marketing Service (USDA-AMS). 2007c. Rules and regulations. At http://www.ams.usda.gov/AMSv1.0/getfile? dDocName=STELPRDC5060710&acct=noprulemaking. Accessed May 2008.

United States Department of Agriculture-Economic Research Service (USDA-ERS). 2005. Food consumption. At http://www.ers.usda.gov/Briefing/Consumption/. Accessed December 2006.

United States Department of Agriculture-Foreign Agricultural Service (USDA-FAS). 2005. U.S. market profile for organic food products. Commodity and Marketing Programs—Processed Products Division, International Strategic Marketing Group. At www.USMarketProfileOrganicFoodFeb2005.pdf. Accessed December 2005.

United States Department of Agriculture-Foreign Agriculture Service (USDA-FAS). 2005 February. U.S. market profile for organic food products. At http://www.fas.usda.gov/agx/organics/USMarketProfileOrganicFoodFeb2005.pdf. Accessed December 2005.

United States Department of Agriculture-Food Safety Inspection Service (USDA-FSIS). 2006a. Progress report on Salmonella testing of raw meat and poultry products,

1998–2005. At http://www.fsis.usda.gov/Science/Progress_Report_Salmonella_Testing/index.asp. Accessed November 2006.

United States Department of Agriculture-Food Safety Inspection Service (USDA-FSIS). 2006b. Food labeling: meat and poultry labeling terms. At http://www.fsis.usda.gov/Fact_Sheets/Meat_&_Poultry_Labeling_Terms/index.asp. Accessed December 2006.

United States Department of Agriculture-Food Safety Inspection Service (USDA-FSIS). 2007. Quarterly progress report on Salmonella testing. At http://www.fsis.usda.gov/Science/Q1_2007_Salmonella_Testing/index.asp. Accessed July 2007.

United States Department of Agriculture-Food Safety Inspection Service (USDA-FSIS) 2008. HACCP and pathogen reduction. At http://www.fsis.usda.gov/Science/Hazard_Analysis_&_Pathogen_Reduction/index.asp. Accessed May 2008.

Weaver, W. D., and R. Meijerhof. 1991. The effect of different levels of relative humidity and air movement on litter conditions, ammonia, levels, growth and carcass quality for broiler chickens. Poult. Sci. 70: 746–755.

Wedderkopp, A., E. Rattenborg, and M. Madsen. 2000. National surveillance of Campylobacter in broilers at slaughter in Denmark in 1998. Avian Dis. 44: 993–999.

Williams, J. E. 1984. Paratyphoid infections. Pages 91–129 in Diseases in poultry. M. S. Hofstad, H. J. Barnes, W. M. Reid, and H. W. Yoder, eds. Iowa Sate University Press, Ames, IA.

∎ 22 ∎

Alternative and Organic Beef Production: Food-Safety Issues

Steven C. Seideman, Todd R. Callaway, Philip G. Crandall,
Steven C. Ricke, and David J. Nisbet

Organic Beef Production and Markets

Beef is the second most popular organic meat item behind poultry. Sales of organic beef have been increasing steadily based on consumers' perception that organic beef is more nutritious, better for the environment, contains minimal levels of pesticides, is produced without growth hormones and antibiotics, is not genetically engineered (GE), and is produced using improved animal welfare practices.

Health, well-being, and nutrition are the drivers in marketing of organic fruits and vegetables. These same drivers certainly play a part in organic beef sales. Claire Williamson (2007) of the British Nutrition Foundation recently reviewed the question, "Is organic food better for our health?" She concluded that there appears to be a perception among many consumers that organic foods are more nutritious and therefore healthier than conventional foods; however, to date there is very limited data to support this view. In addition, Williamson (2007) noted that nutrition is not seen as a major reason why people consume organic food. Concerns about the environment, pesticide levels, food additives, and animal welfare are listed as more important factors by many UK consumers.

In addition to the positive consumer motivations listed previously, the avoidance of growth-promoting hormones (GPH) and antibiotics in conventional cattle production are of even greater concern to some consumers. Since 1988 consumers have read about the European Union ban on anabolic steroids used as growth promoters in beef while the United States continues to permit their use. Recently,

ElAmin (2007) reported that the European Food Safety Agency (EFSA) had just concluded that there is not enough evidence for the EU to remove its ban on GPH in cattle. It is important to note that the amount of growth-promoting hormones given to cattle feedlot animals is quantitatively quite small. The FDA has already set acceptable limits on the six hormones typically used as GPH in cattle.

Other key market drivers are environmental awareness and humane animal rearing conditions. There is a perception by some consumers of organic beef that conventional beef production damages the environment by surface water runoff from feedlots that pollute rivers and streams. These consumers may also support the notion that raising animals on pasture and giving them the opportunity for exercise is a more humane practice.

Issues Specific to Organic Beef Production

The U.S. beef production system will probably never become as vertically integrated as the poultry industry because of underlying financial reasons such as the ever-increasing cost of land, replacement cow costs, and the production of only one offspring per year. Sharp (2005) reported that 79% of all beef producers had herds of fewer than 50 animals. It has been estimated that over 70% of all beef cattle were produced on farms where the farm income was not the main source of income. The high returns these small organic beef production operations receive make it an excellent niche for small farmers. However, when the farmers read all the organic regulations, they may decide the increased costs and restrictions of organic production practices outweigh the increased returns.

The single biggest hurdle most small beef producers face is locating a certified organic, federally inspected slaughter facility that is willing to process their organic beef. Currently, the number of federally inspected beef slaughter facilities is extremely low and the ones that exist are not very profitable. To sell organic beef in interstate commerce it must be processed in a USDA-inspected meat-slaughtering facility. Most of these inspected slaughter facilities cannot afford to accommodate small-volume, consumer-style packaging. Additional problems surround marketing and selling organic beef. Since most small farmers have other full-time jobs, they rarely have the time or skills required to market and sell their own organic beef. There are several large organic production and slaughter facilities in Colorado currently selling all the organic and/or natural beef they can produce, but most small farmers are not large enough to do this unless they join a cooperative or some other arrangement where slaughter, marketing, and sales costs are spread out over numerous producers.

Artisan, Exotic, and "Country Club Beef Production"

The *American Heritage Dictionary* (1981) defines "artisan" as a person manually skilled in making a particular product. "Artisan farming" is farming in harmony with livestock, land, and nature. Artisan farmers use practices that allow animals to eat and grow in ways consistent with their natural physiology and to express their natural behaviors, such as ranging and grazing. This environment eliminates the use of artificial hormones and the routine use of antibiotics, which act as a "crutch" to prevent illness in feedlot animals kept under confinement conditions (http://www.nenbeef.com/fhf_what.html).

Numerous livestock farm-to-market organizations have begun using the term artisan beef to describe their production practices. In most instances, artisan beef is grass-fed, raised without growth hormones, and raised using sustainable agriculture practices and is often labeled as "natural." By subscribing to these practices and marketing beef to consumers some forward-thinking, entrepreneurial beef producers have been able to capture a portion of the consumers' dollar that would normally buy USDA certified organic beef. Artisan producers are focusing on the major issues driving consumers' purchasing decision: their family's health, nutrition, natural production, ecologically friendly, and no hormones. Artisan producers are making better returns without actually having to go through the long and difficult process of becoming USDA certified organic. The following Web site provides an example of a producer-couple who direct market their grass-fed Angus beef to customers in the Plano and Dallas, Texas, metropolitan areas. (http://www.stockmangrassfarmer.net/cgi-bin/page.cgi?id=638).

The customers of artisan beef typically have a high level of food sophistication and demand foods that are fresh, wholesome, and flavorful with a high level of nutrition. Their customers are environmentally aware and can afford and are willing to pay a premium price for artisan, grass-fed beef.

Some of these producer groups belong to the American Grass-fed Association, which claims that grass-fed is better for animals "eating in the pasture is what nature intended, grass-feeding is better for the environment." They claim this production system is in "harmony with the land and the animals and may be better for the farmer/rancher living and working in a healthy, sustainable atmosphere." They also contend that grass-fed beef is lower in saturated fats, higher in essential nutrients, and supports better overall health (http://www.americangrassfed.org/index.html).

Roach (2006) reported that the USDA is concerned about the amount of consumer confusion about the term "grass-fed" and is proposing to carefully define it. The initial USDA draft included "grass" to mean vegetable and forage such as corn. There was no specification that the cattle had to be grazed on pastureland and thus could be raised in feedlots on forage.

Some artisan producers in Texas use Texas Longhorn cattle, which producers claim are well adapted to the region, and the resulting meat is reported to be lower in cholesterol and higher in protein than conventional beef. Their Web site (http://www/banderagrassland.com) claims that their "beef cattle graze lush open pastures and are allowed to grow at their own, natural rate with no antibiotics or growth hormones administered." Typically these cattle grow much longer and larger before their meat begins to lay down intramuscular fat or marbling. They claim that these adaptable cattle have taught them to be better stewards of their land and are advocates for sustainable agriculture: a holistic approach that rewards first the soil, then native plant life, then varieties of domestic and wild animal species, and ultimately the production of wholesome, uncompromised food.

Northeast Artisan Meats is an alliance of organizations dedicated to promoting sustainable livestock agriculture in the Northeast (500 farms) region of the United States that work with livestock family farmers to produce Northeast Artisan Beef. They consider their beef to be all natural and it is marketed through distributors to exclusive restaurants and food-service operations. They are positioned as the only artisan meats in the Northeast region and claim that their meats are free from antibiotics and hormones, exceptionally tender and flavorful, healthier than meat raised on organic grains, and raised entirely on local farms. They promote the use of breeds of cattle from the Heritage Breeds Conservancy (http://www.nehbc.org/) in selecting a breed that fits their ranching operation and promote the farm, not the organization. They also point out significant ecological benefits through reduced consumption of water, fossil fuels, pesticides, and herbicides as well as greatly reduced soil loss due to erosion (http://www.nenbeef.com/fhf_belief.html).

An Added Challenge to Organic Beef Production: Food Safety

Foodborne Illness

Much of the support of the organic movement is based upon the benefits to the health of consumers and the environment. However, another critical moral and ethical responsibility that faces organic beef producers includes food safety (Rollin 2006). The food supply of the United States is one of the safest in the world and is constantly improving; but too many foodborne illnesses linked to beef consumption continue to occur.

Beef products can be contaminated by several of the most common foodborne pathogenic bacteria. *Campylobacter* is the most common foodborne pathogenic bacteria (Mead et al. 1999), and it is commonly isolated from cattle and beef products (Harvey et al. 2004; Besser et al. 2005). Another serious beef-linked foodborne pathogenic bacteria is *Salmonella enterica* (Mead et al. 1999). Collectively, these two bacteria cause over an estimated 3.2 million human illnesses and over 650 deaths and are estimated to cost the U.S. economy more than $3.6 billion each year (Mead et al. 1999; USDA-ERS 2001). However, the most critical pathogen that impacts the beef (traditional and organic) industry remains *Escherichia coli* O157:H7, which causes hemorrhagic colitis in humans. Each year, more than 60 people die and 73,000 people are made ill by *E. coli* O157:H7 in the United States (Mead et al. 1999), and enterohemorrhagic *E. coli* (EHEC) infections cost the American economy approximately $1 billion per year (USDA-ERS, 2001). While a number of human foodborne illnesses have been linked to vegetables and fruits, the most common infection route still remains through foods of animal origin (Braden, 2006). In fact, bovine-derived products have been linked to approximately 75% of *E. coli* O157:H7 outbreaks (Vugia et al. 2007). Repeated outbreaks of hemorrhagic colitis and large-scale recalls of *E. coli* O157:H7 contaminated ground beef as well as other contaminated foods have firmly established the connection between cattle and *E. coli* O157:H7 (Steinmuller et al. 2006; Jay et al. 2007). The well-publicized deaths of children who consumed foods contaminated by exposure to beef products or ruminants have further shaken the confidence of consumers in the wholesomeness and safety of beef.

Pathogen Distribution in Cattle

Pathogenic bacteria are widely distributed in cattle at all stages of the production continuum. Studies have shown that more than one-third of all cattle are asymptomatic carriers of *E. coli* O157:H7 over the course of the year, with seasonal prevalences of >80% being fairly common (Barkocy-Gallagher et al. 2003; Stanford et al. 2005; Callaway et al. 2006). Similar or greater prevalences have been reported for *Salmonella* and *Campylobacter* carriage in cattle as well (Besser et al. 2005; Callaway et al. 2006). This indicates that foodborne pathogenic bacteria are a widespread problem in cattle that constitutes a threat to the safety of the beef industry, both organically and traditionally produced.

Foodborne pathogenic bacteria can live in the intestinal tract of animals without causing illness, yet they are not predominant members of the proportion of the intestinal population (Laven et al. 2003; Tkalcic et al. 2003). The oral cavities of cattle also can contain pathogens (Keen and Elder 2002; Smith et al. 2005), and the hides of cattle represent a reservoir for fecal pathogens to be transferred within a processing plant and to carcasses and finished beef products (Arthur et al. 2007; Reicks et al. 2007).

Where Do We Need to Reduce Pathogens in an Organic Beef System?

In-plant postharvest pathogen reduction strategies do reduce carcass contamination with *E. coli* O157:H7 (Bosilevac et al. 2006; Woerner et al. 2006; Arthur et al. 2007). However, no matter how effective processing-plant strategies become, they are not sufficient to ensure food safety and human health. Foodborne pathogenic bacteria can be frequently found as transient members of the bovine intestinal microbial population that do not cause animal disease (Rasmussen et al. 1993; Besser et al. 2005; Callaway et al. 2006). Thus relying on animals "looking sick" is not an effective intervention strategy to detect and/or eliminate these bacteria. Furthermore, cattle facilities provide alternate routes for human exposure to pathogens, including water runoff from farm feedlots and pastures (Thurston-Enriquez et al. 2005; Soupir et al. 2006) or effluent lagoons (Hill and Sobsey 2003), direct animal (Chapman et al. 2000) and/or fecal contact (Keen et al. 2007). Pathogenic bacteria shed into the environment can linger on a farm for long periods of time, even after animal depopulation (LeJeune et al. 2004; Doane et al. 2007).

Thus because these pathogenic bacteria can live undetected in food animals and still pose significant risks to humans, pathogen control strategies must be developed to work at the on-farm level, during transport and lairage up until entry to the processing plant. Furthermore, these strategies must be tailored to meet the needs of each specific phase of the beef production chain, yet still be applicable to large numbers of animals that are grown in a variety of conditions, from feedlots to pasture to open range. Additionally, in order for solution strategies to be implemented in the organic beef industry, they must meet the pathogen reduction goals while remaining organic in nature and environmentally friendly, as well as being economically feasible.

Pathogen Reduction: Thinking Strategically and Organically

General Concepts

The microbial ecology of cattle gastrointestinal tracts is very complex and fiercely competitive. Foodborne pathogenic bacteria tend to be "utility" bacteria that can survive in many environments and conditions, but because they are generalists, they typically do not become a major component of the microbial ecosystem. Because of this, it means that our approaches must take into consideration the natural ecosystem that is present. Organic and natural beef production can be supported through using this existing microbial population to enhance growth efficiency or to eliminate or reduce the unwanted foodborne pathogens via "natural" mechanisms. Many of these mechanisms are described in greater detail in the accompanying chapter of this book on "On-Farm Interventions to Reduce Epizootic Bacteria in Food-Producing Animals and the Environment" (chapter 5).

One simple way that the foodborne pathogen populations can be controlled is through the use of pro-commensal strategies, which included the use of probiotics, prebiotics, and competitive exclusion. In general, pro-commensal strategies can be categorized into two groups: (1) the introduction of a "normal" (non-pathogen containing) intestinal microbial population (probiotics and/or competitive exclusion); or (2) providing a limiting substrate (a prebiotic) that is not digestible by the host animal but can allow an already existing microbial population to expand its niche in the gastrointestinal population. However, prebiotics are currently not economically viable for use in the cattle industry and will not be discussed further. Because of increased concerns about the issue of antimicrobial/antibiotic resistance, it is expected that the prophylactic use of antibiotics in cattle will decrease in the future, causing "pro-commensal" strategies to become increasingly widely utilized.

Probiotics

Various probiotics have been used in the cattle industry to increase growth rate, milk production, or production efficiency (Fuller 1989; Callaway et al. 2005b). Researchers examined the effects of these products on pathogen populations and found that feeding these probiotics provided neither benefit nor detriment in regards to pathogen levels in cattle (Keen and Elder 2000). Subsequently, researchers have developed probiotics to specifically target pathogen populations in cattle, such as a direct-fed-microbial (DFM) *L. acidophilus* culture derived from a cattle rumen that reduced *E. coli* O157:H7 shedding by more than 50% in feedlot cattle (Brashears and Galyean 2002; Brashears et al. 2003a; 2003b). Refining this DFM by adding *Propionibacterium freudenreichii* (a propionate-producing commensal bacteria) reduced the fecal prevalence of *E. coli* O157:H7 by almost 50% and the prevalence on hides three-fold (Elam et al. 2003; Younts-Dahl et al. 2004). Further work with this DFM has indicated that it reduced fecal and hide *Salmonella* popu - lations (Stephens et al. 2007) DFM product while improving the growth efficiency of cattle fed it. The efficiency increase can balance the cost of inclusion in rations and "pay for" a food-safety improvement.

Competitive Exclusion

Competitive exclusion (CE) is the addition of an exogenous bacterial population (nonpathogenic) to the intestinal tract in order to reduce colonization or decrease existing populations of pathogenic bacteria in the gastrointestinal tract (Fuller 1989; Nurmi et al. 1992; Steer et al. 2000). CE cultures are composed of a single or multiple strains of a single or several bacterial species that compete with one another for available nutrients or physical space, or through the production of antimicrobial substances; the best-adapted species flourishes in each intestinal ecological niche.

Competitive exclusion has been widely used in poultry to prevent *Salmonella* colonization of broilers (Nurmi et al. 1992). Research has also shown that CE can be effective in preventing *Salmonella* and enterotoxigenic *E. coli* colonization of swine (Anderson et al. 1999; Fedorka-Cray et al. 1999; Genovese et al. 2000). Field trials with a porcine-derived commercial CE product for use in growing swine have been highly encouraging (Genovese et al. 2003; Harvey et al. 2003). Historically, CE was not considered a useful technique in cattle due to the large microbial reservoir of the rumen and intestinal tract. Yet CE reduces *E. coli* O157:H7 (as well as *Salmonella*) in cattle by adding a defined population of multiple non-EHEC *E. coli* strains that were isolated from cattle. These researchers found that this generic *E. coli* CE culture could displace an established *E. coli* O157:H7 population from calves (Zhao et al. 1998). To date, this is the only true CE culture for cattle that is being developed as a commercial product, and is currently in the field trial stage.

Bacteriophage, a Natural Antibiotic Solution Strategy for Foodborne Pathogens in Organic Beef

In addition to pro-bacterial strategies, there are also some natural or organic antibacterial strategies that can be utilized in the production of organic beef. Bacteriophages are viruses that specifically kill bacteria and are common members of the intestinal microbial flora of food animals, including cattle (Klieve and Bauchop 1988; Klieve et al. 1991; Klieve and Swain 1993). Bacteriophage are highly specific and can be active against a single strain or several related strains of bacteria (Barrow and Soothill 1997). Dosing bacteriophage in cattle (and other food animals) has been suggested to specifically eliminate pathogens from a mixed microbial population (Merril et al. 1996; Summers 2001). Phage have been used successfully in several *in vivo* research studies examining the effect of phage on diseases that impact production efficiency or health in food animals (Smith and Huggins 1982, 1983; Huff et al. 2002).

Bacteriophage treatment reduced Enteropathogenic *E. coli* (EPEC)-catalyzed diarrhea, and splenic EPEC colonization in calves (Smith and Huggins 1983, 1987), indicating that bacteriophages could be useful in the effort to reduce foodborne pathogenic bacteria entering the food chain from the gut of food animals. Several *E. coli* O157-specific phage have been isolated and reduced *E. coli* O157:H7 populations in experimentally infected animals (Callaway et al. 2008b). In field studies phage that killed *E. coli* O157:H7 were widespread in open-range sheep and feedlot cattle (Callaway et al. 2003a; 2006). Other researchers have isolated phage active against *Salmonella, Campylobacter,* and pathogenic *E. coli* strains (Huff et al. 2002; Higgins et al. 2005; Loc Carrillo et al. 2005). The effectiveness of phage treatment in "real world" conditions has been variable, therefore more basic research needs to be completed before bacteriophage can be considered a truly viable natural/organic method to control foodborne pathogenic bacteria in cattle.

Diet Effects on Pathogen Populations

Specific Food Ingredients

One management factor that can be altered to affect the foodborne pathogen populations in beef cattle is diet. For example, barley feeding has been shown to increase fecal shedding of *E. coli* O157:H7 in cattle (Bach et al. 2002; 2005; Berg et al. 2004). Other researchers have demonstrated that feeding dried distillers' grains can under some situations increase fecal *E. coli* O157 populations (Jacob, Fox, Drouillard et al. 2008; Jacob, Fox, Narayanan et al. 2008; Jacob, Parsons et al. 2008). Thus finding that altering the diet can profoundly change pathogen populations of the gut has great implications for the production of organic beef. Further research is underway to determine which feedstuffs can be used to reduce pathogen populations.

One alternative feedstuff has been shown to impact foodborne pathogen populations in cattle. Orange peel and pulp included at 2% of the total volume have been demonstrated to have anti-*E. coli* O157:H7 and *Salmonella* Typhimurium activity in *in vitro* fermentations (Callaway, Carroll et al. 2005; Fisher and Phillips 2006; Callaway, Carroll et al. 2008). This effect appears to be a result of the antimicrobial action of essential oils (e.g., limonene) found in the peel (Callaway, Carroll et al. 2008). Other research has found that feeding a seaweed (*Ascophyllum nodo - sum*) extract reduced *E. coli* O157:H7 populations in feces by two-thirds and reduced hide samples positive from 85% to less than 50% (Braden et al. 2004). These results underscore that certain feedstuffs can exert potent effects on the microbial population and can be used to control pathogens in certain circumstances and dietary regimens.

Dietary Shifts

In a study that excited a great deal of controversy concerning the effects of dietary shifts on foodborne pathogenic bacteria, cattle fed a 90% corn/soybean meal ration (feedlot-type ration) contained generic *E. coli* populations that were 1,000-fold higher than cattle fed a 100% Timothy hay diet (Diez-Gonzalez et al. 1998; Callaway, Elder et al. 2003). When cattle were abruptly switched from a 90% grain finishing ration to a 100% hay diet, fecal *E. coli* populations declined 1,000-fold, and the population of *E. coli* resistant to an extreme acid shock (similar to the human gastric stomach) declined more than 100,000-fold within 5 days (Diez-Gonzalez et al. 1998). Based on these results the authors suggested that feedlot cattle be switched from high grain diets to hay for five days prior to slaughter to reduce *E. coli* contamination entering the abattoir (Diez-Gonzalez et al. 1998). This study has been used widely by the organic and natural beef industries to support grass or hay feeding to cattle to produce high quality beef. Subsequent research has shown in some cases that a sudden switch to a high forage diet can increase shedding of pathogens, but there are other factors affecting shedding of pathogens that are not understood clearly at this time (Callaway et al. 2003b). However, this research demonstrates that solutions that are natural and/or organic can reduce foodborne pathogenic bacteria populations in cattle and still maintain an organic approach to beef production.

Conclusion

The sales of organic, artisan, natural, grass-fed, and similar terms used for beef will continue to increase due to consumer perceptions. As scientists, the need to safeguard the beef from pathogenic bacteria and loss of consumer confidence is of extreme concern. Further research is needed to determine if the gastrointestinal tract and meat from these organic-type cattle are different from conventional beef and to investigate whether or not the use of probiotics and competitive exclusion techniques are of benefit.

References

American Heritage Dictionary. 1981. P.5.

Anderson, R. C., L. H. Stanker, C. R. Young, S. A. Buckley, K. J. Genovese, R. B. Harvey, J. R. DeLoach, N. K. Keith, and D. J. Nisbet. 1999. Effect of competitive exclusion treatment on colonization of early-weaned pigs by *Salmonella* serovar Cholerasuis. Swine Health Prod. 12:155–160.

Arthur, T. M., J. M. Bosilevac, D. M. Brichta-Harhay, M. N. Guerini, N. Kalchayanand, S. D. Shackelford, T. L. Wheeler, and M. Koohmaraie. 2007. Transportation and lairage environment effects on prevalence, numbers, and diversity of *Escherichia coli* O157:H7 on hides and carcasses of beef cattle at processing. J. Food Prot. 70:280–286.

Bach, S. J., T. A. McAllister, J. Baah, L. J. Yanke, D. M. Veira, V. P. J. Gannon, and R. A. Holley. 2002. Persistence of *Escherichia coli* O157:H7 in barley silage: effect of a bacterial inoculant. J. Appl. Microbiol. 93:288–294.

Bach, S. J., L. J. Selinger, K. Stanford, and T. McAllister. 2005. Effect of supplementing corn- or barley-based feedlot diets with canola oil on faecal shedding of *Escherichia coli* O157:H7 by steers. J. Appl. Microbiol. 98:464–475.

Barkocy-Gallagher, G. A., T. M. Arthur, M. Rivera-Betancourt, X. Nou, S. D. Shackelford, T. L. Wheeler, and M. Koohmaraie. 2003. Seasonal prevalence of shiga toxin-producing *Escherichia coli,* including O157:H7 and non-O157 serotypes, and *Salmonella* in commercial beef processing plants. J. Food Prot. 66:1978–1986.

Barrow, P. A., and J. S. Soothill. 1997. Bacteriophage therapy and prophylaxis: rediscovery and renewed assessment of potential. Trends Microbiol. 5:268–271.

Berg, J. L., T. A. McAllister, S. J. Bach, R. P. Stillborn, D. D. Hancock, and J. T. LeJeune. 2004. *Escherichia coli* O157:H7 excretion by commercial feedlot cattle fed either barley- or corn-based finishing diets. J. Food Prot. 67:666–671.

Besser, T. E., J. T. LeJeune, D. H. Rice, J. Berg, R. P. Stilborn, K. Kaya, W. Bae, and D. D. Hancock. 2005. Increasing prevalence of *Campylobacter jejuni* in feedlot cattle through the feeding period. Appl. Environ. Microbiol. 71:5752–5758.

Bosilevac, J. M., X. Nou, G. A. Barkocy-Gallagher, T. M. Arthur, and M. Koohmaraie. 2006. Treatments using hot water instead of lactic acid reduce levels of aerobic bacteria and enterobacteriaceae and reduce the prevalence of *Escherichia coli* O157:H7 on preevisceration beef carcasses. J. Food Prot. 69:1808–1813.

Braden, C. R. 2006. *Salmonella enterica* serotype Enteritidis and eggs: a national epidemic in the United States. Clin. Infect. Dis. 43:512–517.

Braden, K. W., J. R. Blanton Jr., V. G. Allen, K. R. Pond, and M. F. Miller. 2004. *Ascophyllum nodosum* supplementation: a preharvest intervention for reducing *Escherichia coli* O157:H7 and *Salmonella* spp. in feedlot steers. J. Food Prot. 67:1824–1828.

Brashears, M. M., and M. L. Galyean. 2002. Testing of probiotic bacteria for the elimination of *Escherichia coli* O157:H7 in cattle. Am. Meat Inst. Found. Available at http://www.amif.org/PRProbiotics042302.htm. Accessed April 24, 2007.

Brashears, M. M., M. L. Galyean, G. H. Loneragan, J. E. Mann, and K. Killinger-Mann. 2003. Prevalence of *Escherichia coli* O157:H7 and performance by beef feedlot cattle given *Lactobacillus* direct-fed microbials. J. Food Prot. 66:748–754.

Brashears, M. M., D. Jaroni, and J. Trimble. 2003. Isolation, selection, and characterization of lactic acid bacteria for a competitive exclusion product to reduce shedding of *Escherichia coli* O157:H7 in cattle. J. Food Prot. 66:355–363.

Callaway, T. R., J. A. Carroll, J. D. Arthington, R. C. Anderson, T. S. Edrington, K. J. Genovese, and D. J. Nisbet. 2005. Orange pulp reduces growth of *E. coli* O157:H7 and *Salmonella Typhimurium* in pure culture and in vitro mixed ruminal microorganism fermentation. Page 236 in Am. Soc. Anim. Sci/ Am. Dairy Sci. Assoc./Can. Anim. Sci. Assoc. Ann. Mtg., Cincinnati, OH..

Callaway, T. R., J. A. Carroll, J. D. Arthington, C. Pratt, T. S. Edrington, R. C. Anderson, M. L. Galyean, S. C. Ricke, P. Crandall, and D. J. Nisbet. 2008. Citrus products decrease growth of *E. coli* O157:H7 and *Salmonella* Typhimurium in pure culture and in fermentation with mixed ruminal microorganism *in vitro.* Foodborne Path. Dis. 5:621–627.

Callaway, T. R., K. D. Dunkley, R. C. Anderson, T. S. Edrington, K. J. Genovese, T. L. Poole, R. B. Harvey, and D. J. Nisbet. 2005. Probiotics, vaccines and other intervention strategies. Pages 192–213 in Raw material safety: Meat. J. N. Sofos, ed. Woodhead Pub., Cambridge, UK.

Callaway, T. R., T. S. Edrington, A. D. Brabban, R. C. Anderson, M. L. Rossman, M. J. Engler, M. A. Carr, K. J. Genovese, J. E. Keen, M. L. Looper, E. M. Kutter, and D. J. Nisbet. 2008. Bacteriophage isolated from feedlot cattle can reduce *Escherichia coli* O157:H7 populations in ruminant gastrointestinal tracts. Foodborne Path. Dis. 5:183–192.

Callaway, T. R., T. S. Edrington, A. D. Brabban, J. E. Keen, R. C. Anderson, M. L. Rossman, M. J. Engler, K. J. Genovese, B. L. Gwartney, J. O. Reagan, T. L. Poole, R. B. Harvey, E. M. Kutter, and D. J. Nisbet. 2006. Fecal prevalence of *Escherichia coli* O157, *Salmonella, Listeria,* and bacteriophage infecting *E. coli* O157:H7 in feedlot cattle in the southern plains region of the United States. Foodborne Pathog. Dis. 3:234–244.

Callaway, T. R., T. S. Edrington, P. D. Varey, R. Raya, A. D. Brabban, E. Kutter, Y. S. Jung, K. J. Genovese, R. O. Elder, and D. J. Nisbet. 2003. Isolation of naturally-occuring bacteriophage from sheep that reduce populations of *E. coli* O157:H7 *in vitro* and *in vivo.* Page 25 in Proc. 5th Int. Symp. on Shiga Toxin-Producing *Escherichia coli* Infections, Edinburgh, UK.

Callaway, T. R., R. O. Elder, J. E. Keen, R. C. Anderson, and D. J. Nisbet. 2003. Forage feeding to reduce pre-harvest *E. coli* populations in cattle, a review. J. Dairy. Sci. 86:852–860.

Chapman, P. A., J. Cornell, and C. Green. 2000. Infection with verocytotoxin-producing *Escherichia coli* O157 during a visit to an inner city open farm. Epidemiol. Infect. 125:531–536.

Diez-Gonzalez, F., T. R. Callaway, M. G. Kizoulis, and J. B. Russell. 1998. Grain feeding and the dissemination of acid-resistant *Escherichia coli* from cattle. Science 281:1666–1668.

Doane, C. A., P. Pangloli, H. A. Richards, J. R. Mount, D. A. Golden, and F. A. Draughon. 2007. Occurrence of *Escherichia coli* O157:H7 in diverse farm environments. J. Food Prot. 70:6–10.

Elam, N. A., J. F. Gleghorn, J. D. Rivera, M. L. Galyean, P. J. Defoor, M. M. Brashears, and S. M. Younts-Dahl. 2003. Effects of live cultures of *Lactobacillus acidophilus* (strains np45 and np51) and *Propionibacterium freudenreichii* on performance, carcass, and intestinal characteristics, and *Escherichia coli* strain O157 shedding of finishing beef steers. J. Anim. Sci. 81:2686–2698.

ElAmin A. 2007. Scientific panel advises keeping ban on growth hormones. At http://www.foodnavigator-usa.com/news/ng.asp?id=78282. The full text of the opinion is available on the EFSA Web site atwww.efsa.europa.eu/en/science/contam/contam_opinions/ej510_hormone.html.

Fedorka-Cray, P. J., J. S. Bailey, N. J. Stern, N. A. Cox, S. R. Ladely, and M. Musgrove. 1999. Mucosal competitive exclusion to reduce *Salmonella* in swine. J. Food Prot. 62:1376–1380.

Fisher, K., and C. A. Phillips. 2006. The effect of lemon, orange and bergamot essential oils and their components on the survival of *Campylobacter jejuni, Escherichia coli* O157, *Listeria monocytogenes, Bacillus cereus* and *Staphylococcus aureus* in vitro and in food systems. J. Appl. Microbiol. 101:1232–1240.

Fuller, R. 1989. Probiotics in man and animals. J. Appl. Bacteriol. 66:365–378.

Genovese, K. J., R. C. Anderson, R. B. Harvey, T. R. Callaway, T. L. Poole, T. S. Edrington, P. J. Fedorka-Cray, and D. J. Nisbet. 2003. Competitive exclusion of *Salmonella* from the gut of neonatal and weaned pigs. J. Food Prot. 66:1353–1359.

Genovese, K. J., R. C. Anderson, R. B. Harvey, and D. J. Nisbet. 2000. Competitive exclusion treatment reduces the mortality and fecal shedding associated with enterotoxigenic *Escherichia coli* infection in nursery-raised pigs. Can. J. Vet. Res. 64:204–207.

Harvey, R. B., R. E. Droleskey, C. L. Sheffield, T. S. Edrington, T. R. Callaway, R. C. Anderson, D. L. J. Drinnon, R. L. Ziprin, H. M. Scott, and D. J. Nisbet. 2004. *Campylobacter* prevalence in lactating dairy cows in the United States. J. Food Prot. 67:1476–1479.

Harvey, R. B., R. C. Ebert, C. S. Schmitt, K. Andrews, K. J. Genovese, R. C. Anderson, H. M. Scott, T. R. Callaway, and D. J. Nisbet. 2003. Use of a porcine-derived, defined culture of commensal bacteria as an alternative to antibiotics used to control *E. coli* disease in weaned pigs. Pages 72–74 in 9th Intl. Symp. Dig. Physiol. in Pigs. Banff, AB, Canada.

Higgins, J. P., K. L. Higgins, H. W. Huff, A. M. Donoghue, D. J. Donoghue, and B. M. Hargis. 2005. Use of a specific bacteriophage treatment to reduce *Salmonella* in poultry products. Poult. Sci. 84:1141–1145.

Hill, V. R., and M. D. Sobsey. 2003. Performance of swine waste lagoons for removing *Salmonella* and enteric microbial indicators. Trans. Am. Soc. Agric. Eng. 46:781–788.

Huff, W. E., G. R. Huff, N. C. Rath, J. M. Balog, H. Xie, P. A. Moore, and A. M. Donoghue. 2002. Prevention of *Escherichia coli* respiratory infection in broiler chickens with bacteriophage (spr02). Poult. Sci. 81:437–441.

Jacob, M. E., J. T. Fox, J. S. Drouillard, D. G. Renter, and T. G. Nagaraja. 2008. Effects of dried distillers' grain on fecal prevalence and growth of *Escherichia coli* O157 in batch culture fermentations from cattle. Appl. Environ. Microbiol. 74:38–43.

Jacob, M. E., J. T. Fox, S. K. Narayanan, J. S. Drouillard, D. G. Renter, and T. G. Nagaraja. 2008. Effects of feeding wet corn distiller's grains with solubles with or without monensin and tylosin on the prevalence and antimicrobial susceptibilities of fecal food-borne pathogenic and commensal bacteria in feedlot cattle. J. Anim. Sci. 86:1182–1190.

Jacob, M. E., G. L. Parsons, M. K. Shelor, J. T. Fox, J. S. Drouillard, D. U. Thomson, D. G. Renter, and T. G. Nagaraja. 2008. Feeding supplemental dried distiller's grains increases faecal shedding of *Escherichia coli* O157 in experimentally inoculated calves. Zoon. Pub. Health 55:125–132.

Jay, M. T., M. Cooley, D. Carychao, G. W. Wiscomb, R. A. Sweitzer, L. Crawford-Miksza, J. A. Farrar, D. K. Lau, J. O'Connell, A. Millington, R. V. Asmundson, E. R. Atwill, and R. E. Mandrell. 2007. *Escherichia coli* O157:H7 in feral swine near spinach fields and cattle, central California coast. Emerg. Infect. Dis. Available from http://www.cdc. gov/EID/content/13/12/1908.htm.

Keen, J. E., L. M. Durso, and T. P. Meehan. 2007. Isolation of *Salmonella enterica* and shiga-toxigenic *Escherichia coli* O157 from feces of animals in public contact areas of United States zoological parks. Appl. Environ. Microbiol. 73:362–365.

Keen, J., and R. Elder. 2000. Commercial probiotics are not effective for short-term control of enterohemorrhagic *Escherichia coli* O157 infection in beef cattle. Page 92 in 4th Intl. Symp. Works. Shiga Toxin (Verocytotoxin)-producing *Escherichia coli* Infect., Kyoto, Japan.

Keen, J. E., and R. O. Elder. 2002. Isolation of shiga-toxigenic *Escherichia coli* O157 from hide surfaces and the oral cavity of finished beef feedlot cattle. J. Am. Vet. Med. Assoc. 220:756–763.

Klieve, A. V., and T. Bauchop. 1988. Morphological diversity of ruminal bacteriophages from sheep and cattle. Appl. Environ. Microbiol. 54:1637–1641.

Klieve, A. V., and R. A. Swain. 1993. Estimation of ruminal bacteriophage numbers by pulsed-field electrophoresis and laser densitometry. Appl. Environ. Microbiol. 59:2299–2303.

Klieve, A. V., K. Gregg, and T. Bauchop. 1991. Isolation and characterization of lytic phages from *Bacteroides ruminicola* ss *brevis*. Curr. Microbiol. 23:183–187.

Laven, R. A., A. Ashmore, and C. S. Stewart. 2003. *Escherichia coli* in the rumen and colon of slaughter cattle, with particular reference to *E. coli* O157. Vet. J. 165:78–83.

LeJeune, J. T., T. E. Besser, D. H. Rice, J. L. Berg, R. P. Stillborn, and D. D. Hancock. 2004. Longitudinal study of fecal shedding of *Escherichia coli* O157:H7 in feedlot cattle: predominance and persistence of specific clonal types despite massive cattle population turnover. Appl. Environ. Microbiol. 70:377–385.

Loc Carrillo, C. M., R. J. Atterbury, A. El-Shibiny, P. L. Connerton, E. Dillon, A. Scott, and I. F. Connerton. 2005. Bacteriophage therapy to reduce *Campylobacter jejuni* colonization of broiler chickens. Appl. Environ. Microbiol. 71:6554–6563.

Mead, P. S., L. Slutsker, V. Dietz, L. F. McCraig, J. S. Bresee, C. Shapiro, P. M. Griffin, and R. V. Tauxe. 1999. Food-related illness and death in the United States. Emerg. Infect. Dis. 5:607–625.

Merril, C. R., B. Biswas, R. Carlton, N. C. Jensen, G. J. Creed, S. Zullo, and S. Adhya. 1996. Long-circulating bacteriophage as antibacterial agents. Proc. Natl. Acad. Sci. USA 93:3188–3192.

Nurmi, E., L. Nuotio, and C. Schncitz. 1992. The competitive exclusion concept: development and future. Int. J. Food Microbiol. 15:237–240.

Rasmussen, M. A., W. C. Cray, T. A. Casey, and S. C. Whipp. 1993. Rumen contents as a reservoir of enterohemorrhagic *Escherichia coli*. FEMS Microbiol. Lett. 114:79–84.

Reicks, A. L., M. M. Brashears, K. D. Adams, J. C. Brooks, J. R. Blanton, and M. F. Miller. 2007. Impact of transportation of feedlot cattle to the harvest facility on the prevalence of *Escherichia coli* O157:H7, *Salmonella*, and total aerobic microorganisms on hides. J. Food Prot. 70:17–21.

Roach, S. 2006. USDA proposal could redefine grass-fed meat. At http://www.meatprocess.com/news/ng.asp?id=70409. Accessed September 2006.

Rollin, B. E. 2006. Food safety—who is responsible? Foodborne Pathog. Dis. 3:157–162.

Sharp, G. 2005. Traditions and transitions; the uneven industrialization of the U.S. beef sector. PhD Diss., University of Wisconsin-Madison.

Smith, H. W., and R. B. Huggins. 1982. Successful treatment of experimental *E. coli* infections in mice using phage: its general superiority over antibiotics. J. Gen. Microbiol. 128:307–318.

Smith, H. W., and R. B. Huggins. 1983. Effectiveness of phages in treating experimental *Escherichia coli* diarrhoea in calves, piglets and lambs. J. Gen. Microbiol. 129:2659–2675.

Smith, H. W., and R. B. Huggins. 1987. The control of experimental *E. coli* diarrhea in calves by means of bacteriophage. J. Gen. Microbiol. 133:1111–1126.

Smith, D. R., R. A. Moxley, S. L. Clowser, J. D. Folmer, S. Hinkley, G. E. Erickson, and T. J. Klopfenstein. 2005. Use of rope devices to describe and explain the feedlot ecology of *Escherichia coli* O157:H7 by time and place. Foodborne Path. Dis. 2:50–60.

Soupir, M. L., S. Mostaghimi, E. R. Yagow, C. Hagedorn, and D. H. Vaughan. 2006. Transport of fecal bacteria from poultry litter and cattle manures applied to pastureland. Water, Air, Soil Poll. 169:125–136.

Stanford, K., D. Croy, S. J. Bach, G. L. Wallins, H. Zahiroddini, and T. A. McAllister. 2005. Ecology of *Escherichia coli* O157:H7 in commercial dairies in southern Alberta. J. Dairy Sci. 88:4441.

Steer, T., H. Carpenter, K. Tuohy, and G. R. Gibson. 2000. Perspectives on the role of the human gut microbiota and its modulation by pro and prebiotics. Nutr. Res. Rev. 13:229–254.

Steinmuller, N., L. Demma, J. B. Bender, M. Eidson, and F. J. Angulo. 2006. Outbreaks of enteric disease associated with animal contact: not just a foodborne problem anymore. Clin. Infect. Dis. 43:1596–1602.

Stephens, T. P., G. H. Loneragan, E. Karunasena, and M. M. Brashears. 2007. Reduction of *Escherichia coli* O157 and *Salmonella* in feces and on hides of feedlot cattle using various doses of a direct-fed microbial. J. Food Prot. 70:2386–2391.

Summers, W. C. 2001. Bacteriophage therapy. Ann. Rev. Microbiol. 55:437–451.

Thurston-Enriquez, J. A., J. E. Gilley, and B. Eghball. 2005. Microbial quality of runoff following land application of cattle manure and swine slurry. J. Water Health 3:157–171.

Tkalcic, S., T. Zhao, B. G. Harmon, M. P. Doyle, C. A. Brown, and P. Zhao. 2003. Fecal shedding of enterohemorrhagic *Escherichia coli* in weaned calves following treatment with probiotic *Escherichia coli*. J. Food Prot. 66:1184–1189.

USDA-ERS. 2001. ERS estimates foodborne disease costs at $6.9 billion per year. Economic Research Service-United States Department of Agriculture. Available at: http://www.ers.usda.gov/publications/aer741/aer741.pdf. Accessed October 16, 2007.

Vugia, D., A. Cronquist, J. Hadler, M. Tobin-D'Angelo, D. Blythe, K. Smith, S. Lathrop, D. Morse, P. Cieslak, T. Jones, K. G. Holt, J. J. Guzewich, O. L. Henao, E. Scallan, F. J. Angulo, P. M. Griffin, and R. V. Tauxe. 2007. Preliminary foodnet data on the incidence of infection with aathogens transmitted commonly through food—10 states, 2006. Morbid. Mortal. Weekly Rep. 56:336–339.

Williamson, C. S. 2007. Is organic food better for our health? Nutr. Bulletin 32:104–108.

Woerner, D. R., J. R. Ransom, J. N. Sofos, G. A. Dewell, G. C. Smith, M. D. Salman, and K. E. Belk. 2006. Determining the prevalence of *Escherichia coli* O157 in cattle and beef from the feedlot to the cooler. J. Food Prot. 69:2824–2827.

Younts-Dahl, S. M., M. L. Galyean, G. H. Loneragan, N. A. Elam, and M. M. Brashears. 2004. Dietary supplementation with *Lactobacillus*- and *Propionibacterium*-based direct-fed microbials and prevalence of *Escherichia coli* O157 in beef feedlot cattle and on hides at harvest. J. Food Prot. 67:889–893.

Zhao, T., M. P. Doyle, B. G. Harmon, C. A. Brown, P. O. E. Mueller, and A. H. Parks. 1998. Reduction of carriage of enterohemorrhagic *Escherichia coli* O157:H7 in cattle by inoculation with probiotic bacteria. J. Clin. Microbiol. 36:641–647.

■ 23 ■

The Consumers' Perspective
on the Safety of Organic Foods:
An Opportunity for Future Research

Jenna Anding and Philip G. Crandall

Introduction

Organic food production and sales continue to be one of the fastest-growing areas of U.S. agriculture. Once limited to farmer's markets and health food stores, organic foods are now part of mainstream supermarkets and club stores (Organic Trade Association 2006a). All 50 states in the United States have some amount of certified organic farmland, although the total amount of certified organic cropland is less than 1% (USDA-ERS 2007a). Over the past decade, the industry has sustained an impressive rate of growth in sales ranging from 15 to 21% annually since 1997 (Figure 23.1) (Organic Trade Association 2006b; 2007). More than one-half of all Americans have reported trying organic foods (Whole Foods Market 2003; 2004) and the percentage of adults who identify themselves as regular organic foods purchasers ranges from 11 to 66% (Dimitri and Greene 2002). Produce, fruits and vegetables, continues to be the leading category in organic food sales accounting for 43% of U.S. organic food sales in 2002, followed by breads/cereals, 13%, packaged/prepared foods, dairy foods, and beverages at 11% each. Other organic foods such as meat and soy foods comprise less than 10% of organic food sales (USDA-ERS 2007b).

Beginning October 2002 the USDA National Organic Program was implemented to assure consumers that the organic foods they purchased were produced according to established standards (USDA-AMS 2008a). As a result, foods marketed as organic must be "produced without using most conventional pesticides;

fertilizers made with synthetic ingredients or sewage sludge; bioengineering (GMO or GE); or ionizing radiation." Additionally, those foods are subjected to labeling requirements (USDA-AMS 2008b) that reflect the percentage of the organic ingredients in the product. Products labeled "100 percent organic" contain only ingredients (with the exception of salt and water) that were organically produced. Those with at least 95% organic ingredients can legally be labeled as "organic," but not 100% organic, while those with at least 70% organic ingredients can include the phrase "made with organic ingredients" and list those organic ingredients on the label. Products that contain less than 70% organic ingredients are not allowed to use the term "organic" on the display panel but may identify those ingredients that are organic in the ingredients section of the label. Only those labeled "100% Organic" and "Organic" are allowed to include the USDA organic seal on the label. Preliminary research suggests that the USDA organic seal is being recognized by consumers (Raab and Grobe 2005), but it is unclear if they fully understand the definition of the term "organic" (Batte et al. 2003).

For individuals who buy organic, the willingness to pay a premium for organic foods ranges from 10% to 25% (University of Nebraska 2001; Govindasamy and Italia 1999), which can significantly increase a household's food costs. For example, Brown and Sperow (2005) analyzed the cost of an all-organic diet based on food lists and recipes associated with the USDA's Thrifty Food Plan. Comparing the costs of organic and nonorganic foods, the authors noted that an all-organic diet would increase the food bill by as much as 49%. In their study, the average premiums for organic foods ranged from 15% for vegetables to a maximum of 122% for fats and oils.

Who Buys Organic and Why?

Consumer studies (Govindasamy and Italia 1999; Onyango et al. 2007; Loureiro and Hine 2002; Govindasamy et al. 2001; Loureiro et al. 2001 and Thompson and Kidwell 1998) have attempted to identify the characteristics of individuals and their reasons for purchasing organic foods. Female gender, small household size, and the presence of young children in the household are a few of the demographics that have been used to describe organic food consumers for the past 10 years (Table 23.1). High income, a higher level of education, as well as younger consumers have been positively correlated to organic food purchases but not consistently across published studies.

In addition to the usual demographic characteristics, organic consumers' attitudes and beliefs help shape the image of an organic food consumer. Magnusson et al. (2003) examined the self-reported purchases of organic foods among Swedish adults and found that more than 70% of those responding believed that their

Table 23.1. Reported Demographic Characteristics of Organic Food Consumers

Author	Year	Sample	Methodology	Characteristics supporting the purchase of organic food
Onyango et al	2007	1,185 adults; national survey	Telephone survey to assess organic food preferences	Young age (18–32 years) Female gender Moderately religious College education or more
Loureiro and Hine	2002	437 adults randomly selected in grocery stores across Colorado	In-store survey to assess willingness to pay for local, organic, and GMO-free potatoes	High income Highly educated
Godvindasamy et al	2001	606 adults randomly selected in New Jersey, New York, and Pennsylvania	Mail survey	Small household size Households with few children Suburban location Female gender Young age Highly education Not married
Loureiro et al	2001	285 randomly selected adults shopping in one of two Portland, OR, retail food outlets	In-store survey to assess consumer preference of eco-labeled, organic, and conventionally grown apples	Households with children under 18 years of age Concern about food safety Concern about environment vs. jobs
Govindasamy and Italia	1999	291 adults shopping in one of 5 retail grocery stores in New Jersey	Survey distributed in store; returned by mail. Survey assessed consumers' willingness to pay for organic produce	Young age (less than 36 years) Female gender High income (> $70,000/year) Small household size
Thompson and Kidwell	1998	340 adults; 2 retail food outlets in Tucson, AZ, that sold both organic and conventionally grown produce	In-store interviews	Households with children under 18 years of age

choice of organic foods would improve the environment. Additionally, more than half (52%) perceived that organic foods improved either their health or the health of their family. These findings echo those conducted in Norway (Wandel and Bugge 1997), The Netherlands (Schifferstein and Ophuis 1998), and in Northern Ireland (Davies et al. 1995). In the United States, consumers are buying organic foods not only in support of their own health and the health of the environment (Williams and Hammitt 2000) but also to support local farmers (Raab and Grobe 2005; University of Nebraska 2001; Lohr 2001) and ensure that the foods purchased (especially produce) are as fresh as possible (Raab and Grobe 2005; University of Nebraska 2001 Davies 1995). The idea that the nutrition, taste, and quality of organic foods rank superior to conventionally grown/produced are additional beliefs widely held by consumers of organic foods (Raab and Grobe 2005; Batte et al. 2003; University of Nebraska 2001; Wandel and Bugge 1997; Saba and Messina 2003; Williams and Hammitt 2001).

Food Safety and Organic Food

Another factor driving the organic industry is the consumer's continued concern about food safety (Batte et al. 2003; Govindasamy and Italia 1999; Huang 1996). While most Americans trust the nation's food supply, the percentage of consumers who characterize themselves as either "completely confident" or "mostly confident" has dropped from 83% in 1996 to 74% in 2000 (USDA-ERS 2002). In 2006, a Web-based survey (IFIC 2006) of nearly 500 adults found that 72% of consumers expressed confidence in the safety of the food supply. Confidence in the food supply, however, may depend on the type of consumer. A Boston study (Williams and Hammitt 2000) found that consumers of organic produce had significantly less trust in government agencies charged with protecting the nation's food supply compared with those who purchase those foods that are conventionally grown.

When it comes to safety of our nation's food supply consumer concerns about pesticide residues are well documented (Williams and Hammitt 2000; USDA-ERS 2002; Bruhn et al. 1992; Misra et al. 1991) and are a major reason why people choose organic food. Although the percentage of consumers expressing concern about pesticide residues has decreased (USDA-EARS 2002), some continue to see pesticides as a serious health threat. However, that "threat" may be more perception than reality. For example, Bourn and Prescott (2002) and Magkos et al. (2006) examined whether or not pesticide residues were actually lower in organically produced foods compared to those conventionally. Both papers noted that reliable data were limited and extremely difficult to collect; however, research by Baker et al. (2002) suggests that synthetic pesticide residues are less prevalent in food items that are organically produced. That certainly does not suggest that organic foods are void of all pesticides. In fact, pesticide contamination can occur if the soil is contaminated, if the water used to irrigate the crops contains pesticides, or if chemicals used on a neighboring farm drift onto a field that is practicing organic food production (Magkos et al. 2006). Pesticides approved for organic production may also carry health risks (Magkos et al. 2006).

Since 1995, the percentage of consumers who identify bacteria in food as a serious risk has grown from an estimated 80% to nearly 90% (USDA-ERS 2002). This may be due, in part, to the increased media attention that has been given to outbreaks of foodborne illness. In fact, produce-related foodborne-disease outbreaks have been steadily increasing over the past three decades (Lynch et al. 2006; Sivapalasingam et al. 2004). Surveillance data from the Centers for Disease Control and Prevention (CDC) report that the number of foodborne-disease outbreaks linked to fruits, vegetables, and nuts totaled 44 in 1998, peaked at 62 outbreaks in 1999 and 2000, and dropped to 53 in 2002 (Lynch et al. 2006). A foodborne-disease outbreak (FBDO) is defined by the CDC as "an incident in which two or more persons experience a similar illness resulting from the ingestion of a common food" (CDC 2000).

Since the 1990s produce has been linked to 6% of all foodborne-disease outbreaks, a significant increase from 0.7% in the 1970s (Sivapalasingam et al. 2004). Whether this increase is due to better reporting systems or a true increase in produce-linked foodborne disease is unclear but definitely worth consideration. The reported number of individuals who contract a produce-linked foodborne illness each year also has climbed from an average of 708 in the 1970s to more than 8,800 in the 1990s. Produce items most often implicated in foodborne-disease outbreaks during this period include lettuce (25 outbreaks), melons (13), sprouts, apple and orange juice (11) and berries (11). Of the 190 produce-related outbreaks reported by 22 states between 1973 and 1997, *Salmonella* was implicated in the largest number of outbreaks (30) followed by *E. coli* O157:H7 (13 outbreaks), Hepatitis A (12 outbreaks), Norovirus (9 outbreaks), and *Cyclospora cayetanensis* (8 outbreaks). Other pathogens include *Campylobacter, Bacillus cereus, Yersinia enterocolitica,* and *Staphylococcus aureus.*

Both conventional and organic farmers have the option of using well-composted animal manure as a source of fertilizer, although the practice is believed to be more common among organic farmers, since synthetic fertilizers are not allowed (Albihn 2001). Certified organic farms must follow strict guidelines when manure is used since it can be a source of *E. coli* along with other pathogens. There have been concerns that organically grown crops fertilized with animal manure could have a higher amount of microbial contaminants compared to those fertilized with conventional fertilizers (Avery 1998). However, this concern may be unfounded. Mukherjee et al. (2004) analyzed organic and conventionally grown produce for the presence of coliform bacteria (including *E. coli*), *E. coli* O157:H7, and *Salmonella*. Produce varieties included tomatoes, leafy greens, cucumbers, zucchini, apples, and strawberries. Participating organic farms included those who were certified and those who were not USDA certified but self-reported that they adhered to organic farming practices (noncertified). The prevalence of *E. coli* (noted as a percentage of produce samples testing positive for the microorganism) was more than six times higher in the organic (9.7% of samples) compared to those conventionally grown produce (1.6%). However, when compared to non-certified produce there was even a greater contamination by microorganisms (11.4%). The percentage of samples of certified organic produce with *E. coli* was 4.3%, which was higher but not significantly different from those conventionally grown. Two samples from the uncertified produce tested positive for *Salmonella*; however, none of the certified or conventionally grown samples tested positive for *E. coli* O157:H7 or *Salmonella*. A followup study comparing the microbial quality of preharvest produce from certified, uncertified, and conventional farms over two growing seasons yielded similar results (Mukherjee et al. 2006). *E. coli* has been documented in samples of organically grown lettuce in Norway but, like the Mukherjee studies, *E. coli* O157:H7 and *Salmonella* were not detected. *Listeria monocytogenes* was detected in two out of the 179 samples tested (Locarevic et al. 2005). Assessment of spring mix (mesclun) by Phillips and Harrison (2005) also reported the presence of *E. coli* in both organic and conventionally grown samples; however, *Salmonella* and *Listeria monocytogenes* was not detected in any samples. Based on these and other studies, it is likely that a number of microorganisms, including *E. coli*, are present in both organic and conventionally grown produce. However, the presence of generic *E. coli* does not automatically suggest a presence of *E. coli* O157:H7. Clearly where contamination with pathogenic microorganisms is concerned, there is not enough evidence to conclude that one method of farming (i.e., organic or conventional) is safer than another.

Public Opinions about Food Safety, and Organic Food: Knowledge, Attitudes, and Questions Identified at the Tri-State Food Safety Consortium

There is a need for additional research to provide consumers, regulators, growers, and manufacturers baseline data on their current practices and to point out areas where good agricultural practices (GAP) could further reduce the risks of food-borne illness. Additional consumer and microbiological research is needed to estimate how much of these risks are real and the extent to which they are due to difference in perceptions. Are there differences in handling practices among consumers who tend to buy organic foods?

To begin identifying research as well as possible outreach/education activities, individuals attending the 2006 Tri-State Food Safety Consortium in Fayetteville, Arkansas, were invited to participate in a focus group discussion led by a professor of food science and a Cooperative Extension nutrition specialist. The two-hour discussion featured questions targeting the following areas: (1) characteristics of organic food consumers; (2) safety of organic food (including perceived risk of contracting foodborne disease from organic produce); and (3) the extent to which the term "organic" is identifiable and understood. Attendees (n = 21) included university researchers, Cooperative Extension faculty, consumers, producers, and retailers of organic food. The discussion group ranged from consumers who regularly or exclusively purchased organic foods to two who did not purchase organic foods because they did not believe that organic foods were any better than conventional. Most participants agreed that organic foods carried a price premium.

Characteristics of organic food consumers: The demographic characteristics of organic food consumers identified by the respondents were consistent with existing research (e.g., female, often single, well educated, above average in disposable income). An individual reported on a survey of their natural foods cooperative and found that many of their consumers had completed some college, typically had more disposable income, and were either between the ages of 40 and 65 or in their early twenties. Most of their customers were female and some had indicated that they or a member of their family had had a history of "health problems." Another participant noted that in their east coast state, organic food consumers were often between 25 and 40, single, and had a higher income. They were also perceived as being part of the "microbrew crowd, more hip." Participants also noted that some young mothers felt that it was important enough to budget for organic foods to buy healthy/nutritious foods for their children. The desire to improve health and concern for the environment were common factors behind the decision that led at least one participant to choose organic foods over those grown conventionally.

Several respondents commented that the increasing availability of organic foods will probably impact these organic consumer characteristics. One participant noted that one of the largest retailers had recently begun carrying organic produce targeting the "younger, selective shopper" and "soccer mom" demographics. However, it was also mentioned that only a small percentage of their produce is organic due to low demand and logistic difficulties including keeping organic and nonorganic foods separate. This seemed to prompt others to identify additional challenges related to acquiring organic foods. At least one participant (a retailer) reported that their business only stocked USDA certified organic foods (no conventional or transitional) while another identified that their state's short growing season was a significant limiting factor. Lack of organic farms and auditors necessary to certify farms as organic, challenges controlling disease and insect pests, and the higher costs that consumers must pay for organic foods were also identified as factors that may be limiting the growth of the organic market.

Perceived safety of organic food: Similar to those findings reported in the published research, participants in this focus group overwhelmingly tended to relate pesticide residues with food safety. Probing with additional questions related to concerns about microbes or the need to wash produce to reduce pathogens did not seem to elicit a measurable response from the group.

Understanding of the term "organic": This part of the discussion produced some interesting responses that may indicate a lack of understanding of the term among consumers, as the term "organic" was often linked with "natural." While the USDA has a definition of the term "natural" the term is not synonymous with organic and is currently limited to meat and poultry products (USDA-FSIS 2005). The Food and Drug Administration, which is charged with the implementation and enforcement of food labeling laws, does not have a legal definition of the term "natural." Whether or not respondents understood the terms "100% organic," "organic," and "made with organic ingredients" also was unclear.

Summary

The documented growth and public interest in organic food production supports the need for continued research from both a food-safety and consumer perspective. Current research, while limited, suggests that fruits and vegetables could become contaminated with microbial pathogens regardless of how they are grown. However, the available research does not appear to firmly legitimize concerns that contamination by food pathogens is more likely to occur among organically grown produce. Additional studies, particularly those that compare the microbial quality among organic, uncertified organic, and conventionally grown produce can benefit the industry's supporters, justify the need for a having a national certification policy, and help calm the fears of organic food critics.

The implementation of the USDA National Organic Program assured consumers that foods labeled "organic" are indeed organically grown. However, it is unclear whether the public understands the different categories of certified organic foods. Consumer studies addressing this question may be beneficial. Outreach efforts that educate consumers about differences between the terms "organic" and "natural" may also be in order.

Finally, it is acknowledged that while a large percentage of individual consumers are confident in the safety of the U.S. food supply, some concerns remain. How do these concerns differ between consumers of organic foods and those who choose those conventionally produced? It has been suggested that consumers of organic food may be less confident in the conventional food supply compared to other consumers. If this is accurate, what are the factors behind that reduced confidence? Furthermore, are food-handling practices between the two groups different? Answers to these questions may be helpful to educators as they develop and conduct food-safety outreach programs to different clientele groups.

References

Albihn, A. 2001. Recycling biowaste—human and animal health problems. Acta Vet. Scand. 95:69–75.

Avery, D. T. 1998. The hidden dangers in organic food. American Outlook fall 1998. At http://ao.hudson.org/index.cfm?fuseaction=article_detail&id=1196. Accessed September 11, 2007.

Baker, B. P., C. M. Benbrook, E. Groth, and L. Benbrook. 2002. Pesticide residues in conventional, integrated pest management (IPM)-grown and organic foods. Insights from three US data sets. Food Additives and Contaminants 19:427–446.

Batte, M. T., J. Beaverson, and N. Hooker. 2003. Organic food labels: a customer intercept survey of central Ohio food shoppers. Ohio State University Report AEDE-RP-0038–03. At http://aede.osu.edu/Programs/VanBuren/PDF/AEDE-RP-0038–03.pdf. Accessed September 10, 2007.

Bourn, D., and J. Prescott. 2002. A comparison of the nutritional value, sensory quality, and food safety of organically and conventionally produced foods. Crit. Rev. Food Sci. Nutr. 42:1–34.

Brown, C., and M. Sperow. 2005. Examining the cost of an all-organic diet. J. Food Distr. Res. 36:20–26.

Bruhn, C. M., K. Diaz-Knauf, N. Feldman, J. Harwood, G. Ho, E. Ivans, L. Kubin, C. Lamp, M. Marshall, S. Osaki, G. Stanford, Y. Steinbring, I. Valdez, E. Williamson, and E. Wunderlich. 1992. Consumer food safety concerns and interest in pesticides. J. Food Safety 12:253–262.

Centers for Disease Control and Prevention. 2000. Guidelines for Confirmation of Foodborne-Disease Outbreaks. MMWR Surveillance Summaries 49(SS01):54–62. At http://www.cdc.gov/mmwr/preview/mmwrhtml/ss4901a3.htm. Accessed September 18, 2007.

Davies, A., A. J. Titterington, and C. Cochrane. 1995. Who buys organic food? Br. Food J. 97:17–23.

Dimitri, C., and C. Greene. 2002 Recent growth patterns in the U.S. organic foods market. Economic Research Service/USDA. At http://www.ers.usda.gov/Publications/AIB777/. Accessed June 4, 2007.

Govindasamy, R., M. DeCongelio, J. Italia, B. Barbour, and K. Anderson. 2001. Empirically evaluating consumer characteristics and satisfaction with organic products. New Jersey Agricultural Experiment Station P-02139–1-01. At http://aesop.rutgers.edu/~agecon/pub/organicproduction.pdf. Accessed August 14, 2007.

Govindasamy, R., and J. Italia. 1999. Predicting willingness-to-pay a premium for organically grown fresh produce. J. Food Distr. Res. 30: 44–53.

Huang, C. 1996. Consumer preferences and attitudes toward organically grown produce. European Rev. Agric. Econ. 29:43–53.

International Food Information Council (IFIC). 2006. Food biotechnology: A study of U.S. consumer attitudinal trends, 2006 report. At http://www.ific.org/reserch/biotechres.cfm#PDF. Accessed August 14, 2007.

Locarevic, S., G. S. Johannessen, and L. M. Rorvik. 2005. Bacteriological quality of organically grown leaf lettuce in Norway. Lett. Appl. Microbiol. 41:186–189.

Lohr, L. 2001. Factors affecting international demand and trade in organic food products. USDA Economic Research Service. In Changing structure of global food consumption and trade. WRS No. (WRS01–1) May 2001. At http://www.ers.usda.gov/publications/wrs011/. Accessed September 14, 2007.

Loureiro, M. L., and S. Hine. 2002. Discovering niche markets: a comparison of consumer willingness to pay for local (Colorado grown), organic, and GMO-free products. J. Agric. Appl. Econ. 34:477–487.

Loureiro, M. L., J. J. McCluskey, and R. C. Mittelhammer. 2001. Assessing consumer preferences or organic, eco-labeled, and regular apples. J. Agric. Res. Econ. 26:404–416.

Lynch, M., J. Painter, R. Woodruff, and C. Braden. 2006. Centers for Disease Control and Prevention. Surveillance for foodborne-disease outbreaks—United States, 1998–2002. MMWR Surveillance Summary 55:1142.

Magkos, F., F. Arvaniti, and A. Zampelas. 2006. Organic food: buying more safety or just peace of mind? A critical review of the literature. Crit. Rev. Food Sci. Nutr. 46:23–56.

Magnusson, M. K., A. Arvola, U. K. K. Hursti, L. Aberg, and P. Sjoden. 2003. Choice of organic foods is related to perceived consequences for human health and to environmentally friendly behavior. Appetite 40:109–117

Misra, S. K., C. L. Huang, and S. L. Oh. 1991. Consumer willingness to pay for pesticide-free fresh produce. Western J. Agric. Econ. 16:218–227.

Mukherjee, A., D. Speh, E. Dyck, and F. Diez-Gonzalez. 2004. Preharvest evaluation of coliforms, *Escherichi coli, Salmonella,* and *Escherichi coli* O157:H7 in organic and conventional produce grown by Minnesota farmers. J. Food Prot. 67:894–900.

Mukherjee, A., D. Speh, A. T. Jones, K. M. Buesing, and F. Diez-Gonzalez. 2006. Longitudinal microbiological survey of fresh produce grown by farmers in the upper Midwest. J. Food Prot. 69:1928–1936.

Onyango, B. M., W. K. Kallman, and A. C. Bellows. 2007. Purchasing organic food in US food systems: a study of attitudes and practice. Br. Food J. 109(5):399–411.

Organic Trade Association. 2006a. U.S. organic industry overview. At http://www.ota.com/pics/documents/short%20overview%20MMS.pdf. Accessed June 4, 2007.

Organic Trade Association. 2006b. Manufacturer survey press release. At http://www.organicnewsroom.com/2006/05/organic_sales_continue_to_grow.html. Accessed June 4, 2007.

Organic Trade Association. 2007. Manufacturer survey press release. At http://www.organicnewsroom.com/2007/05/us_organic_sales_show_substant_1.html. Accessed June 4, 2007.

Phillips, C. A., and M. A. Harrison. 2005. Comparison of the microflora on organically and conventionally grown spring mix from a California processor. J. Food Prot. 68:1143–1146.

Raab, C., and D. Grobe. 2005. Consumer knowledge and perceptions about organic food. J. Extension, 43 (4), Article Number 4RIB3. At http://www.joe.org/joe/2005august/rb3.shtml. Accessed August 4, 2007.

Saba, A., and F. Messina. 2003. Attitudes towards organic foods and risk/benefit perception associated with pesticides. Food Qual. Pref. 14:637–645.

Schifferstein, H. J., and O. Ophuis. 1998. Health-related determinants of organic food consumption in the Netherlands. Food Qual. Pref. 9:119–133.

Sivapalasingam, S., C. R. Friedman, L. Cohen, and R. V. Tauxe. 2004. Fresh produce: a growing cause of foodborne illness in the United States, 1973–1997. J. Food Prot. 67:2342–2353.

Thompson, G. D., and J. Kidwell. 1998. Explaining the choice of organic product: cosmetic defects, prices, and consumer preferences. Am. J. Agric. Econ. 80:277–287.

USDA-Agricultural Marketing Service. 2008a. National Organic Program standards. At http://www.ams.usda.gov/nop/NOP/standards/FullText.pdf. Accessed May 23, 2008.

USDA-Agricultural Marketing Service. 2008b. National organic program: organic production and handling standards. At http://www.ams.usda.gov/AMSv1.0/getfile?dDocName=STELDEV3004445&acct=nopgeninfo. Accessed May 23, 2008.

USDA Economic Research Service. 2002. Consumer food safety behavior: consumer concernsAt http://www.ers.usda.gov/briefing/consumerfoodsafety/consumerconcerns.htm. Accessed August 14, 2007.

USDA Economic Research Service. 2007a. Organic production data sets. At http://www.ers.usda.gov/Data/Organic, 2007. Accessed June 4, 2007.

USDA Economic Research Service. 2007b. Organic farming and marketing: questions and answers. At http://www.ers.usda.gov/Briefing/Organic/Questions/orgqa5.htm.March 19, 2007/. Accessed March 19, 2007.

USDA Food Safety and Inspection Service. 2005. Food standards and labeling policy book. At http://www.fsis.usda.gov/OPPDE/larc/Policies/Labeling_Policy_Book_082005.pdf. Accessed September 18, 2007.

University of Nebraska-Lincoln Food Processing Center, Institute of Agriculture and Natural Resources. 2001. Attracting consumers with locally grown products.At http://www.foodmap.unl.edu/report_files/Locally_Grown_Consumer_Survey_Report.pdf. Accessed August 14, 2007.

Wandel, M., and A. Bugge. 1997. Environmental concern in consumer evaluation of food quality. Food Qual. Pref. 8:19–26.

Whole Foods Market. 2003. One year after USDA organic standards enacted more Americans are consuming organic food. At http://www.wholefoodsmarket.com/company/pr_10–14–03.html. Accessed June 4, 2007.

Whole Foods Market. 2004. Organic foods continue to grow in popularity according to Whole Foods Market survey. At http://www.organicconsumers.org/organic/popularity102204.cfm. Accessed June 4, 2007.

Williams, P. R. D., and J. K. Hammitt. 2000. A comparison of organic and conventional fresh produce buyers in the Boston area. Risk Analysis 20:735–746.

Williams, P. R. D., and J. K. Hammitt. 2001. Perceived risks of conventional and organic produce: pesticides, pathogens, and natural toxins. Risk Analysis 21:319–330.

▌24▐

Future Prospects for Advancing Food-Safety Research in Food Animals

Steven C. Ricke

Introduction

Although great strides have been made by the scientific community in understanding the biology and dissemination of foodborne pathogens, recent media headlines publicizing pathogenic bacterial contamination of foods such as spinach and peanut butter indicate that foodborne disease is still a high-profile issue for the consumer. Therefore, food safety continues to be a primary focus for the food industry. The USDA Economic Research Service estimates that foodborne illness costs about $6.9 billion per year on account of five foodborne pathogens alone— *Campylobacter, Salmonella* (nontyphoidal serotypes only), *Escherichia coli* O157:H7 and non-O157 STEC, and *Listeria monocytogenes* (USDA ERS 2004).

The increasing complexity of the food-safety issues associated with most agricultural commodities requires a multidisciplinary effort across academic institutions and public health agencies. Specific issues that have been identified within the past few years by consumer groups, federal agencies, and the food animal industry include preharvest control of *Salmonella* spp. and *Campylobacter,* and post harvest control of *Listeria* (Jackson et al. 2007; Lianou and Sofos 2007; White et al. 2007). Within the food animal research arena, a substantial investment has been made in food-safety research that would immediately benefit from input of environmental and ecological tracking of foodborne pathogens. Foodborne pathogen transmission and survival, transfer of genetic material, and control of nutrient cycling in these ecosystems not only represent agricultural issues but will become public health issues as the boundaries between municipalities and farms become more blurred. This will come to a head as more zero discharge rules for production agriculture are promoted. Given the continued prevalence of foodborne illness in a time of limited resources it becomes necessary to find cross-cutting directions for research applications. More important, these issues must be addressed in the context of ongoing and anticipated changes in the nature and production of the food supply.

Advances and Opportunities in Food-Safety Research

Antimicrobials and the Development of Alternative Antimicrobial Systems

A medical health crisis is looming due to the rapid spread of antibiotic resistance in human and animal pathogens coupled with a decreasing commercial interest in developing and/or discovering new antibiotics (Amábile-Cuevas 2003). Gram-positive microorganisms, particularly the methicillin-resistant *Staphylococcus*, have received considerable news media coverage, but antibiotic-resistant Gram-negative bacteria that can cause problems for the elderly, young, and those already sick are also emerging (Jarvis 2008). This has obvious major implications for the medical community but also impacts the food animal industry as well. For example, in broiler production antibiotics were traditionally included in the diets of healthy birds not only for prevention of diseases such as *Clostridium* spp. induced necrotic enteritis but also for increased weight gain and feed efficiency (Chapman and Johnson 2003; Amábile-Cuevas 2003). However, in response to consumer and political pressure, the European Union has banned several growth-promoting antibiotics over their concern that antibiotic resistance could be transferred to humans (Casewell et al. 2003).

Finding alternatives to antimicrobials has become more attractive as the agricultural use of antibiotics in the United States continues to wane. Historically, a number of chemicals, physical methods, and natural antimicrobials have been implemented in food production and processing to reduce or eliminate microbial contamination in meats (Davidson and Taylor 2007; Farkas 2007; Montville and Chikindas 2007; Ricke 2003; Ricke et al. 2005; Fisher and Phillips 2008). Most of these interventions represent fairly broad-spectrum antimicrobial activities that would influence the host gut microflora and not just the pathogens. More recently, a number of biological compounds have emerged as potential agents for specifically targeting foodborne pathogens. Bacteriophages, bacteriocins, and antibodies have all been researched for control of foodborne pathogens at both the preharvest and postharvest phase of food animal production (Joerger 2003; Berghman et al. 2005; Hagens and Offerhaus 2008). These biological agents can be targeted against a particular foodborne pathogen without having activity against other microflora. Although several of these biological compounds look promising in the laboratory much remains to be done to make their implementation in the rigorous commercial environment a reality.

Given the expense and risk of overexposure when antimicrobial compounds are used at high concentrations, delivery systems that minimize the concentration at the targeted intervention site are needed to achieve complete efficacy and become more attractive to the food industry. Combinations of antimicrobial substances as part of multiple-hurdle technologies can provide added barriers against the growth of foodborne pathogens in meat products. By combining several antimicrobial compounds for simultaneous exposure to the pathogen the opportunities for synergisms are possible that would lessen the concentrations of individual antimicrobials used.

However, conventional applications of antimicrobials using delivery systems such as spraying represent inefficient delivery systems for the application of multiple antimicrobials, which in turn can lead to inconsistent efficacy as well as inconsistent long-term control during storage. Incorporating compounds such as bacteriocins into plastic films for packaging of foods has been explored as a means for delivering such compounds in a stabilized form (La Storia et al. 2008). A more versatile approach for antimicrobials is in the form of polymer-nano-composites (nanotechnology) that would serve as a means to deliver any antimicrobial combination as a single application and could potentially allow for use of lower concentrations of individual antimicrobial compounds. Nanotechnology involves the process of constructing particles that are in the 1 to 100 nm range either by creating complex nanomaterial containing individual antimicrobials or by transforming large materials into smaller ones (Nickols-Richardson 2007; Uskokovi 2007). There are already commercial food and nutrition products containing nanoscale additives and the nanofood market is expected to grow to $20 billion by 2010 (Siegrist et al. 2007). Nanotechnology has been touted as having considerable potential in the form of food packaging bio-nanocomposites to extend the shelf life of food products (Sorrentino et al. 2007). It is anticipated that nanotechnology may even become attractive in the future for reducing foodborne pathogens. For example, the construction of an immunonanoparticle has been shown to enhance antimicrobial activity of lysozyme against *Listeria monocytogenes* (Yang et al. 2007). However, several regulatory issues and public concerns over safety of incorporating nanoparticles may cloud the near future for the food industry before adoption of these technologies become more commonplace in foods (Chau et al. 2007; Dobrovolskaia and McNeil 2007; Uskokovi 2007)

Preharvest Food Safety and Gastrointestinal Microbiology

Gastrointestinal microbial ecology has become an important component of preharvest food safety encompassing foodborne pathogen colonization of animals (e.g., companion, food, and exotic) and the control or treatment of those pathogens before, during, and after establishing a gut microbiota using antimicrobial agents, prebiotics, and probiotics, among others. Even though there has been a tremendous emphasis of academic research and commercial products particularly for prebiotics and probiotics, consistent and predictable effectiveness still remains an unattained target. Much of this status is due to the fact that although much is known about colonization mechanisms of foodborne pathogens in the gut, the indigenous gut microbiota are still poorly understood. Historically, the rumen of ruminant animals has been the most extensively studied gastrointestinal site, and much less is known about monogastric gastrointestinal tract microbial ecology. To better understand gut microbial ecology requires expertise for working with strict anaerobes including methanogens as well as an understanding of anaerobic microbial physiology and gut ecology. The more recent application of molecular techniques, not only to identify individual gut microbes but also to quantitatively profile entire gut populations, has the potential to enable a more comprehensive understanding of the dynamics that impacts the gut microbiota responses to changes in diet, age of the animal, and the presence of feed additives (Wise and Siragusa 2007; Gong et al. 2007). Such knowledge will strengthen the opportunities for better understanding of identifying gut microorganisms that have important probiotic implications and interface with immune response and disease resistance in food animals. Research issues are expected to include the reduction in reliance on antibiotics for controlling gut microbial populations; the development of strategies to select for better GI tract microorganisms to promote animal health, performance, and well-being; and the characterization of cellular mechanisms through which nutraceuticals modulate the impact of environmental stressors and improve disease resistance.

A better understanding of the gut microbiota extends beyond simply focusing on foodborne pathogen control. An in-depth understanding of gut microbiota and gut function will also have implications for human health and well-being. As more cooperation and coordination of research is fostered among gastrointestinal microbiology and physiology, researchers involved in animal physiology, human and animal nutrition, and food science can be more directed toward normal human gut microflora function. This topic is one of the major remaining frontiers in nutrition research, and its importance has been highlighted by previous and more recent findings on the influence of the human gut microbiota on obesity and immune function (Bajzer and Seeley 2006; Blumberg and Strober 2001; Ley et al. 2006; Strober et al. 2002; Turnbaugh et al. 2006). However, additional anaerobic molecular microbiology research is needed to better understand human and animal gut microbiology and to complement the ongoing programs in conventional food-safety and microbiological research.

Molecular and Genomic Approaches

Food-safety research is also beginning to benefit from progress made with functional genomics tools to facilitate an understanding of virulence mechanisms of animal pathogens, mechanisms of interactions between microbes or between microbes and their hosts, and mechanisms used by these organisms to survive. The increased understanding on the underlying molecular processes will become a foundation for development of more efficient and cost-effective prevention and control of animal pathogens.

Opportunities to employ gene function studies on pathogens, to identify roles of individual microorganisms in complex microbial communities, and to utilize metabolomics to quantify microbial metabolic pools will provide answers to fundamental questions on community structure and function and pathogen behavior in agricultural environments (ASM 2007). Innovation of molecular techniques and increased through-put of samples have allowed rapid development of large data sets that have increased the need for computational biologists. Since understanding and valid interpretation of data remain critical and complex, better management of these large data sets will facilitate greater understanding and, ultimately, use of the molecular data sets to develop the necessary understanding for application to animal health and food-safety issues. Likewise, development of DNA- and RNA-based analyses has greatly facilitated detection of foodborne microbial species but effective sample preparation is still problematic (Batt 2007). In addition it is becoming more apparent that it is important to distinguish biologically active individual bacteria from naked DNA or inactive cells within a microbial community to better assess their contributions to microbial diversity, phenotype, and intragenomic heterogeneity (Klein 2007).

Challenges and Emerging Issues in Food-Safety Research

Greening/Organics of the U.S. National Food Supply

Organic animal production is defined as meat and animal products generated without the use of synthetic substances and has become increasingly more prominent in the past few years (Berg et al. 2001). In general natural and organic foods are increasing in popularity due to a consumer perception that they are safer and more nutritious as well as being produced under more sustainable and ethical conditions (Doyle 2006). While "natural" foods remain yet to be comprehensively defined by the federal regulatory agencies, the USDA has defined the 100% organic label to be applied solely to foods that contain only certified organic ingredients and are produced only with certified organic processing aids (Doyle 2006). U.S. organic foods purchases appear to be influenced by both socioeconomic factors as well as food attributes such as production location, naturalness, and product familiarity (Onyango et al. 2008). Long-term growth in this sector of the food industry will remain positive as long as U.S. household incomes enable flexible spending habits and human population dynamics continue to evolve toward shifts in food spending to embrace emerging food trends (Henderson 2007).

For food animals, eggs and poultry are considered among the most rapidly growing food products in the U.S. organic food industry (Oberholtzer et al. 2006). Increase in consumer demand is anticipated to drive expansion of organic poultry and egg products into mass-market grocery stores and will probably lead to additional introduction of organic brands by conventional production entities (van der Sluis 2007). Organic and naturally raised poultry can involve a wide variety of production systems including free-range habitats, pasture with portable house or pens that can be moved periodically, and permanent housing that provides access to the outside (Fanatico 1998; Oberholtzer et al. 2006). Such systems, particularly the free-range and pasture operations, present new biosecurity and food-safety challenges. The opportunity for exposure to foodborne pathogens such as *Salmonella* or *Campylobacter* is increased especially from contact with potential carriers such as wild birds and rodents (Berg 2001). This situation becomes a concern not just from the standpoint of increased contact with the pathogens but the possibility of exposure to more diverse serovars and strains of these pathogens that may present unpredictable detection and control challenges. In addition, organic animal production is prohibited from using synthetic antibiotics and, although current research results are relatively inconsistent, there is some thought that removal of these antibiotics may lead to increased pathogen levels and elevated microbial safety risks although reported research is relatively inconsistent (Winter and Davis 2006). The continued expansion of organic poultry meat and egg production will require natural antimicrobials and preservatives not only at the processing stages but development of similar compounds for use in the preharvest grower stages of production. Accomplishing the goal of a safe organic food supply will require more characterization and identification of any environmental factors that differ between conventional and organic production that may be critical control points unique to organic production. Likewise characterization of physiological, behavioral, and gastrointestinal tract differences will be needed to design better preharvest control measures for organic production.

Globalization and Integration of Food Animal Production

Just as animal food production has become more vertically integrated, so have the food-safety issues associated with meat production. More focus is being placed on understanding all facets of foodborne pathogen prevalence from farm to fork but food-safety knowledge deficiencies persist in several of the animal production steps. For example, preharvest food-safety research has substantially progressed over the past 20 years, but as discussed previously could still benefit from molecular microbial ecology methodologies on the gut microflora to further investigate the microbial communities that control fermentation and attachment sites on the gut lining. Other issues intersect environmental and food safety such as fate and transport of emerging pathogens, which can encompass confined animal housing, animal populations, and their subsequent impact on bacterial communities in the surrounding watershed. However, in order to conduct research in these associated food-safety areas, systems oriented research is needed that involves several areas of agricultural systems that impact food safety and animal production.

If strategic choices are made that maximize fostering of interaction across research disciplines then application of advanced research tools can be included to generate more comprehensive data on the relationship between environment and prevalence of foodborne pathogens. Assessment of food-safety research needs and addressing opportunities for the infusion of new technological tools into these research programs will have to include a comprehensive evaluation of the optimal use of resources, facilities, and wise choices of problems to solve. For agriculture in general, more emphasis will need to be placed on sustainability and development goals in small-scale agricultural production sectors (Kiers et al. 2008). This emphasis certainly holds true for food-safety future research as well. Even though many of the current food-safety control measures are directed toward large-scale vertical animal product and processing, new paradigms with more diffuse critical control points will need to be considered as local agricultural markets become more popular. At the same time international trade is becoming more integrated with implementation of several international trade agreements among countries (Kali and Reyes 2007). Consequently, food safety of internationally traded food commodities both as a public health concern and as a political leverage issue is becoming an important contributor to driving food-safety research into new technological requirements.

Impact of Non-Food-Safety Factors

There is a need to foster further consideration of the non-food-safety agricultural production and processing sectors about policy and legislative actions that will impact food safety. With finite land, water, and air resources, in conjunction with the emerging biofuels industry, production agricultural personnel will soon be facing increased expenses associated with the total costs of food animal production/processing and accompanying food-safety challenges.

One of the key changes in U.S. agriculture in the past five years has been the rapid expansion of biofuel production. In particular, the production of ethanol from corn has risen dramatically, nearly tripling by 2006 (Fischer 2007). By 2005–2006,

4.5 billion gallons of ethanol were being produced by corn grain-to-ethanol refineries (Somerville 2006). It is anticipated that unless corn prices decrease below profitability, corn-produced ethanol will continue to expand (Cassman et al. 2006). However, by 2004 predictions were already indicating that the U.S. corn supply-demand balance was changing from excess production to one of limited supply (Wisner and Baumel 2004). Although concerns regarding corn as a source of human food have received considerable attention, about half of U.S.-produced corn actually goes to animal feed, 20% is exported, 20% converted to ethanol, and 10% is used for industrial and food-related products (Johnson 2007). Governmental calls for ethanol production to account for 10% of gasoline use by 2010 could, based on existing corn production levels, translate to over 50% of the annual crop being devoted to fuel production (Mayday 2007).

Although the immediate impact on food safety is not explicit, the changes caused by shifting crop production toward corn and away from other row crops such as soybeans, which serve as a primary protein source, will create ripple effects in the animal feed industry. One shift already occurring is the feeding of more co-products of ethanol production such as wet and dry distillers' grains with cattle producers being able to use as much as 35 to 45% dry matter in beef cattle and 30% in dairy cattle (Mayday 2007). However, research has indicated that there is a positive association between the prevalence of the foodborne pathogen *Escherichia coli* O157:H7 and feeding dried distillers' grain (Depenbusch et al. 2008; Jacob et al. 2008). Whether this type of feed is also a potential factor in the occurrence of foodborne pathogens in other farm animal species fed distillers' grains remains to be determined. Other possible consequences of reducing soybeans as a source of protein to nonruminants and switching to alternative ingredients could be an increase in *Salmonella* infection as protein supplements such as meat and bone meal and other animal byproducts are being used at higher rates. Historically, specific *Salmonella* serovars have been linked to single-source animal byproducts such as fishmeal (Maciorowski et al. 2004, 2006). As these ingredients become more prominent further development of control measures will need to be implemented to limit contamination during animal feed processing and milling.

Agricultural and Food Bioterrorism

Traditionally unintentional contamination of the U.S. food supply was perceived as the primary objective for food-safety research and policy. However, now in the twenty-first century, intentional food contamination also has to be addressed. The threat from food bioterrorism necessitates some of the same research needs as that from unintentional contamination. Rapid detection of pathogens, access to extensive databases, and comprehensive interpretation are required to minimize or eliminate the threat from both. In other aspects, such as anticipated human behavior and origin of contamination, the needs may differ substantially.

The mere threat potential of bioterrorism leads to a myriad of consequences some of which directly impact food-safety research. For example, when the concept of biological agents as instruments of terror became more probable, the openness of scientific research and sharing of methods, databases, and strains came under scrutiny (Omenn et al. 2006). Consequently, the idea that some types of research should not be publicized becomes a more acceptable concept (Omenn et al. 2006). Although foodborne pathogen research has remained fairly accessible, tighter restrictions and more accountability of cataloging and sharing of strains between institutions has become more commonplace. Part of the difficulty in confronting bioterrorism lies in the potential for employment of a multiple-point attack on a food supply chain at rural, industrial, and urban interfaces (Norton 2003). Such a disruption would preclude the conventional linear cause-and-effect scenarios currently used to assess foodborne-disease outbreaks. In addition the economic consequences for the food industry may very well resonate for a longer period of time than those of a conventional foodborne outbreak. Not only are the economic losses associated with decreased sales and public avoidance a factor, but the potential for loss of employment in the respective food industry becomes significant (Norton 2003). This scenario becomes more likely as food-industry mergers continue to increase, resulting in the consolidation of meat and dairy industries with fewer firms accounting for more of the meat animal slaughter capacity (Woteki and Kineman 2003). In addition, as these mergers occur, they are often followed by construction of larger processing facilities that allow production of larger quantities for greater distribution (Woteki and Kineman 2003). This consolidation amplifies the relative impact of an act of terror because of the sheer quantity of food produced that is affected as well as the economics associated with the food product (Crutchley et al. 2007). In addition, more importation of foods will further complicate the capacity to effectively limit potential attacks.

Food-safety and other scientific research can also be impacted indirectly by human activism based on the concept of animal rights. Attempts to disrupt the food supply via deliberate contamination may represent a variety of motives but attacks on animal research result from highly focused sustained efforts to limit all forms of animal research and eventually animal agriculture production in general. Animal welfare is an important concern that must always be factored into experimental design but assessment of risk of infection and host responses still requires some animal models to account for all the complexities. As extremist activities continue to escalate, more problems for animal research facilities are created (Miller 2007; Conn and Parker 2008). Experimental animal protocols have evolved to address many of the issues that have been raised but concerns remain. Methods that employ non-animal means to evaluate foodborne pathogen infection mechanisms such as tissue culture can supplement some of the experimental information needed, but they cannot provide data on host defenses when microbial risks have to be evaluated (Coleman et al. 2007).

Education and Training

The training and mentorship of students in advanced degree programs that fit food-safety research programs at the respective universities and subsequent placement of these students into well-suited industry, academic, and government positions will have an impact on food-safety nationally and internationally for years to come. Students from academic programs in the biological sciences who have experience in various aspects of microbial genomics are an important resource for meeting the demands of currently funded food-safety research. Many of the food-safety programs focused on foodborne pathogens and epidemiologically related disease issues now require routine in-depth molecular and genomic analyses. Teaching food safety has become more challenging at the undergraduate level due to the information explosion and ready access to large quantities of information. In 2007 it was estimated that approximately 20% of the world's population used the Web and the large majority of these individuals use Web-based search engines (Henzinger 2007). As students who grew up in this environment enter the food-safety workforce they will be more likely to embrace technological advances for acquiring information but they may have difficulties in filtering relevant data from this wealth of information. Consequently, the focal point of the academic instructor may very well be not so much on retrieving information but rather on teaching students how to evaluate the quality of the information they obtain. Advances such as electronic virtual worlds may provide scientific potential for interactively evaluating scientific methodology across networks (Bainbridge 2007); however, hands-on training at the bench will still remain vital to develop and assess methodology.

Funding/Resources and Shifting Political Emphasis

For the past 10 to 15 years there has been considerable success in the development of food-safety research with fairly immediate applications, but funding has at times been unpredictable and sometimes problematic. This is becoming more of a pressing issue as the climate for federal support becomes increasingly more competitive without continued growth in the respective competitive federal grants programs (Bazer 2007; Morrissey 2007). This has translated into more researchers pursuing fewer federal research dollars. Traditionally, food-safety funding has originated from the USDA National Research Initiative (NRI) grants (recently renamed Agricultural and Food Research Initiative or AFRI) in the form of traditional food-safety research programs as well as more specialized focus areas such as epidemiology and the integrated multidisciplinary programs. Other funding contributors included congressional earmarks and industry and foundations to support specific targeted projects on high impact needs (Scanes 2007). Both the U.S. Food and Drug Administration and the USDA Agricultural Research Service have federally funded laboratories devoted to investigation of food-safety issues and solutions. In addition, endowments from private sources for research centers, departments, chairs, and fellowships have added some long-term stability to available resources.

Despite the diversity of funding sources, research funding is becoming more difficult to secure since all of these sources are undergoing economic and financial pressures and/or have received calls for decreased emphasis on food-safety research. Meanwhile, local and state industries face rising costs, limiting their ability to support further development of core research capacities. The more recent federal research and development budgets have traditionally offered little increase in most of the major science programs (Ember et al. 2007; Morrissey et al. 2007), although there are now some increases in certain programs with the changes in the political landscape. In addition, political debate continues to push the concept of a "mega food-safety agency," a concept that becomes more popular each time a new outbreak receives significant publicity (Shire 2007). The debate over such agency mergers becomes more intense as food consumed in the United States increasingly originates from overseas (Fox 2007). Research funding from earmarks and endowments have also received their share of scrutiny, with questions being asked regarding justification for their targeted focus as well as restrictions on uses of the funds. However, such sources are often the only means for multidisciplinary research teams to tackle the key critical issues that require long-term solutions (Scanes 2007).

Conclusions

Food-safety challenges are currently being addressed by development of probiotic cultures and biological compounds to limit *Campylobacter* and *Salmonella* colonization in poultry flocks and by isolation and characterization of *Listeria* antimicrobials from organic sources and their incorporation into food-packaging systems. Rapid-detection systems continue to be developed based on new generation molecular approaches and innovative sensors for foodborne pathogens in food matrices. Molecular and genomic tools are being used to better understand foodborne pathogens in food and animal production settings and to increase understanding of how management practices alter microbial community structure and function in the environment. To more completely control the pathogens in the food supply, the need remains to better understand how these pathogenic organisms interact with commensal organisms *in vivo* and *in situ* on plants and in animals in order to identify effective interventions and control strategies.

Numerous unsolved problems related to foodborne disease persist, among them how foodborne pathogens such as *E. coli* O157:H7, *Salmonella* Enteritidis, and *Campylobacter* spread among animals and how to prevent this occurrence. Other critical concerns include determining the microbial cause of outbreaks in which no pathogen can be identified by current methods; how to assure that the food and water that animals consume is safe; control strategies in the slaughter plant to reduce the contamination of meat; and proper and effective utilization and disposal of waste products in the environment. Another challenging question involves whether human food that receives no cooking before consumption can be effectively protected from microbial contamination at the preharvest stage. The requirements of new standards for safer food supplies include improvement of existing methods as well as introducing new technologies.

Given the continued importance of food contamination as a food-processing and potential bioterrorism threat, the need for food-safety and environmental science research programs will not diminish. The integration of new technologies into the food-safety research programs will also aid in the recruitment of the best students from both national and international sources with strong foundations in contemporary microbial genomics.

Meeting these challenges in land-grant institutions requires strategic and economical resource management of academic assets. This includes adequate facilities for fostering successful microbial research support for collaboration across microbiological food and animal science programs. This strategy will strengthen all programs and discourage isolationism that only leads to minimal opportunities for integration and eventual dilution of already limited resources. In addition, highly qualified students must be recruited to provide continuity to microbiological core research beyond routine data generation. This influx will enhance physical and intellectual capacity to develop the widest spectrum of food-safety applications possible.

Acknowledgments

This review was supported by a USDA Food Safety Consortium grant.

References

Amábile-Cuevas, C. F. 2003. New antibiotics and new resistance. Am. Scientist 91:138–149.

American Society for Microbiology. 2007. Basic research on bacteria: the essential frontier—Report on the American Society for Microbiology, National Institutes of Health Workshop on basic bacterial research. ASM Press, Washington, DC.

Bainbridge, W. S. 2007. The scientific research potential of virtual worlds. Science 317:472–476.

Bajzer, M., and R. J. Seeley. 2006. Obesity and gut flora. Nature 444:1009–1010.

Batt, C. A. 2007. Food pathogen detection. Science 316:1579–1580.

Bazer, F. W. 2007. How to boost agricultural research—US land-grant universities need a radical rethink of their priorities. The Scientist 21:29.

Berg, C. 2001. Health and welfare in organic poultry production. Acta Vet. Scand. 95 (Suppl.):37–45.

Berghman, L. R., D. Abi-Ghanem, S. D. Waghela, and S. C. Ricke. 2005. Antibodies: an alternative to antibiotics? Poult. Sci. 84:660–666.

Blumberg, R. S., and W. Strober. 2001. Prospects for research in inflammatory bowel disease. JAMA 285:643–647.

Casewell, M., C. Friis, E. Marco, P. McMullin, and I. Phillips. 2003. The European ban on growth-promoting antibiotics and emerging consequences for human and animal health. J. Antimicrob. Chemo. 52:159–161.

Cassman, K., V. Eidman, and E. Simpson. 2006. Convergence of agriculture and energy: implications for research and policy. CAST Commentary QTA2006–3:1–12.

Chapman, H. D., and Z. B. Johnson. 2002. Use of antibiotics and roxarsone in broiler chickens in the USA: analysis for the years 1995 to 2000. Poult. Sci. 81:356–364.

Chau, C.-F., S.-H. Wu, and G.-C. Yen. 2007. The development of regulations for food nanotechnology. Trends Food Sci. Technol. 18:269–280.

Coleman, M. E., B. K. Hope, H. G. Claycamp, and J. T. Cohen. 2007. Microbial risk assessment scenarios, causality, and uncertainty. Microbe 2:13–17.

Conn, P. M., and J. V. Parker. 2008. Winners and losers in the animal-research war. Am. Scientist 96:184–186.

Crutchley, T. M., J. B. Rodgers, H. P. Whiteside Jr., M. Vanier, and T. E. Terndrup. 2007. Agroterrorism: where are we in the ongoing war on terrorism? J. Food Prot. 70:791–804.

Davidson, P. M., and T. M. Taylor. 2007. Chap. 33. Chemical preservatives and natural antimicrobial compounds. Pages 713–745 in Food microbiology—fundamentals and frontiers. 3rd ed. M. P. Doyle, L. R. Beuchat, and T. J. Montville, eds. Am. Soc. Microbiol. Press, Washington, DC.

Depenbusch, B. E., T. G. Nagaraja, J. M. Sargeant, J. S. Drouillard, E. R. Loe, and M. E. Corrigan. 2008. Influence of processed grains on fecal pH, starch concentration, and shedding of Escherichia coli O157 in feedlot cattle. J. Anim. Sci. 86:632–639.

Dobrovolskaia, M. A., and S. E. McNeil. 2007. Immunological properties of engineered nanomaterials. Nature Nanotechnol. 2:469–478.

Doyle, M. E. 2006. Natural and organic foods: safety considerations: a brief review of the literature. UW-FRI Briefings 1–9.

Ember, L. R., D. J. Hanson, B. Hileman, C. Hogue, J. Johnson, and S. R. Morrissey. 2007. 2008 R&D budget holds no surprises. Chem. Eng. News 85:25–30, 41.

Fanatico, A. 1998. ATTRA livestock production guide: sustainable chicken production. USDA-ATTRA. 12p.

Farkas, J. 2007. Physical methods of food preservation. Pages 685–712 in Food microbiology—fundamentals and frontiers. 3rd ed. M. P. Doyle, L. R. Beuchat, and T. J. Montville, eds. Am. Soc. Microbiol. Press, Washington, DC.

Fischer, J. R. 2007. Building a prosperous future in which agriculture uses and produces energy efficiently and effectively. NABC Report 19. Pages 27–40 in Agricultural biofuels: technology, sustainability, and profitability. A. Eaglesham and R. W. F. Hardy, eds. Natl. Agric. Biotech. Council, Ithaca, NY.

Fisher, K., and C. Phillips. 2008. Potential antimicrobial uses of essential oils in food: is citrus the answer? Trends in Food Sci. Technol. 19:156–164.

Fox, J. L. 2007. Concerns over food safety could be "coming to a head." Microbe 2:326–327.

Gong, J., W. Si, R. J. Forster, R. Huang, H. Yu, Y. Yin, C. Yang, and Y. Han. 2007. 16S rRNA gene-based analysis of mucosa-associated bacterial community and phylogeny in the chicken gastrointestinal tracts: from crops to ceca. FEMS Microbial Ecol. 59:147–157.

Hagens, S., and M. L. Offerhaus. 2008. Bacteriophages—new weapons for food safety. Food Technol. 62:46–54.

Henderson, G. 2007. America's wealth will shape future food trends. Meat & Poultry 53:2.

Henzinger, M. 2007. Search technologies for the internet. Science 317:468–471.

Jackson, C. R., P. J. Fedorka-Cray, N. Wineland, J. D. Tankson, J. B. Barrett, A. Douris, C. P. Gresham, C. Jackson-Hall, B. M. McGlinchey, and M. V. Price. 2007. Introduction to United States Department of Agriculture VetNet: status of Salmonella and Campylobacter databases from 2004 through 2005. Foodborne Path. Dis. 4:241–248.

Jacob, M. E., J. T. Fox, J. S. Drouillard, D. G. Renter, and T. G. Nagaraja. 2008. Effects of dried distillers' grain on fecal prevalence and growth of Escherichia coli O157 in batch culture fermentations from cattle. Appl. Environ. Microbiol. 74:38–43.

Jarvis, L. M. 2008. Imminent threat. Chem. Eng. News 86:21–24.

Joerger, R. D. 2003. Alternatives to antibiotics: bacteriocins, antimicrobial peptides and bacteriophages. Poult. Sci. 82:640–647.

Johnson, J. 2007. Ethanol—is it worth it? Chem. Eng. News 85:19–21.

Kali, R., and J. Reyes. 2007. The architecture of globalization: a network approach to international economic integration. J. Intl. Business 38:595–620.

Kiers, E. T., R. R. B. Leakey, A.-M. Izac, J. A. Heinemann, E. Rosenthal, D. Nathan, and J. Jiggins. 2008. Agriculture at a crossroads. Science 320:320–321.

Klein, D. A. 2007. Microbial communities in nature: a postgenomic perspective. Microbe 2:591–595.

La Storia, A., D. Ercolini, F. Marinello, and G. Mauriello. 2008. Characterization of bacteriocin-coated antimicrobial polyethylene films by atomic force microscopy. J. Food Sci. 73:T48–T54.

Ley, R. E., P. J. Turnbaugh, S. Klein, and J. I. Gordon. 2006. Human gut microbes associated with obesity. Nature 444:1022–1023.

Lianou, A., and J. N. Sofos. 2007. A review of the incidence and transmission of Listeria monocytogenes in ready-to-eat products in retail and food service environments. J. Food Prot. 70:2172–2198.

Maciorowski, K. G., P. Herrera, M. M. Kundinger, and S. C. Ricke. 2006. Animal production and contamination by foodborne Salmonella. J. Consumer Prot. Food Safety 1:197–209.

Maciorowski, K. G., F. T. Jones, S. D. Pillai, and S. C. Ricke. 2004. Incidence and control of food-borne Salmonella spp. in poultry feeds—a review. World's Poult. Sci. J. 60:446–457.

Mayday, J. 2007. Food, feed, or fuel: ethanol boom reverberates throughout the food system. Meat & Poultry 53:10–12.

Miller, G. 2007. Animal extremists get personal. Science 318:1856–1858.

Montville, T. J., K. Chikindas, and M. L. Chikindas 2007. Biopreservation of foods. Pages 647–764 in Food microbiology—fundamentals and frontiers. 3rd ed. M. P. Doyle, L. R. Beuchat, and T. J. Montville, eds. Am. Soc. Microbiol. Press, Washington, DC.

Morrissey, S. R. 2007. Supporting science—AAAS science and technology policy forum focuses on R&D funding and science advocacy. Chem. Eng. News 85:29–30.

Nickols-Richardson, S. M. 2007. Nanotechnology: implications for food and nutrition professionals. J. Am. Diet. Assoc. 107:1494–1497.

Norton, R. A. 2003. Symposium: Agro-terrorism: biological threats and security measures. Food security issues—a potential comprehensive plan. Poult. Sci. 82:958–963.

Oberholtzer, L., C. Greene, and E. Lopez. 2006. Organic poultry and eggs capture high price premiums and growing share of specialty markets. Outlook from the Economic Research Service—USDA LDP-M-150–01, pp. 1–18.

Omenn, G. S. 2006. Grand challenges and great opportunities in science, technology, and public policy. Science 314:1696–1704.

Onyango, B. M., W. K. Hallman, and A. C. Bellows. 2007. Purchasing organic food in U.S. food systems: a study of attitudes and practice. Br. Food J. 109:399–411.

Ricke, S. C. 2003. Perspectives on the use of organic acids and short chain fatty acids as antimicrobials. Poult. Sci. 82:632–639.

Ricke, S. C., M. M. Kundinger, D. R. Miller, and J. T. Keeton. 2005. Alternatives to antibiotics: chemical and physical antimicrobial interventions and foodborne pathogen response. Poult. Sci. 84:667–675.

Scanes, C. G. 2007. The case for funding agricultural research. Poult. Sci. 86:2483–2484.

Siegrist, M., M.-E. Cousin, H. Kastenholz, and A. Wiek. 2007. Public acceptance of nanotechnology food and food packaging: the influence of affect and trust. Appetite 49:459–466.

Shire, B. 2007. Single food safety agency discussion resurfaces. Meat & Poultry 53:24–25.

Somerville, C. 2006. The billion-ton biofuels vision. Science 312:1277.

Sorrentino, A., G. Goarrasi, and V. Vittoria. 2007. Potential perspectives of bio-nanocomposites for food packaging applications. Trends Food Sci. Technol. 18:84–95.

Strober, W., I. J. Fuss, and R. S. Blumberg. 2002. The immunology of mucosal models of inflammation. Annu. Rev. Immunol. 20:495–549.

Turnbaugh, P. J., R. E. Ley, M. A. Mahowald, V. Magrini, E. R. Mardis, and J. I. Gordon. 2006. An obesity-associated gut microbiome with increased capacity for energy harvest. Nature 444:1027–1031.

U.S. Department of Agriculture Economic Research Service. 2004. Briefing room: economics of foodborne disease. At http://www.ers.usda.gov/Briefing/FoodborneDisease/.

Uskokovi , V. 2007. Nanotechnologies: what we do not know. Technol. Soc. 29:43–61.

van der Sluis, W. 2007. More organic poultry meat and eggs in the U.S. World Poult. 23:26–28.

White, P. L., A. L. Naugle, C. R. Jackson, P. J. Fedorka-Cray, B. E. Rose, K. M. Pritchard, P. Levine, P. K. Saini, C. M. Schroeder, M. S. Dreyfuss, R. Tan, K. G. Holt, J. Harman, and S. Buchanan. 2007. *Salmonella* Enteritidis in meat, poultry, and pasteurized egg products regulated by the U.S. Food Safety and Inspection Service, 1998 through 2003. J. Food Prot. 70:582–591.

Winter, C. K., and S. F. Davis. 2006. IFT scientific status summary: organic foods. J. Food Sci. 71:R117–R124.

Wise, M. G., and G. R. Siragusa. 2007. Quantitative analysis of the intestinal bacterial community in one- to three-week-old commercially reared broiler chickens fed conventional or antibiotic-free vegetable-based diets. J. Appl. Microbiol. 102:1138–1149.

Wisner, R. N., and C. P. Baumel. 2004. Ethanol, exports and livestock: will there be enough corn to supply future needs? Feedstuffs 30:1, 20.

Woteki, C. E., and B. D. Kineman. 2003. Challenges and approaches to reducing foodborne illness. Annu. Rev. Nutr. 23:315–344.

Yang, H., L. Qu, A. Wimbrow, X. Jiang, and Y.-P. Sun. 2007. Enhancing antimicrobial activity of lysozyme against *Listeria monocytogenes* using immunonanoparticles. J. Food Prot. 70:1844–1849.

Contributors

Robin Anderson, research microbiologist, USDA Agricultural Research Service, College Station, Texas.

Jenna D. Anding, associate professor of nutrition and food science and Extension program leader, Texas A&M University.

Bradley L. Bearson, microbiologist, USDA Agricultural Research Service, Ames, Iowa.

Shawn M. D. Bearson, microbiologist, USDA Agricultural Research Service, Ames, Iowa.

Dhruva Bhattacharya, Department of Poultry Science, University of Arkansas.

Elizabeth A. Bihn, Good Agricultural Practices program coordinator, Department of Food Science, Cornell University.

Sheridan L. Booher, research associate, Veterinary Medical Research Institute, Iowa State University.

Aleksandar Božic, University of Novi Sad, Serbia.

Christine Bruhn, consumer food marketing specialist, Department of Food Science, University of California, Davis.

J. Allen Byrd, research microbiologist, USDA Agricultural Research Service, College Station, Texas.

Todd R. Callaway, research microbiologist, USDA Agricultural Research Service, College Station, Texas.

Vesela I. Chalova, assistant professor, University of Food Technologies, Plovdiv, Bulgaria.

Nancy A. Cornick, associate professor of veterinary microbiology and preventive medicine, Iowa State University.

Philip G. Crandall, professor of food science, University of Arkansas.

James H. Denton, professor emeritus of poultry science, University of Arkansas.

Ann M. Donoghue, research leader, USDA Agricultural Research Service, Fayetteville, Arkansas.

Dan J. Donoghue, professor of poultry science, University of Arkansas.

Scot E. Dowd, Research and Testing Laboratory, Lubbock, Texas.

Thomas S. Edrington, research animal scientist, USDA Agricultural Research Service, College Station, Texas.

Satchi Eswaranandam, postdoctoral associate, Department of Food Science, University of Arkansas.

Anne C. Fanatico, USDA Agricultural Research Service, Fayetteville, Arkansas.

Daniel Fung, professor of animal sciences and industry, Kansas State University.

Kenneth J. Genovese, research microbiologist, USDA Agricultural Research Service, College Station, Texas.

Michael Gregory, Ozark Mountain Poultry, Rogers, Arkansas.

Irene Hanning, postdoctoral associate, Department of Food Science, University of Arkansas.

Roger B. Harvey, veterinary medical officer, USDA Agricultural Research Service, College Station, Texas.

Navam S. Hettiarachchy, professor of food science, University of Arkansas.

Michael G. Johnson, professor emeritus of food science, University of Arkansas.

Frank T. Jones, Extension poultry specialist emeritus, University of Arkansas.

Dianna M. Jordan, assistant professor of veterinary diagnostic and production animal medicine, Iowa State University.

Yong S. Jung, USDA Agricultural Research Service, College Station, Texas.

Woo-Kyun Kim, Department of Medicine, David Geffen School of Medicine, University of California, Los Angeles.

Young Min Kwon, associate professor of poultry science, University of Arkansas.

Bwalya Lungu, postdoctoral associate, Department of Food Science, University of Georgia, Athens.

John Marcy, professor, Center of Excellence for Poultry Science, University of Arkansas.

James L. Marsden, distinguished professor of animal sciences and industry, Kansas State University.

Teresa Maurer, sustainable agriculture and rural development program manager, National Center for Appropriate Technology, Fayetteville, Arkansas.

Jackson L. McReynolds, research microbiologist, USDA Agricultural Research Service, College Station, Texas.

Harley W. Moon, professor emeritus of veterinary pathology, Iowa State University.

Arunachalam Muthaiyan, postdoctoral associate, Department of Food Science, University of Arkansas.

Ramakrishna Nannapaneni, assistant professor of immunology and molecular food microbiology, Mississippi State University.

David J. Nisbet, supervisory microbiologist, USDA Agricultural Research Service, College Station, Texas.

Lisa M. Norberg, senior research assistant, University of Texas–M. D. Anderson Cancer Center, Houston, Texas.

Corliss A. O'Bryan, postdoctoral associate, Department of Food Science, University of Arkansas.

Ken Over, graduate student, Department of Food Science, University of Arkansas.

Omar A. Oyarzabal, assistant professor of poultry science, Auburn University.

Taha M. Rababah, graduate student, Department of Food Science, University of Arkansas.

Ron Rainey, Extension economist, University of Arkansas.

Ixchel Reyes-Herrera, Doctoral Academy Fellow, Department of Poultry Science, University of Arkansas.

Steven C. Ricke, Donald "Buddy" Wray Food Safety Endowed Chair and director of the Center for Food Safety, University of Arkansas.

Scott Russell, associate professor of poultry science, University of Georgia.

Orhan Sahin, assistant scientist, Department of Veterinary Microbiology and Preventive Medicine, Iowa State University.

Steven C. Seideman, Extension food processing specialist, University of Arkansas.

Gregory R. Siragusa, Danisco Animal Nutrition, Waukesha, Wisconsin.

Sujata A. Sirsat, doctoral student, Department of Poultry Science, University of Arkansas.

Theivendran Sivarooban, senior microbiologist, Water Security Corporation, Sparks, Nevada.

Michael Slavik, professor of poultry science, University of Arkansas.

Qijing Zhang, associate professor of veterinary microbiology and preventive medicine, Iowa State University.

Index

prevalence, 235–43, 311; beef production, 307–16; regulatory guidelines, 236, 291–92, 295–96
organic poultry: bacterial pathogens, 340; market growth rate, 294–95; market share, 294; popularity, 294, 340; processing guidelines, 296–99; regulatory guidelines, 295–96
Organic Trade Association (OTA), 289
ovary regression, 65–67
oxytetracycline, 240, 241
ozone-based treatments, 5, 125

Paenibacillus polymyxa, 17
Papillibacter, 228
Pathatrix system, 158–59
Pathogen Modeling Program (PMP), 100
Pathogen Reduction-Hazard Analysis Critical Control Point Final Rule, 95, 122
peanut butter, 196, 335
pediocin, 7
penicillin, 240
personal food-handling practices, 277–78
pesticide residues, 283, 326–27
pet food contamination, 273
phagocytic cells, 41
Phebus, Randy, 4
phenolic compounds, 265–66
pigs: *see* pork; swine
pinocytosis, 197
pirlimycin, 240
plant extracts, 260–62, 265–66
polyclonal antibodies (PAb), 179–80, 183, 186
polymerase chain reaction (PCR) testing, 158–60, 170–71, 201, 202, 203–5, 228–31
polymorphic-tag-lengths-transposon-mutagenesis (PTTM), 170–71
pork: irradiation pasteurization, 122; *Listeria monocytogenes* contamination, 136; market share, 295; popularity, 294; *see also* swine
postal surveys, 274
postharvest control strategies: beef contamination, 119–26, 312; *Listeria monocytogenes*, 139–44; poultry, 296; processing plant contamination, 87–93; ready-to-eat foods, 130–32, 134, 140, 144, 178; research areas, 335–36; *Salmonella* contamination, 195–210; small plant validation processes, 95–103; *see also* food-processing environments
potassium sorbate, 141
poultry: antibiotic growth promotants (AGP), 225–26, 231; antimicrobial resistance rates, 241–42; bacterial prevalence, 238–39; breeder chickens, 88; broiler chickens, 89–91, 225–32, 247, 294–302; *Campylobacter jejuni*, 11–21, 105, 108–13, 230, 238–39, 247–52, 298–99; contaminated

product recalls, 260; decontamination strategies, 196; egg-laying hens, 63–75; *Escherichia coli* O157:H7 contamination, 228; hatchery contamination, 88–89; intestinal microbiota profiling, 225–32; irradiation pasteurization, 264–66; *Listeria monocytogenes* contamination, 132, 135–36, 178; organic poultry, 294–302, 340; organic versus conventional livestock operations, 238–39, 241–42; plant extract treatments, 265–66; preharvest control strategies, 11–21, 55, 296–99, 314; processing plant contamination, 87–93, 135, 297–301; raw poultry, 110–11; *Salmonella* contamination, 55, 63–64, 68–75, 87–93, 140, 195–96, 238–39, 297–301
poultry water treatment (PWT), 90–91
preharvest control strategies: beef, 120–22, 125; cattle, 55; competitive exclusion, 54–56, 88, 313–14; egg-laying hens, 63–75; epizootic bacteria reduction efforts, 49–50; experimental chlorate product (ECP) treatment, 50–53; lauric acid/Lauricidin, 54; poultry, 11–21, 55, 296–99, 314; research areas, 335–36; short-chain nitroalkanes, 53; swine, 27–31, 35–44, 55–56, 314
probiotic culture supplementation strategies, 14–17, 72–75, 313
processing plants: *see* food-processing environments
produce, 293, 327–28
Propionibacterium freudenreichii, 313
Propionibacterium spp., 51
Pseudomonas aeriginosa, 109–10, 113
Pseudomonas fragi, 112
Pseudomonas spp., 53, 228
pulsed field gel electrophoresis (PFGE) testing, 160, 187

quorum sensing, 37, 111

radiation treatments, 196, 199
rapid identification methods: aerobic, anaerobic, and "real-time" viable cell count advances, 157; air and surface sampling, 157; antibody-based technologies, 185–87; background information, 155; biomass measurements, 159; biosensor, microchip, and biochip advances, 160–61; food-based bioterrorism, 342–43; future directions, 162–63; genetic testing advances, 159–60; immunological testing advances, 158–59, 185–87, 337; instrumentation advances, 159; miniaturization advances, 157–58; molecular data sets, 339; poultry product contamination, 252; research areas, 4;